Make: The Annotated Build-It-Yourself Science Laboratory

Raymond Barrett and Windell Oskay

MAKER MEDIA

SAN FRANCISCO, CA

Make: The Annotated Build-It-Yourself Science Laboratory

by Raymond Barrett and Windell Oskay

Based upon: Build-It-Yourself Science Laboratory by Raymond E. Barrett. Illustrated by Joan Metcalf. Originally published by Doubleday & Company, Inc.

Printed in the United States of America.

Published by Maker Media, Inc., 1160 Battery Street East, Suite 125, San Francisco, California 94111.

Maker Media books may be purchased for educational, business, or sales promotional use. Online editions are also available for most titles (*http://safaribooksonline.com*). For more information, contact our corporate/institutional sales department: 800-998-9938 or corporate@oreilly.com.

Editor: Brian Jepson	**Indexer:** WordCo Indexing Services
Production Editor: Nicole Shelby	**Interior Designer:** David Futato
Proofreader: Marta Justak	**Cover Designer:** Riley Wilkinson
	Illustrator: Joan Metcalf

May 2015:   First Edition

Revision History for the First Edition
2015-04-24:   First Release

See *http://oreilly.com/catalog/errata.csp?isbn=9781457186899* for release details.

978-1-457-18689-9

[LSI]

Praise for The Annotated Build-It-Yourself Science Laboratory

My father's purpose in life was to see the excitement of discovery from children experimenting. He was an innovator believing that students learned best by independent hands-on experiences with everyday items. It's wonderful seeing Windell Oskay bringing this book back to stimulate the next generation of children.

—Stephen Barrett

At first glance, projects in this book (such as building a carbon arc furnace or a hydrogen generator) may seem intimidating, even dangerous, and that is exactly the point! The science practices and skills explored through the real experiences in this publication will build the critical thinking and careful observation skills needed to support teachers and students to develop a real understanding of science. As relevant today as when it was first published, this book will support curious people of all ages to engage in serious fun—the starting points for falling in love with science all over again!

—Mike Petrich, Tinkering Studio—Exploratorium

Much, much more than a DIY Lab. It's really a fairly full course in experimental science.

—Forrest M. Mims III, Author and Amateur Scientist

Table of Contents

PART I.   Chemistry

PART II.   Physics

PART III. Biology

8. General Biology Equipment 209

PART IV.   Appendix

Foreword

In this book, you will learn how to make some amazing things: a carbon arc furnace, cloud chamber, mechanical stroboscope, radiometer, optical micrometer, electromagnet, microtome, spectroscope, and so many others. You will blow glass, catch bugs, and cut the ends off of power cords. You will learn how acids and alkalis taste, what kinds of things live in a drop of water, and how your lungs draw in air. You will measure mass, density, volume, pressure, temperature, time, humidity, cosmic rays, conductivity, and optical polarization. You will isolate hydrogen, build an electric motor, grow rock candy crystals, and literally *burn* a record of the day's weather using a Campbell–Stokes sunlight recorder.

While there are a lot of neat things to build, not everything is about making equipment. To look at the apparatus alone would be to miss the point; to not see the forest for the trees. At its heart, this is a *science book*. Every project comes with a set of questions for you to investigate, frequently challenging you, asking *Can you work like a scientist?* Beyond these is something yet more: one of the most extraordinary collections of "science fair" research project ideas ever put to paper, with over 1,600 open-ended questions for investigation, spanning the fields of chemistry, physics, biology, and geology.

All considered, this is one of the finest hands-on science project books ever written. Originally published in 1963, it has held up quite well, especially when you consider the pace of scientific and technical progress over the last half century. One reason is that it feels *authentic*, rather than dumbed down or bowdlerized: There are a fair number of deliciously real (i.e., potentially dangerous) projects that would never be allowed in young-adult science books today, yet were perfectly acceptable in a less litigious age. We understand many hazards better today, but as surely as night follows day, *nothing in this book is any more dangerous than it was when the book was first published*.

My own personal experience with this book began when I was 10 years old, in 1984, at Ainsworth Elementary School in Portland, Oregon. My fifth grade classroom made regular trips to the school library. It was at one of our regular trips that the librarian spoke to us all about the Dewey Decimal System, and how the library was organized. While I was already an avid reader, I had always simply browsed (as kids of a certain age do) among the set-out books on display whenever I visited the library. Learning about the Dewey Decimal System changed all that. The science books were in the 500s, and since I already knew that *I was going to be a sci-*

entist when I grew up, the 500s were where *I should be spending my time.*

The science book section in a school library is apt to be inhabited by all kinds of titles, including the abstract, the esoteric, the dull, and (hopefully) the amazing. Perhaps it is little wonder that this one caught my attention, with its bold and inviting title: *BUILD-IT-YOURSELF SCIENCE LABORATORY*. Because, well, that was exactly what I wanted to do.

What I don't know is how many other people had this kind of experience growing up. In a sense, simply by attending school in Portland, I was (unknowingly) growing up within the author's local sphere of influence. Raymond E. Barrett was a teacher in the Portland school district for seven years before he was hired in 1959 as the education director of OMSI, the Oregon Museum of Science and Industry—a post where he remained for 22 years. At OMSI, Barrett developed new hands-on, experiential approaches to teaching science. He broadened the appeal with classes, workshops, and camps. He provided leadership for science education both in the Pacific Northwest and across the nation, teaching teachers better ways to teach science.

During his years as a teacher, and later in his first few years at OMSI, Barrett began to develop a set of lesson plans for do-it-yourself science projects targeted at middle and high school students. The plans were designed to stimulate interest in the sciences, invoking Galileo, Newton, and Faraday, who (as the story goes) constructed their laboratories from the simplest possible materials. Through the plans, one could build or improvise some 200 pieces of laboratory equipment from mostly household materials, and use them in over 2,000 experiments.

The early 1960s were in so many ways a different time. There was the Sputnik crisis, still lingering. America's cold war adversaries were smart and technological; we had to compete.

The space race was on. The United States had a credible human spaceflight program, and putting earthlings on the moon was a realistic priority. Science education was booming. And people were hungry for better ways to teach science.

For all of these reasons—plus the fact that it was simply good—Barrett's "build it yourself science" program became so popular that individuals and institutions across the US ordered more than 4,000 sets of his mimeographed lesson plans. It even led to Barrett having his own local television show, teaching science with home-built equipment. The program's fame eventually attracted the attention of the Doubleday company, which contracted Barrett to collect his lesson plans into book form. Barrett refined and expanded his plans, and the results are here in the book that you have before you, illustrated by OMSI staff artist Joan Metcalf.

By the time that I had come across the book in the mid 1980s, the book was already 20 years old, and Barrett had already retired from OMSI.

As a child, I remember being particularly delighted at one little "discovery" that I made while working on a project from the book. I had been looking at tiny protists in drops of pond water through my school-grade microscope, but found them hard to see, since they were small, fast, and *transparent*. A project in the book talked about using crossed polarizers in a microscope to look at crystals or a fish tail, but it seemed like they would be able to solve the problem with the microbes. I modified my microscope to have polarized filters above and below the sample ("Polarized Light Filters" on page 215), having scrounged the filters from a set of improvised 3D glasses. Through crossed polarizers, you can only see things between them that rotate the polarization of light; everything else will simply be black. And wow, what an effect: the protists were still small and fast, but now they were *glowing white* on a black background. I had (re)discovered a primitive form of dark-field microscopy, and it was amazing.

In modern times, our contemporary maker and maker education movements have helped to rekindle our cultural interest in hands-on education, especially in the STEM (science, technology, engineering, and mathematics) and STEAM (STEM with art) fields, in a way that hasn't been seen since the 1960s.

In some ways, it is an uphill battle. We live in an era now where zero tolerance guidelines mean that kids routinely get suspended or expelled from school for possession of "dangerous" items like *scissors*. A professor friend of mine relayed to me an anecdote about a parent who recently called in to complain that their child was being placed at risk of infection from *sharing needles* because their youth e-textile class did not routinely sterilize their sewing needles. You might find that it's not trivial to draw blood with fat kid-friendly sewing needles, let alone puncture two different students such that you would have a credible risk of infection. Did they simply confuse this "sharing needles" with the very high risk of sharing hollow hypodermic needles among intravenous drug-users?

There are signs of hope. Gever Tulley and Julie Spiegler's book *50 Dangerous Things (you should let your children do)* was a great reminder that well-planned but potentially risky activities are an incredible teaching tool for learning how to do things safely and well. And makerspaces and maker education are helping people to recognize that children and young adults need time to *play*, tinker, explore, make things with their hands, and learn on their own. Learning how to build things is an important part of developing physical intuition, learning about how the world really works, and helping to hone critical thinking skills.

And for me personally, *this* is the book that taught me how to make things.

—Windell H. Oskay, October 2014

Getting Started

On the Nature of This Annotated Edition

This edition of *Build-it-yourself Science Laboratory* has been adapted from the 1960s original in order to make it more accessible and useful to modern readership, while at the same time preserving its general character and set of experiments.

We have left the original text uncensored and "as-is" to the extent that it is practical. Rather than directly editing the text, we have in most cases added footnotes[1] or extended notes in a new Appendix E where appropriate.

Exceptions to this rule include:

- Minor technical and typographical errors.

- Minor wording and punctuation changes for clarity.

- Specific postal addresses for mail-order supplies, which have been removed.

- Specific dollar price estimates, which have been updated to current values.

- Chemical formulas for minerals, which have been updated with current notation.

- The wind chill table, which has been updated with current values.

- The original illustrations, which have been cleaned up and retouched for clarity and layout reasons.

- One missing illustration referred to in the text, which has been synthesized.

- One project, based around a (now) particularly rare neon bulb, has been replaced with a modern LED-based apparatus that serves the same function. The text describing the original apparatus is preserved in the appendix.

- The foreword and new prefaces that have been added: this section and the others up through and including "Conventions Used in This Book" on page 19.

- Appendix A, which describes sources for chemicals and other materials, has been rewritten with modern sources.

---

1 There were no footnotes in the original text; all footnotes are added annotations.

- The data table appendices (Appendix B, Appendix C, and Appendix D), which have been replaced with new versions.

Finding Materials

The supplies and chemicals specified within the experiments were, of course, intended to be a set of "easily obtainable" materials. However, the set of materials that might be described that way has changed greatly over the last half century. You might understand if a few of them are simply (or practically) unavailable in the modern era. One of the motivations for this annotated edition is to make the book more *usable* by filling in these gaps with modern sources and (in some cases) modern substitutes.

We might broadly group the hard-to-find materials into a few categories:

- Materials that were once in widespread use, but have fallen out of favor for technological reasons. For example, ditto fluid (once used as widely for school handouts as laser printers are today) or the chemicals used for developing photographic film. By and large, these chemicals are still available from specialty sources or on the Internet— just no longer from your corner drugstore.

- Materials that are genuinely hazardous and rarely used (by individuals) for reasons of safety. Substances such as mercury and asbestos are still legal and in widespread industrial use, but have either been banned or fallen out of favor in many of their former applications. Kitchen potholders are no longer lined with asbestos, and oral thermometers are more likely to be digital than mercury-filled. *And that is not a bad thing.* In the one place where asbestos actually comes up—an asbestos tile to pro-

tect your desk from hot glassware— we'll suggest that you use a regular kitchen trivet instead.

- Materials that are no longer produced, for example, vacuum tubes for radios.

- Materials that are hard to purchase without appropriate credentials. Many chemical supply companies will only sell to institutions or approved companies. As an independent or amateur scientist, nothing could be more frustrating. We will endeavor to provide alternate sources or substitutes when barriers like these might become issues.

For all of these and additional material needs, please see the annotations throughout the text and in Appendix A. You may also want to look on our website, *http://biyscience.com*, for an online list of sources for materials, complete with purchasing links.

Making Versus Buying Versus Printing

There are many places in this book where you might reasonably judge it to be more prudent, quick, or cost-effective to simply purchase a given piece of equipment, rather than fabricating it from the given instructions. And, if you have the resources and inclination to do so, there is no shame whatsoever in doing exactly that.

But acknowledging that fact, why should you even consider making such basic things as your own directional compass or thermometer, when you could simply buy them—and shiny new digital versions, at that? The answer, of course, is that there is often great value both in learning how to make things and in learning how things work. In making something, you learn about its principles of operation, about mechanisms and fabrication, and you develop your own physical intuition. As an experimental scientist, you will inevitably need to make some *new* kind of apparatus in your work; something that you can't simply purchase *because it has never existed before*. Experience with making things and developing that physical intuition—your sense of what will and will not work when you build things—can give you a tremendous head start.

There are also any number of cases where it may be helpful (or just fun!) to take advantage of computerized tools, such as 3D printers or laser cutters, to help fabricate your equipment. Doing so may save you time, improve the quality of your output, or allow customization that isn't otherwise straightforward. You might have your own 3D printer or laser at home, or access to one at your school, local library, hackerspace, or makerspace. If not, online services such as Shapeways and Ponoko offer easy access to 3D printing and laser cutting (respectively) at a modest price.

If you do have easy access to a 3D printer, it may be tempting to just print every piece of apparatus that you might need. However, some consideration should be given to what materials things actually need to be made out of. Most common, low-cost 3D printers make objects out of plastic, either by melting filament or polymerizing a resin. Thus, as compared with laboratory glassware, 3D printed objects will typically have a low maximum service temperature (because of melting), be flammable (because they are plastic), and be more chemically reactive. For general-purpose applications, like making models and mounting optics, almost any material may be used, but there are cases where more robust or inert materials like glass or metal are required. This can be a little trickier than it sounds. For example, a test tube rack—one of those "obvious" things that people like to print in 3D—really needs to be able to handle hot test tubes, right from an alcohol burner.

Safety

There are both era-appropriate and updated safety tips throughout the text, but it is important for you to understand that modern safety practice demands additional attention beyond simply reading a few lines of warning text. It is not so much the case that a given experiment is "safe" or "unsafe"—rather, it is the human element, *you, the experimenter*, that renders a situation safe or unsafe. Some of the experiments and fabrication procedures explained in this book carry a risk of serious injury, death, or severe property damage. And yet *all of them* can be performed safely when carried out with diligence and care.

Use a consistent approach to every experiment and step of fabrication: onsider the safety implications of what you will be doing, make sure that you are (along with anyone else in the vicinity) aware of the potential hazards, and take appropriate precautions. If you aren't *certain* that you understand the safety implications, then it is your unwavering obligation to seek and obtain outside assistance before proceeding.

Although we cannot anticipate every situation, here are some of the types of things that you will need to think about: If something could potentially fly out at you, make sure that you're wearing approved safety glasses with side protection. If there's fire or extreme heat, have a charged fire extinguisher on hand and a working phone—just in case you need to call the fire department. If there's exposed line (wall/mains) voltage, make sure that you have access to a circuit breaker to shut it off. If you're working with a knife or scissors, be careful not to cut yourself. And so on.

Electrical safety is a topic of particular concern, since there are projects that involve exposed electric wiring. A common rule of thumb is that electronics are safe to touch below 25 V AC or 60 V DC. However, there is credible evidence to suggest that there is *no level* of voltage that it is

100% safe to touch. You can minimize risk when working with line voltage by using a fused isolation transformer and having an easily accessible power switch (such as a wall switch or power strip) that is upstream and separate from any power switch on your project.

If you are a young person, some of the projects will require adult supervision. Discuss your projects with an adult to figure out which projects. Regardless of your age, good safety practice requires that you have another responsible human being nearby when working with power tools or any project that could produce potentially lethal hazards such as fire or exposed line voltage.

If we could give one "soundbite" of advice for staying safe, it would be this: Pay careful attention to what you are doing, and approach every new situation with patience and above all, common sense.

International Power and Units

The projects in this book that involve line voltage are designed for use with the power grid in the US, 117 V AC, at 60 Hz. If you live in an area with different wall power, do not assume that a project involving line voltage can be built without accounting for the change.

Most projects in this book use inches and other US units, rather than metric units. Please see Appendix B for a list of unit conversions.

On Independent Thinking

The many projects and questions in this book are designed to improve your critical thinking skills and provide less hand holding than you may be used to in other contexts. Here are some things that you may want to consider as you approach new problems.

- Questions do not always have answers.
- Some questions that the book asks will require research to answer, and not just of the experimental sort. (Who *was* von Jolly, and what was his method for measuring the mass of the earth? You'll need to find that out first, before trying to reproduce his results.)

- Some ingenuity will occasionally be required. If a procedure calls for rubber tubing and glass tubing that will be connected together, the author has assumed that you can make it work, even if the parts don't fit together precisely when you first sit down and try. (Do you need to get different size tubing? Make an adapter? Shim it with tape?)

- Some degree of responsibility is always required. In terms of safety practice, care with chemicals, care with animals, and so on. If something *could potentially* go terribly wrong, stop and re-evaluate the situation and your approach. You are not yet properly prepared.

Cultural Influences

Little shows the age of the original book better than two places where the author tacitly assumed that a scientist or the director of a research laboratory would be male. That may have been culturally acceptable in the 1960s, but it has no place any longer. Today, the director of the National Science Foundation is an astrophysicist, Dr. France A. Córdova. She is not the first woman to direct the NSF. Anyone with the drive to do so can be a scientist; assumptions to the contrary are universally harmful to our society.

Cigar boxes and cigarettes find their way into a few of the projects. Much like asbestos and mercury, the health risks of tobacco are better understood today. While there is little good that might be said about cigarettes, cigar boxes are an excellent class of project box that there isn't really a good replacement for today. Fortunately, it's possible to purchase "cigar boxes" that never had cigars in them—just a box for the sake of being a box.

Casual experiments with animals—like putting a rat through a maze—were once commonplace, but (for good reason) are now heavily scrutinized. Some other experiments with animals that appear in this book would be very unlikely to pass muster from a review committee in the present age. I urge you to think twice before involving animals, and to follow international "science fair" guidelines (*including* the use of a review committee) when doing so. See Note 1 for further discussion.

Acknowledgments

Thanks to Steven Herzberg for extensive assistance in the preparation of this book. In particular, for his many long hours proofreading and formatting the text and cleaning up the illustrations.

Thanks to Alan Yates for allowing us to use his LED electroscope project.

Thanks to Jonathan Foote (*http://rotor-mind.com*) for help with the neon bulb "oscilloscope" project.

Thanks to Stephen Barrett for reviewing the Foreword and his kind words of encouragement.

Thanks to Eric Bulmer and Brian Jepson for many helpful comments on the annotations.

Thanks to my partner Lenore Edman for patience, encouragement, and many helpful suggestions.

Additional Image and Data Sources

The periodic table in Appendix D is a work of the US government (NIST), and is in the public domain. www.nist.gov/pml/data/

The graphic for the Morse code table ("International Code" on page 137) was adapted from a public domain source at wikimedia.org.

Conventions Used in This Book

The following conventions are used in this book:

Safety Tips

The "Safety Tips" sections throughout the text are the author's original guidance on safety practice. These tend to be a little *understated*, and should be taken quite seriously. When the safety tips say something like "Don't touch the bare wires," that may be because a potentially lethal(!) shock could result. (So *really* don't touch the bare wires!)

Modern Safety Practice

The "Modern Safety Practice" sections throughout the text are new notes about contemporary safety practice, added where necessary in the annotated edition. These notes frequently refer to extended documentation in the appendix.

Safari® Books Online

 Safari Books Online is an on-demand digital library that delivers expert content in both book and video form from the world's leading authors in technology and business.

Technology professionals, software developers, web designers, and business and creative professionals use Safari Books Online as their primary resource for research, problem solving, learning, and certification training.

Safari Books Online offers a range of plans and pricing for enterprise, government, education, and individuals.

Members have access to thousands of books, training videos, and prepublication manuscripts in one fully searchable database from publishers like O'Reilly Media, Prentice Hall Professional, Addison-Wesley Professional, Microsoft Press, Sams, Que, Peachpit Press, Focal Press, Cisco Press, John Wiley & Sons, Syngress, Morgan Kaufmann, IBM Redbooks, Packt, Adobe Press, FT Press, Apress, Manning, New Riders, McGraw-Hill, Jones & Bartlett, Course Technology, and hundreds more. For more information about Safari Books Online, please visit us online.

How to Contact Us

Please address comments and questions concerning this book to the publisher:

> Make:
> 1160 Battery Street East, Suite 125
> San Francisco, CA 94111
> 877-306-6253 (in the United States or Canada)
> 707-829-0515 (international or local)

Make: unites, inspires, informs, and entertains a growing community of resourceful people who undertake amazing projects in their backyards, basements, and garages. Make: celebrates your right to tweak, hack, and bend any technology to your will. The Make: audience continues to be a growing culture and community that believes in bettering ourselves, our environment, our educational system—our entire world. This is much more than an audience, it's a worldwide movement that Make: is leading—we call it the Maker Movement.

For more information about Make:, visit us online:

> Make: magazine: *http://makezine.com/magazine/*
> Maker Faire: *http://makerfaire.com*
> Makezine.com: *http://makezine.com*
> Maker Shed: *http://makershed.com/*

We have a web page for this book, where we list errata, examples, and any additional information. You can access this page at *http://bit.ly/biy-lab*. You can also visit *http://www.biyscience.com* for an active list of resources and references.

To comment or ask technical questions about this book, send email to *bookquestions@oreilly.com*.

Original Preface

You as a student or teacher can begin your *Real* understanding of science in the same way as Newton, Galileo, and Faraday did hundreds of years ago. These early scientists had none of the modern tools of science, and yet from simple materials they were able to make the great findings that are the basis of much scientific investigation today.

Each of these scientists built his own laboratory and his own equipment. You, *as a pioneer of the twentieth century*, can work much the same way. You can design and build your own experimental equipment and use this equipment to find out things for yourself. While building and working with this equipment, you will open up many new questions and problems. These questions can start you on an exciting journey into the real world of science.

The materials needed to build the basic tools of science described in this book come from your home, the garage, the dime store, or the hardware store. In all cases, the cost is very low. Whenever a basic material is needed that is not easily available, suggestions are made as to where the material may be obtained.

The basic idea behind the book is to encourage students and teachers to build their own science laboratories in the home and at school. With this homemade equipment the young scientist can experience the thrill of creativity, and the desire for and satisfaction of personal discovery. Ideas or problems are suggested after many of the instructions to open part way the doors to the many paths that lead to an understanding of the universe.

Where do you start? It really does not matter. As in exploring a new country, there is no certain place to begin or to end. Each time you travel the path, you will make new observations and arouse new curiosity. Start where your interest lies. A practical suggestion might be to build the basic pieces of science equipment that are starred in the table of contents. These are tools to build tools, and with these you are well equipped to start down the path of your choice. You will find that many paths cross again and again. It's fun to explore. Each path traveled will give you greater understanding of this adventureland we call science.

Original Introduction

Making the Classroom a Research Laboratory

The most efficient learning of our day comes in the research laboratory. It matters little whether this laboratory is sponsored by industry or is a part of an institution of higher learning. The organization that produces this efficiency in learning is much the same in both places. The research approach to learning can be applied from kindergarten through college.

Organization of the Research Laboratory

A key figure in the research laboratory is the director. This person does not know the answer to all questions and problems, but he does have a good background in the particular area in which his group is working. Because of greater experience, he can channel the efforts of his fellow workers into what is seemingly the most profitable direction.

The research team is comprised of people with various talents and abilities. Some are specialists, while others have a broader background. There are those who head up smaller teams and direct operations on a particular aspect of the larger problem. There also are those whose duties are to provide services so that the group as a whole can function more efficiently.

Organization of the Classroom Research Laboratory

The teacher occupies the role of the director in the classroom. The teacher, like the research director, does not know the answer to all questions and problems that will arise. In the case of the research director in the laboratory, it would be foolish to work on a problem to which the answer is already known. In the classroom, where the primary purpose is education, the problems need not be original so long as they are not known by the student. The teacher should not provide the answers. Rather, the answers should come from the student's research and experimentation. The teacher's role is that of giving counsel, pointing out areas that might be profitable to explore, and co-ordinating the knowledge gained by student groups so that all may share the knowledge and have greater understanding of the basic problem, or "big idea."

The class as a whole, or the teacher director, selects a major problem. An example of such a major problem might be: What is the nature of magnetism? The teacher's role in the selection of the problem is to be sure the students are properly motivated to want to find out the answers to the problem suggested. Motivation may come from other students, for instance by the bringing of magnets to the classroom. The teacher is not completely at the mercy of chance happenings, however. The good teacher knows how to arouse interest in an area that should be studied. A teacher starting a unit on magnetism might arouse interest by passing around a box containing a mystery item. Each student, by shaking the box, could try to identify the hidden object. Eventually the "discovery" would be made that the object in the box attracts pins or paper clips. Interest is then aroused to find out more about this mysterious force that penetrates the box.

Problems can come from the students. However, the director again should decide the order in which the problems are attacked and perhaps the most efficient method of attack. The director should have in mind problems whose solution will insure that the students will gain the understanding the teacher feels the students should get from the unit.

The director divides his workers into committees. A chairman is selected for each committee. This chairman meets with the director and the other chairmen regularly in order to share findings and problems that arise.

Smaller aspects of the major problem are assigned to each chairman. The groups then meet together with their individual chairmen and plan the experiment or method of attack. Students should be encouraged to attempt to seek the answer first by experimentation. If the solution is possible by experimental effort, the students should plan, set up, carry out, and finally arrive at their own conclusions. Sometimes references are needed in order to enrich upon the experiment or to provide background information so the students can plan the experiment to test their theory and hypothesis.

This is the place for the classroom science library.

Obviously, the experiments and the findings of the groups should be shared among the class as a whole. This can be done through reports, projects, classroom experimentation, artwork, stories, dramatics, and many other methods not used in other subjects.

The class should be encouraged to challenge findings and point out areas in which the findings are not valid. In other words, the students should be conditioned to the scientific method.

Chance for Pure Research

While the research laboratory as a whole attacks a major problem, there is always a specialist group carrying on research and experimentation in areas of personal interest. These areas mayor may not be related to the work of the larger group; yet from this undirected research comes many future "big problems."

The classroom should contain this feature of the research laboratory. While the group as a whole is working in one particular area, some students with a strong interest and ability in another field should be encouraged to carry on research and experimentation apart from the others. It is through this pure research group that the gifted or academically excelled youngster can realize his full potential. It is through this method that the average youngster with unusually strong interest in one field can develop his full potential and gain learning in depth.

All students should have the opportunity to work with the pure research group sometime during the year in order to gain the experience of working individually on a project of their own interest.

Laboratory Tools and Equipment

In a research laboratory many pieces of equipment are specially designed and built by technicians in order to facilitate the work of the scientist. In any classroom there are many students who are handy with their hands and can be employed in the technician role. These students can collect, design, and build up classroom laboratory equipment. This book will help those students to collect the materials that will be needed as the year progresses.

Storage will be a problem. A room in which students participate in learning is always a greater problem to keep neat and clean than a room in which the teacher imparts knowledge and the students assume the passive role.

All students should assume responsibility for the equipment they use, but the over-all job of maintenance of· supplies and equipment should fall to a technician or service group. Small kit boxes for related items (shoe boxes) make it possible to keep the jumble of odds and ends needed in the classroom and yet make them easily accessible. As much as possible, there should be a place for everything.

Reference librarians will be needed to order the books and collect magazine articles and other materials likely to be needed by the group. Files built up by these student librarians during each year thus benefit the classroom research facilities for following years. These librarians should come from students specially qualified for this role in the research laboratory. Reference material is as much a part of the tools of learning as the microscope, and as much care and effort should be given to this phase of the entire program as any other.

Selecting the Problem

At the end of each chapter is a section containing common problems whose solutions are necessary for an understanding of the whole area. Problems are listed in a developmental order. Some are designated for primary grades *(P)*. This is the grade level at which they can be introduced, but not the only level where they can be taught. Those problems designated with an *(I)* can be introduced in the intermediate grades and then retaught in the upper grades. The problems designated with a *(U)* should probably not be introduced before the seventh grade. These are challenging activities and generally require students with high abilities.

Each area of science is broken down into major problems which can be a source for a unit. Listed under the major problem are the minor problems whose investigation will shed light on the study as a whole.

Answers to Problems

There is seldom any one answer to a problem in science. If you asked the question, "What does the horizon look like?" you would find as many answers as people answering. Your view of the horizon depends on where you are standing. In the same way, your answer to a problem in science depends on what you observe and conclude.

There are no listed answers in this book as there are no listed answers for the scientist. You must make the investigations and plan the experimentation. You must do the thinking, observing, recording, and concluding. Your findings represent your answer. The findings of others represent their answer. You are free to challenge the findings of others as they are free to challenge yours. It is in this way that science progresses.

PART I

Chemistry

General Laboratory Equipment

Science Laboratory Workbench

Purpose: This is a simple science all-purpose worktable that can be easily built in any home or schoolroom. The laboratory contains a power supply, sink, and water source.

Material: Old table (or you may build one), 1" × 12" boards for shelves, two gallon jugs or large oil cans, glass tubing, rubber tubing, a bucket, and a funnel.

What to Do: Any old table can be converted into a science workbench. Cover the table with oilcloth[1] (tack or glue down) or a piece of ⅛" masonite (cost about $5.00). Cut a hole just a little smaller than the diameter of a gallon jug in the top of the table as shown. This is for your sink. Cut the bottom 3" off the jug with your bottle cutter. Smooth the edges with a file or emery paper (see "Bottle Cutter" on page 18). Insert a #6½ one-hole stopper[2] into the neck of the bottle. Connect a rubber hose to the stopper with a short piece of glass tubing. This is

your drain hose, and it should run into a bucket.

The wash bottle serves as a supply of water. The supply of water is controlled by a clothespin that serves as a stopcock, as shown on the next page.

Make your shelves and place them on top of the table. Nail them to the table through the side strips at the bottom. The shelves should have a plywood or masonite backing. The shelves should be spaced to store your science equipment, so plan the sizes with this in mind. You can store your chemicals and glass tubing in the drawer if your table has one. If not, plan your shelves so you don't waste space and yet can store materials safely.

Mount your power supply (see "AC or DC Power Supply" on page 126) under the table with the wires coming through holes in the table. Attach your Fahnestock clips[3] to the top of the table. From here you can tap off either direct or alternating current.

---

1 "Oilcloth" means linoleum in this context. Vinyl flooring is a good substitute.

2 See Appendix A for a list of material sources.

3 Electrical connectors used on the power supply.

Your worktable should be placed so you can plug your power supply into a regular outlet plug.

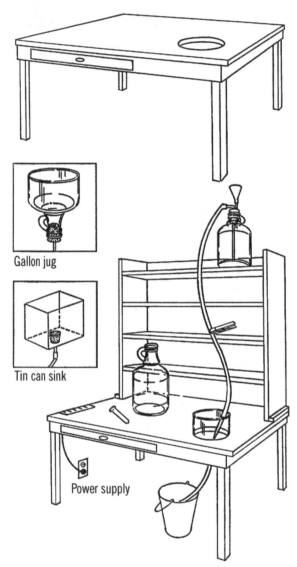

Gallon jug

Tin can sink

Power supply

Figure 1-1 Shelving: Above is a possible arrangement of shelves for your worktable. You can build the shelves with 1" × 12" pine boards as shown in the drawing.

Gravity Wash Bottle

Purpose: This provides a steady supply of water, and the container can be easily refilled.

Material: Gallon jug, #6½ stopper (two-hole), glass tubing, plastic funnel, rubber tubing, and a clothespin.

What to Do: Insert your funnel and glass tubing through one hole in the stopper (see "Thistle Tube" on page 33). Bend the other piece of tubing in your alcohol burner. Insert the long end through the rubber stopper. Slip a piece of rubber tubing over the glass tubing and use a clothespin to stop the flow of water.

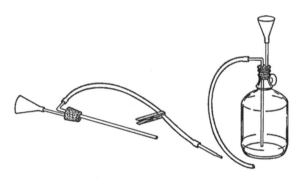

Operation of Equipment: Place the bottle in position. Fill the bottle through the funnel. Suck on the end of the rubber tubing to start the water flowing (siphon) and then clamp the tubing with a clothespin. You can also start the

Modern Safety Practice

Never "prime" (start) a siphon by mouth on anything that you do not actually intend to drink. You can easily end up with a mouthful! See "Modern Safety Practice" on page 10 for more about the hazards of using your mouth in the laboratory and "Wash Bottle" on page 28 for a solution that doesn't require you to blow into the funnel.

water to flow by blowing into the funnel. The bottle should be placed so the hose is directly over the gallon jug sink.

Light Bulb Chemistry Flask

Purpose: To make a glass flask that can be heated and can be used with stoppers.

Material: Burned-out light bulb[4] of any size or shape.

Helpful Hints for Building: Bend back the soft metal tip on the end of the bulb with a pair of pliers or your fingernail. Twist the metal piece so that it breaks off. You will see a hole in the top of the bulb. Use a pointed file or a small screwdriver to break the black material around the hole. A pair of diagonal pliers can be used to break the black substance away. After the black material is broken into pieces, turn the bulb over and shake it. Stick a file through the hole and break the wire holding the filament or center part in the bulb. Shake this out. If you have bent the edge of the top of your flask, you can make it round again by turning it on the end of any piece of round wood such as a broom handle.

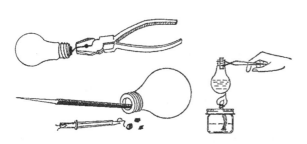

Operation of Equipment: If the light bulb is round on the bottom, use a coffee can tripod (see "Tripod and Adjustable Rings" on page 14) to support the bulb. The heated flask may be handled by a pair of tongs (see "Ring Support for Support Stand and Test Tube Holder" on

page 16). The stoppers may be purchased from any scientific supply house or can be made by drilling corks (see "X Connector" on page 37).

(see "Tripod and Adjustable Rings" on page 14), (see "Ring Support for Support Stand and Test Tube Holder" on page 16), (see "X Connector" on page 37)

Safety Tips

1. Be careful the black material in the neck of the bulb does not fly out.

2. Be careful not to work so fast that you break the bulb. Don't shove the file through the bottom of the bulb.

3. Do not try to use a fluorescent bulb. The material inside is harmful.

Modern Safety Practice

See Note 2 in Appendix E for safety guidelines when working with glass. Chiefly, wear safety glasses and take care to avoid cutting yourself with pieces of glass that may break.

Can You Work Like a Scientist?

1. What is the black material? Can you test it to see if it is an electrical conductor?

2. Is the metal tip at the top of the bulb made out of iron? How can you test this?

3. What is the difference between a bulb that is burned out and one that is not? Try both types and see.

4. What does a fluorescent bulb have in it that makes it different from a regular bulb? *Do not try to open a fluorescent bulb.*

5. Is there a type of bulb that has a wide neck? Look at floor lamp bulbs and

---

4 *Incandescent* light bulb, not fluorescent or LED.

large bulbs that are used to light class-rooms and gymnasiums.

6. Could you use a soda straw as glass tubing for your flask? If you didn't have a stopper, how would you seal the opening in the flask?

Cutting Glass Tubing

Purpose: Glass tubing comes in lengths of four feet from scientific supply houses or in longer lengths from neon sign companies. Ordinary glass tubing costs about $1 - $2 a foot. It is very helpful to be able to cut tubing to the exact length needed.

Materials: Glass tubing (any length) and a file.

What to Do: Place the piece of glass tubing on a smooth surface. Draw the sharp edge of the file across the tubing at the place you want broken. Just one firm stroke in one direction will do. Then place your thumb on the opposite side of the tubing from the mark. Press with your thumbs, and the tubing should snap easily.

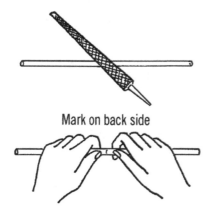

Mark on back side

Operation of Equipment: Dip the tubing in soapy water before trying to insert it into a tight hole in a rubber stopper. If you have drilled holes in a cork for your rubber stopper, seal the holes around the tubing with wax (either from a candle or paraffin).

Modern Safety Practice

See Note 2 in Appendix E for safety guidelines when working with glass.

Bending Glass Tubing

Purpose: Many experiments call for glass tubing bent in different shapes. To buy such bent tubing is very expensive. Tubing can be bent with any good source of heat such as a bunsen burner, propane torch, or alcohol lamp.

Materials: Glass tubing, source of heat (alcohol lamp, propane torch, or bunsen burner). Most alcohol lamps take much longer to bend tubing than do the other two heat sources. A propane torch is not expensive and is of great help in a science laboratory.

What to Do: Turn the heat up quite high on your torch or burner. Turn the piece of glass tubing in the flame. Slowly bring the tubing nearer the tip of the flame. As the glass heats, it bends easily. Bend the tubing to the shape you want, and then hold it in position away from the flame until the glass cools enough to set (about 10 seconds[5]). In order to bend perfect curves, it is necessary to put an attachment on the burner so that the flame will reach two or three inches of the tubing at the same time.

---

5 Careful! It is not *cool enough to touch* after 10 seconds!

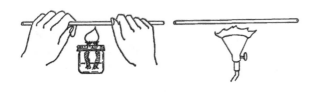

Safety Tips

1. Be very careful of the open flame. Make sure the torch is firmly supported.

2. Don't touch the part of the tubing that was heated. Glass will retain heat for many minutes.

Modern Safety Practice

In addition to working with room temperature glass (Note 2 in Appendix E), this project also involves open flames and hot glass.

See Note 3 in Appendix E for basic safety practice when working with open flames.

See Note 4 in Appendix E for basic safety practice when working with hot glass.

Graduated Cylinder and Chemistry Flask

Purpose: A graduated cylinder[6] is used to measure accurately amounts of a liquid. It has many purposes in chemistry, such as measuring amounts of chemicals and figuring volumes of solids, such as rocks.

Materials: Baby bottle[7] or other straight-sided bottle, medicine dropper with measuring marks.

What to Do: The baby bottle is a wonderful source of science equipment. It is calibrated in cubic centimeters and also in ounces. The standard size is 240 cc. It can be heated with the alcohol burner, and when used with a size #6½ stopper, it makes an excellent flask. Also, because of its shape, it makes a very good test tube.

You can make your own graduated cylinder by using a narrow straight-sided bottle and a medicine dropper. Place a piece of tape down one side of the bottle. The medicine dropper is figured in cubic centimeters. Fill your bottle with water by using the medicine dropper. Add up the number of cc's as you go along and mark these on the piece of tape you are using for a scale.

Can You Work Like a Scientist?

1. Fill your graduated cylinder half full. Note the reading on your scale. Now drop a small rock into the bottle. Is the level of water higher or lower? What did you add to the water besides the rock?

2. If all you did was add the rock to the water, could you find the volume (how much space it fills) of the rock from the height of the water in the graduated cylinder?

---

6 Originally "graduate" here. This terminology has fallen out of favor. Use the term "graduated cylinder" for clarity.

7 See Note 5 in Appendix E about baby bottles.

3. Take rocks of almost the same size. Can you predict accurately which is larger? Measure them in the graduated cylinder. Were you right in your prediction?

4. Can you get two rocks that weigh the same? (Use your gram scale.) Will two rocks that weigh the same always have the same volume?

5. If a rock weighed 10 grams and its volume measured in your graduated cylinder 50 cc, what is its weight per cc? Could you change this fraction into a decimal?

6. Try other types of rocks and figure their weight per cubic centimeter. Do the same kinds of rocks always have the same weight per cubic centimeter?

7. Could you test rocks and classify them as to type by finding their weight per cubic centimeter?

8. How does temperature affect the rate of evaporation? Can you use your graduated cylinder to measure this?

9. How could you find the volume of some object that cannot be placed in water?

10. By using glass tubing, a two-hole rubber stopper, and your graduated cylinder, could you measure the pressure of gases such as the pressure of air in a balloon?

Alcohol Burner

Purpose: To provide a safe source of dependable heat for most experiments and heat for bending and sealing glass tubing.

Materials: Small bottle with a metal lid (ink, salad dressing, or similar bottle) and a piece of clothesline rope[8] about 4 inches long.

What to Do: Punch a small hole in the lid of the jar. The hole should be small enough that you have to force the end of the clothesline rope through the hole. Force the rope through the top side of the lid. Have about ½" of rope stick through the hole. Fill the bottle with alcohol and screw on the lid.

Operation of Equipment: The best source of fluid for your burner is rubbing alcohol which is found in most homes and drugstores.[9] For school use, the alcohol used in the duplicator (ditto machine) found in the school office[10] is an excellent fuel. The clothesline rope should be the type that has many soft cotton fibers on the inside. To start your burner, turn the bottle upside down until the part of the wick outside the bottle is moist. Light the burner with a match.

Can You Work Like a Scientist?

1. What happens if you place a few drops of water in a jar lid and let these drops set for a few hours? Could you time

---

8 Use pure cotton, *not plastic* (e.g., polyester), rope for this application. If you happen to have access to it, fiberglass wick material is a better choice than cotton.

9 Denatured alcohol is generally a better choice, and "rubbing alcohol" can mean different things. See Note 6 in Appendix E for more information.

10 *Formerly* found in an office! See Note 7 in Appendix E about ditto fluid.

Safety Tips

1. It is dangerous to light the burner if the wick does not fit tightly in the hole.

2. Matches should not be used without proper instruction by parents or teacher. Younger students should NEVER use them unless an adult is present.

3. Don't get alcohol on the outside of the bottle, for when you light the wick, this alcohol will burn.

4. Be careful of clothes and your skin when working with this heat source.

5. Never have paper and dry rags near the flame.

6. Don't spill alcohol on the floor. Alcohol will discolor floor tile.

Modern Safety Practice

1. Read Note 3 in Appendix E about working with open flames.

2. Take very seriously the risk of alcohol burning on your skin or clothing. Do not lift or tilt an alcohol burner once it is lit.

3. Have a plan for putting out the burner before you light it. The flame can generally be blown out, but you may want a candle snuffer on hand as well.

about how long it takes for the water to evaporate?

2. What happens if you place the same amount of alcohol in the jar lid?

3. Wet the back of your hand with water. Now blow over the surface. Do the same with alcohol. What is the difference?

4. What would happen to your alcohol if you let your burner sit for a few days?

5. Could you use another lid to solve this problem?

6. Fill a bottle with water. Place an empty jar alongside. Place one end of a piece of cloth in the bottle with water and the other end in the empty bottle. Make sure the cloth touches the bottoms of the jars. Let this set overnight. What happens? Does this explain how the alcohol burner works?

7. Cut the end of a daffodil stem. Place the stem in colored water. Use food coloring or ink. Let the flower soak in the water for several days. Is this the way plants and trees get water from the soil?

Broad Flame Alcohol Burner

Purpose: This alcohol burner provides a broad flame with more heat so that glass tubing can be bent easily into all shapes. This flame enables students to do their own glass blowing.

Materials: Wide-mouth pint jar and lid, two wicks about 5" long (clothesline rope[11] with soft cotton inside).

What to Do: Punch two holes in the lid about one inch apart. These holes should be just a little smaller than the rope wick in diameter. Put the wicks through the holes as shown with about an inch of wick sticking out. Fill your burner with rubbing alcohol or ditto fluid. Screw on the lid.

Operation of Equipment: Let the alcohol climb the rope wick (capillary action). Bend the wicks together so they will make one broad flame and light them with a match. Hold your

---

11 Again: pure cotton, *not plastic*, rope.

glass tubing in the flame, and when it becomes soft, bend it gradually into any shape.

Modern Safety Practice

Read the safety notes on the alcohol burner in the previous section; they all apply equally well here.

Pipette

Purpose: A pipette is used to pick out from a culture and other liquids small plants, insects, and micro-organisms. The pipette is used in chemistry to select and move small amounts of liquid chemicals as well as pick out solid materials from a liquid.

What to Do: Heat the glass tubing over the alcohol burner. When the glass is soft, pull and stretch it. Break it in the middle once cool.

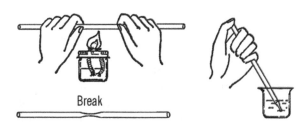

Break

Operation of Equipment: Hold your thumb over the large end of the tube. Move the small end of the tube over the object you want to remove from the water. Lift your thumb off. The liquid will rush into the tube. You can control the amount with your thumb. Why does the liquid go up the tube?

Modern Safety Practice

1. This project involves cutting (cold) glass, open flames, and hot glass. Follow the safety guidelines in Note 2, Note 3, and Note 4 of Appendix E, respectively.

2. The back edge of of your pipette (where you put your thumb) may be sharp. You can correct this by heating the back end over the flame until the glass there softens and then allow it to cool.

Mouth Pipette

Purpose: The mouth pipette serves the same purpose as the hand pipette except that the amount of liquid is controlled by a rubber tube in your mouth instead of by your thumb. You have better control with the mouth pipette.

What to Do: Make your pipette the same as above. Attach a piece of rubber tubing to the large end of the glass tubing. In order to use, hold your tongue over the end of the rubber. When you have located your micro-organism, remove your tongue from the opening. You may bring liquid into the tube with a slight sucking.

Blowtorch Type of Alcohol Burner

Purpose: This type of burner provides a broad flame with a large amount of heat. It is very useful for glass blowing.

What to Do: Make a broad flame alcohol burner. Use your burner to make a long pipette by stretching the glass tubing.

Operation of Equipment: Start your broad flame burner. Hold the glass tubing in your mouth. Have the small end of the tubing near the flame. Blow with a steady pressure. The stream of air will turn your burner into a blowtorch.

Safety Tips

1. Don't touch hot glass. It burns. Glass takes a long time to cool.

2. Don't inhale on the glass tubing. You will suck up hot gas.

Can You Work Like a Scientist?

1. What part of the flame is the hottest? How can you test this?

2. Why does the fine pipette tube work better than just a plain piece of glass tubing?

3. Can you think of a way you can have a steady source of air without blowing? Such a thing would be a compressor.

4. A clothespin fastened to the side of the jar with a rubber band can serve as a holder for your glass tubing. You can blow through a piece of rubber tubing attached to the glass tubing.

Large Pipette-Glass Blowing

Purpose: With this pipette you can move a large amount of water from a large container. You can select the water from any part of the liquid. You can use the glass-blowing ideas to make many different pieces of glassware.

Material: Glass tubing about two feet long (less if you desire).

What to Do: Heat the middle of the glass tubing with your broad flame burner. Be sure to turn the tubing as you heat it. Remove the tubing from the flame and hold your finger over the end. Blow into the tubing. The glass should bulge out at the spot that is heated. Heat the tube again near one end. Pull the heated end into a fine tube. Break the tube at this spot after the glass has cooled.

Operation of Equipment: Hold the pipette in the liquid. Keep your thumb over the end. To fill, remove your thumb. Blow on one end to empty.[12]

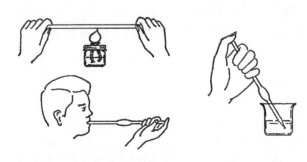

Modern Safety Practice

1. This project involves cutting (cold) glass, open flames, and hot glass. Follow the safety guidelines in Note 2, Note 3, and Note 4 of Appendix E, respectively.

Can You Work Like a Scientist?

1. Try blowing other pieces of equipment. *Remember… don't inhale or touch the hot glass.*

2. Can you find out how glass bottles are made?

Burette Clamp and Test Tube Holder

Purpose: This clamp is used in chemistry. It is attached to a ring stand by one clamp while a second clamp is used to hold a test tube or other small objects.

Materials: Two snap clothespins, tape.

What to Do: Slide one leg of a clothespin into the wire clamp of the second clothespin. Fasten them together by wrapping with tape. One leg of the second pin slides into the end of the cur-

---

12 Blowing on the end is an example of pipetting by mouth, and it should be strictly avoided. See "**Modern Safety Practice**" on page 10.

tain rod ring stand[13]. When the free leg of the first pin is pressed, the clamp opens for a test tube or other small object. This clamp will also hold slides, cardboard, and diffraction gratings. The clamp by itself is a good test tube holder.

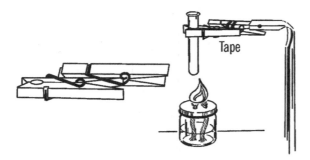

Mason Jar Chemistry Flask

Purpose: Mason jars are made of Pyrex glass and can be heated over alcohol burners. They make an excellent large-size flask. The normal sizes of quart, pint, and half-pint offer a good range for all purposes. They have a second advantage in that they are easier to clean than regular flasks.

Materials: Jar lid for mason jar, drill or short piece of pipe (¾" or 1"×3"), file.

What to Do: Drill a hole in the center of the jar lid large enough for a rubber stopper. This could be through the regular canning lid or through a mayonnaise lid.[14] In either case, drill from the outside in order to keep the outside smooth. If you don't have the right size drill, use a short piece of ¾" or 1" pipe. File or grind one end of the pipe until it is quite sharp. Place the jar lid on a solid piece of wood and cut the hole by pounding the pipe with a hammer as shown.

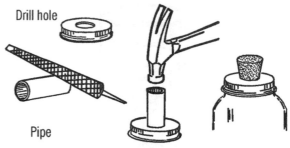

Operation of Equipment: In order to use a mason jar for a flask, tighten the lid on the jar. Insert the rubber stopper.

Funnel

Purpose: A funnel is used for pouring mixtures and for many science experiments.

Material: Top of a gallon jug or the top of any bottle that has a small opening.

What to Do: Cut off the top of the jug with the nichrome wire bottle cutter ("Bottle Cutter" on page 18). Smooth with wet or dry emery paper under water. You can make a smaller opening by using a rubber stopper and a piece of glass tubing.

---

13 See "Support Stand" on page 16.

14 Use a *metal*, not plastic lid.

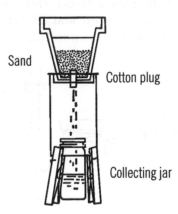

tom. Set the pot over the tin can tripod.[15] Use a collection bottle underneath the cotton plug.

Sand

Cotton plug

Collecting jar

String Filter

Purpose: To filter a liquid without filter paper.

Materials: Two-foot length of cotton rope, two bottles.

What to Do: Place the bottles as shown. Put one end of the rope in the liquid to be filtered. Put the other end in the collecting bottle. This process will take several hours. Why does the liquid climb the rope?

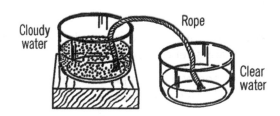

Cloudy water

Rope

Clear water

Plant Pot Filter

Purpose: To strain or filter out materials from a liquid.

Materials: Clay planter pot, sand, cotton plug.

What to Do: Place a cotton plug in the bottom of the pot. Use a few inches of sand in the bot-

Can You Work Like a Scientist?

1. Filters strain out bits of materials suspended in liquids. Try to make such suspensions with liquids and chemicals or other materials. Try your filters out and see if you can strain out the bits of materials in the liquid.

2. Could you make a filter out of your gallon jug funnel?

3. Could you make a gravel, sand, and soil filter by putting small amounts of these in the small-necked bottle funnel and using a cotton plug?

Tripod and Adjustable Rings

Purpose: The tripod is used to hold chemistry flasks or bottles while they are heated by a gas or alcohol burner. The adjustable rings are used to adapt the diameter (distance across opening) to the size of the bottle or flask.

---

15 That is, the tripod described in the next project.

Materials: Tin cans (sizes 300 and 303),[16] Kerr screw lid rings (wide mouth, standard, and "63" [small] sizes[17]).

What to Do: Cut the top and bottom off the cans with a can opener. Place the can on a block of wood as shown and punch holes with a large nail about one inch from one end of the can. You should have at least six. Then take a screwdriver or file and enlarge the holes. Be sure to drive the nails in from the outside of the can as shown.

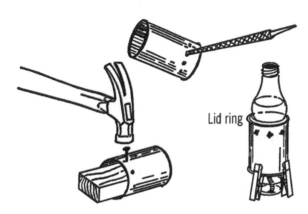

Lid ring

Operation of Equipment: The large ring fits the larger 303 can. The standard ring just fits the size 300 can. (It can slip partly into the can.) The "63" size ring will fit into the standard ring for small bottles. The tin can tripod fits over the alcohol burner. A burner made out of a small bottle is best. The sides of the tin can should be raised by three clothespins as shown. The tip of the wick should be at least two inches below the top of the can. The tin can tripod works perfectly for light bulb flask. The small size bulb fits in the 300 size can. The larger 150 watt bulbs fit the 303 size can and rings. A pound coffee can makes an excellent large tripod.

Safety Tips

1. Always be careful when using the alcohol burner.

2. Remove the wrapper from the can so the paper won't catch on fire.

3. Don't touch the tin can tripod while heating with the burner. The can is very hot. However, the bottom of the can is usually only warm.

4. Use a hot-pad holder or a heavy rag to remove the flask or object from the tripod. Don't touch the bottle until it cools. You may want to use tweezers. The baby bottle tweezers sold at most variety stores are ideal.[18]

Modern Safety Practice

1. When hammering the nails and enlarging the holes, be sure to wear safety glasses and be careful with the tools. Deburr (smooth out) all of the cut metal surfaces to avoid sharp edges.

2. Since wooden clothespins could potentially catch fire, be sure that you are working on a non-flammable surface, and be doubly careful about other flammable materials in your vicinity. Can you design an alternate version made from non-flammable materials, perhaps from steel wire?

Can You Work Like a Scientist?

1. Try not punching holes in the can and see what happens.

---

16 See Note 8 in Appendix E for more about tin cans and their sizes.

17 #63 jars are no longer made. See Note 9 in Appendix E.

18 "Tongs" not tweezers, and they are no longer this common. Either the coat-hanger ring support described next or a test tube clamp ($1–$3 online) may be a better choice.

2. Turn the can over and put the holes at the bottom. What happens?

3. Hold a match near the holes and at the bottom of the can. Can you tell in which direction the air is moving?

4. Why is the can warmer at the top than at the bottom?

5. Put a little water in the bottom of the bulb. Look through the bottom of the bulb directly up at a light. Is this a type of lens?

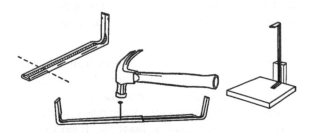

Support Stand

Purpose: The support stand is used to support test tubes or flasks for many experiments.

Materials: Curtain rod,[19] block of wood for base, block of wood for support.

What to Do: Shorten the curtain rods by cutting off part of the straight section with a hack saw. The remaining rods should each be about a foot long. Slide the two pieces together. Turn the rods over and slide them so they are just joined. Take a hammer and a nail and punch a dimple in the metal from the inside. Don't go all the way through. Now slide the two rods together half an inch at a time. Each time lightly tap the same hole on the inside rod. You should have only one dimple on the inside rod and many on the outside. Support the stand by nailing one end to the base and using a block of wood for an upright support.

Operation of Equipment: In order to increase the height of the support stand, pull the rods apart until the dimples in each rod match at the desired height. Fasten the flask or test tube to the stand by means of a clamp as described below.

Modern Safety Practice

Wear safety glasses and be careful with sharp edges.

Ring Support for Support Stand and Test Tube Holder

Purpose: The ring support is used to hold flasks and test tubes. It is attached to the support stand and adjusts on the stand for the desired height.

Materials: Coat hanger[20], pliers.

What to Do: Cut the end off the coat hanger. The piece should be about four inches long. Bend the end with a pair of pliers so it is about the same shape and size of the object (light bulb flask, etc.) to be held. The two end pieces should be spread as shown.

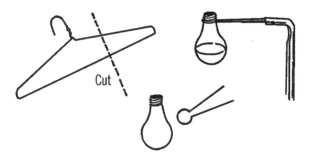

Cut

---

19 Specifically, a "lockseam" type curtain rod.

20 A solid wire type coat hanger, not one made of plastic. If you don't have coat hangers like this, obtain steel "music wire" rod or brass rod from a hardware or home improvement store, roughly 3/32" in diameter.

Operation of Equipment: Slip the flask into your clamp. Squeeze the ends of the support clamp together and insert into the end of the curtain rod support stand. The clamp used by itself is a test tube holder.

Bottle Etcher

Purpose: The bottle etcher works with your nichrome wire bottle cutter. It etches, or scratches, a straight mark all the way around the bottle or gallon jug so that the bottle or jug will break in a smooth and even manner.

Materials: Glass cutter (costs about 3 dollars in a hardware store), wood to make the base as shown, and two wood screws.

What to Do: Nail the two supporting pieces of wood, about 4½" wide, to the base. Cut a notch about ⅛" wide and ½" deep in one of the supporting pieces. Tip: the width of a power saw blade is about the right size. This notch should be about 6½" above the base. Insert the end of your glass cutter as shown. Hold the glass cutter in place by a strip of wood or metal and two screws.

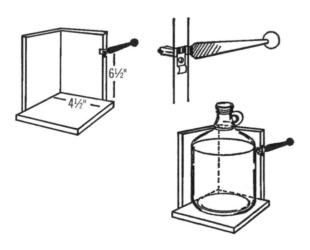

Operation of Equipment: After removing the label, hold the jug firmly against the supporting piece and the base and slowly turn it. The cutter will make a firm scratch on the jug all the

way around. Don't go over the scratch mark again as this dulls the cutter. Always turn the bottle counterclockwise so you don't dull the cutter. Now cut the jug with your nichrome wire cutter.

Can You Work Like a Scientist?

You can use your bottle etcher to cut gallon jugs at any height. The bottom makes excellent shallow dishes. Can you think of a way to change the height of the cutter without removing it from the board?

Glass Cutter

Purpose: Your glass cutter will cut old panes of glass to any desired size. From this glass you can make any number of pieces of science equipment.

Materials: Glass cutter, plastic ruler, and pane of glass.

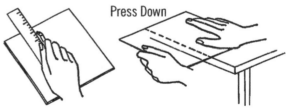

Press Down

What to Do: Lay the glass on some newspaper on a smooth table. Lay your ruler on the glass where you want to cut it. Draw the glass cutter toward you, using the ruler as a straight edge. Make one firm hard line. Don't go back and forth, as it dulls the cutter. Place the cut mark over the edge of the table with the mark up. Hold the glass down firmly and press. The glass should break on the mark. In case the glass is very thick, tap it gently with the ball at the other end of the cutter. This should help the glass break along the line. Any spots that don't break smoothly can be broken off by using the notch on the glass cutter. You can smooth the edges to a fine finish by using wet or dry emery paper. Place the glass in a bucket of water and rub the edges with the emery paper. The water keeps the glass from flying.

Modern Safety Practice

The edges of a pane of glass—before and after you cut it—are very sharp. Handle with extreme caution. Use tough (welding or gardening) gloves to protect your hands when handling large or heavy pieces.

See Note 2 in Appendix E for additional safety guidelines.

Bottle Cutter

Purpose: The bottle cutter is a valuable tool in your science laboratory. It can be used to cut the tops or bottoms off any size bottle. Many pieces of equipment can then be made from the parts of the bottle or jar.

Materials: Wood to make U-shaped frame as shown in drawing, two bolts with nuts and washers, piece of nichrome wire (size 20)[21] about 18 inches in length, and a six-foot extension cord. Nichrome wire may be purchased from any electrical shop or supply house.

What to Do: Build the U-frame. Drill two holes for the bolts as shown. Put the bolts through the holes, twist the ends of the nichrome wire around the heads of the bolts, then tighten the nuts. Connect one wire of the extension cord to the bolt. The other end of the cord connects on the salt water rheostat (see "Salt Water Rheostat" on page 19).

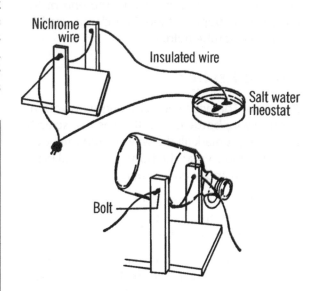

Operation of Equipment: After the bottle is scratched (etched by glass cutter—see "Glass Cutter" on page 17), plug in the cutter and then add salt to the water in the rheostat. Slowly move the two wires in the rheostat closer together until the nichrome wire glows brightly. Your cutter is now ready. Hold the bottle as shown so that it touches the wire along the scratched line. Turn the bottle slowly as the glass pings. The top or bottom should come off in less than a minute. Smooth the edge by rubbing it with wet or dry emery paper under water.

Can You Work Like a Scientist?

1. Why did you use nichrome wire? Try other wire to see what will happen. Be careful in doing this.

2. Why did you use a salt water rheostat? Why didn't you connect up the cutter

21 "20 gauge"; see Appendix A for sourcing information.

Safety Tips

1. Don't touch the hot nichrome wire.

2. Don't touch the bare wires in the rheostat. You will get a severe shock.

3. Younger students: don't try your bottle cutter out unless your teacher or parents are with you.

4. Be careful not to cut yourself on the glass. Be sure to smooth the edge under water.

Modern Safety Practice

This type of bottle cutter can be a highly effective tool. It also presents very real hazards that you need to take seriously and be prepared for.

1. Strictly heed the warnings about touching the exposed wiring: a shock from mains power (AKA line voltage or household wiring) is potentially lethal. Read Note 10 in Appendix E for safety practice around it.

2. The red-hot nichrome wire is capable of causing severe burns or starting a fire. Follow the same set of fire-safety procedures that you would around an open flame (Note 3 in Appendix E).

3. You will also be working with hot (and cold) glass. Hot glass can easily burn you, and cold glass can shatter or cut you. See Note 2 and Note 4 in Appendix E.

4. The salt water rheostat presents its own unique hazards that you need to read about in the next project (see "Modern Safety Practice" on page 20). If you plan to cut bottles on a regular basis, it would be prudent to consider a safer and more permanent solution, such as the "isolated variac" described there.

directly with house current? Be careful here. Your wire may snap.

3. Why did you put salt in the rheostat?

4. What else might work in the rheostat besides salt? How about milk?

5. What would happen if the two wires in the rheostat touched?

6. What causes the water to get hot when electricity passes through it?

Salt Water Rheostat

Purpose: A rheostat changes the resistance to the flow of electricity and thus controls the amount of electricity. The salt water rheostat is a convenient way of reducing the normal house current (117 volts) to a much smaller current to operate such equipment as the bottle cutter and the carbon arc furnace.[22]

Materials: Pyrex pie dish or gallon jug aquarium, two fishing sinkers, extension cord, salt.

What to Do: Cut the socket end off the extension cord. Divide the wires. Remove the insulation from the end of each wire. One wire goes directly to such equipment as a bottle cutter. A wire then goes from the piece of electrical equipment to one of the fishing sinkers. The sinker is placed in the dish, and the dish is filled with water. The electricity passes through the water to the second fishing sinker. A wire is

---

22 For the carbon arc furnace in particular, see Note 11 in Appendix E for an alternative.

then attached to the second sinker, and the wire goes back to the plug.

Safety Tips

1. The water and the dish get quite hot. Place a plywood board under the hot dish so that the dish won't burn the counter.

2. Don't touch bare wires. Don't reach into the water or in any way touch the sinkers. You will receive an electrical shock.

3. Never work with electricity near a sink or water pipe. A wet basement floor is just as bad. When there is a direct path through you to ground, electricity can travel through your body, causing a severe shock.[23]

4. Young experimenters: Never work with the rheostat unless your parents or teacher are present and give you permission.

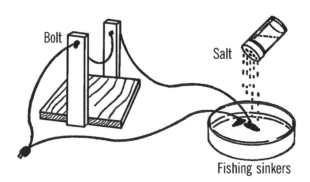

Bolt

Salt

Fishing sinkers

Operation of Equipment: Wire the bottle cutter or carbon arc as shown. Fill the dish with water. Plug the rheostat into house current.

Modern Safety Practice

A salt water rheostat is a classic and functional type of variable resistor. While it is possible to build and use the salt water rheostat safely, it does leave a lot of wiring exposed at line voltage, not to mention a bowl of salt water that can be knocked over. There is no room for error; mistakes with mains voltage can be deadly. Read Note 10 in Appendix E for safety practice around exposed wiring.

There is some value in (carefully) building and testing the salt water rheostat for its own sake, but there is less value in actually using it as part of other projects. For example, if you are building the carbon arc furnace or a bottle cutter, it is likely wiser not to have the added complication of the rheostat's exposed wiring.

In most modern applications of this sort (where you desire a variable amount of voltage from the wall plug), a neatly packaged variable autotransformer or "variac" is the right tool for the job. From a safety standpoint, it is also highly desirable to use an isolation transformer to "dereference" the AC with respect to ground. The best of both worlds is an isolated variable ac power supply or "isolated variac," which combines the two elements, usually along with a fuse (for additional protection). Perhaps it goes without saying, but if you do wish to build and test the salt water rheostat, the safest way to do so would be to power it through an isolated variac.

You will notice that plain water doesn't carry electricity very well. Slowly add salt to the liquid and move the fishing sinkers about two inches apart. *CAUTION: Do NOT touch the bare wires or the sinkers.* Touch only the insulated wires. As you add salt, you will notice the water start to bubble as the electricity moves through

---

23 The original phrasing here was "In the event of a shock, you are perfectly grounded, and the electricity will pass through you." A reminder that "perfectly grounded" is not always a good thing!

the water. The salt is dissolving into the water, and the salt water is becoming electrically charged (ions are formed). The more salt that is dissolved, the more ions to carry the electricity. Now to vary the electricity, move the sinkers either farther apart or closer together. *CAUTION: Do NOT let the two sinkers touch.* This is a direct short, and the full voltage will go through your piece of equipment and burn it out.

Carbon Arc Furnace

Purpose: The arc furnace is a source of brilliant light and produces a very high temperature at the tips of the carbon rod. The temperature is so great that you can melt some metals.

Materials: Wood as shown, a brick, a clay flower pot, two old flashlight batteries,[24] a round curtain rod, and an extension cord.

What to Do: Cut open an old flashlight battery. Cut off both ends with a hack saw. The paste inside contains acid, so wash thoroughly and don't get any on your clothes or in your eyes.[25] There is a black rod in the center of the battery. This is a carbon rod. You need two of these rods.

Build a wooden frame as shown. Drill a hole in each upright just large enough for the round curtain rod. The holes should be high enough to strike the middle of the clay pot when the pot is sitting on the brick. Cut the curtain rod in half. Insert a carbon rod in one end of each half of the curtain rod. The ends of the curtain rods should be crushed with a pair of pliers so firm contact is made with the carbon rods. The cur-

tain rods are then slipped through the uprights. Work a hole in each side of the clay pot with a screwdriver or file. The holes should be large enough for the curtain rod sections. Now slip the carbon rods through the holes and into the pot. Wire as shown. Be sure to tape the curtain rods near the uprights.

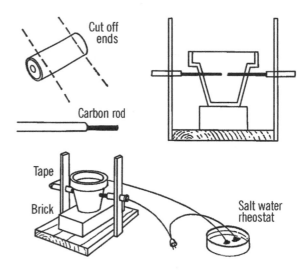

Operation of Equipment: Check your wiring completely before you plug in the extension cord. Be sure your carbon arc furnace goes through a rheostat.[26] *Do not look directly at the ends of the carbon rods. The light might blind you.* Use a good pair of dark glasses when working with your arc furnace.[27] The rods should be spaced so they are separated by about a quarter of an inch. Move (bend) the uprights together. When the rods touch, sparks will fly. Slowly move the rods away from each other, and you can strike a good arc. If you cover your pot, you have a furnace. In order to melt

---

24 A specific type of battery—one with a carbon rod—is needed. Please see Note 12 in Appendix E. You can also purchase carbon rods alone; see Appendix A.

25 Don't touch this material directly. Wear protective rubber gloves while cleaning off the rods and wash your hands thoroughly with soap and water when you are done. Read Note 17 in Appendix E for additional discussion about safety around chemicals.

26 See the safety notes in the previous section and Note 11 in Appendix E about safety when using any alternatives to the salt water rheostat.

27 The carbon arc furnace is as bright as the sun: Use a welding shade approved for carbon arc, *not* just a set of sunglasses. See the *Modern Safety Practice* notes on this project and also Note 13 in Appendix E.

metals, say, a nail, either place directly in the arc, or place in a ceramic dish in the furnace. You can operate the arc furnace without the pot. However, no one in the room should look at the arc.

Safety Tips

1. Don't touch the curtain rods while the furnace is plugged in. Always take care when working with the salt water rheostat.

2. Younger students: never use the arc furnace unless your teacher or parent is present and gives permission. Never work near water pipes, a sink, or on a damp basement floor.

3. Protect your eyes by wearing dark glasses and touch only wood—NOT the rods, pot, wire, curtain rods, or salt water rheostat.

Modern Safety Practice

With the exception of having blinding light and an electric arc instead of hot glass and a red-hot wire, the hazards presented in this project are essentially the same as in the bottle cutter project. The safety guidance there ("**Modern Safety Practice**" on page 18) about wires, fire safety, having another person present, and working with the rheostat still applies.

It is imperative to protect your eyes from the bright-as-the-sun light emitted by the carbon arc. Neither dark sunglasses nor most welding helmets are sufficient protection; you need a welding shade designed specifically for carbon arcs; what is called a "#14" or darker shade. Note 13 in Appendix E discusses eye safety when directly looking at the sun (which is equally demanding) and some alternative types of solar filters.

Adjustable Glass Bottle Etcher

Purpose: To make a bottle etcher that will fit any size bottle.

Materials: Glass cutter, wood for base, two pieces of ¾" plywood for side pieces.

What to Do: Cut ⅛" notches along the edge of one of the side pieces. The notches should be about ¼" deep. These notches can be cut with a hand saw, but a power saw is ideal. The width of the blade of a power saw is just the right width for the notches.

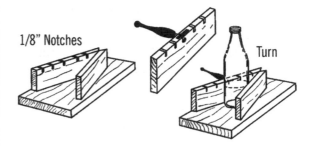

Nail the two side pieces to the base in the form of a V. The glass cutter is then inserted into the proper slot according to the size of the bottle. The cutter should just touch the side of the bottle when the bottle is held firmly against the wood sides. The cutter can be held in position by a short piece of metal, such as tin. The metal is fastened over the cutter and held in place by screws.

Operation of Equipment: Place the bottle firmly against the wood sides and turn the bottle steadily. The cutter will etch a straight line around the bottle. The height of the cut can be varied by placing blocks of wood underneath the bottle. The bottle is cut by placing the scratch or etch mark over a hot nichrome wire (see "Bottle Cutter" on page 18).

Added Suggestion: A second type of adjustable bottle cutter can be made by making a series of bottle cutter boards. The width of the side pieces should vary, thus allowing smaller bottles to be cut. The height can again be varied by using blocks of wood. You can use the same base, and place different width side pieces around the four corners of the base.

Modern Safety Practice

See Note 2 in Appendix E for safety guidelines when etching and cutting glass.

Test Tubes

Purpose: A test tube is used for carrying on chemical experiments, growing bacteria and simple plants, air and water experiments, and experiments with heat. The tube should be made of material that can be heated.

Materials: Burned-out light bulbs (the clear type is best).

Helpful Hints for Building: Remove the element from inside the bulb. (See "Light Bulb Chemistry Flask" on page 5.)

Additional Suggestions: Plastic toothbrush containers, olive or cherry bottles, and perfume vials can serve as test tubes. However, they cannot be heated. Baby bottles[28] are a good substitute for test tubes and they can be heated! Regular test tubes can be bought at any scientific supply house for about fifty cents each. Stoppers can be made from corks by drilling holes or may be bought for about one dollar each.

Test Tube Racks

Purpose: The rack is used to hold unused test tubes or to hold groups of test tubes for experiments, such as growing molds.

Materials: A piece of 2×4 for type A, or pieces of wood for type B. In both cases a drill is needed that is a little larger than the tube.

Helpful Hints for Building: For type A, drill holes in a block of 2×4. If you feel the holes should be deeper, nail two 2×4's together. For type B, nail the boards together as shown and then drill the holes through the top board and almost through the bottom board.

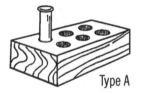
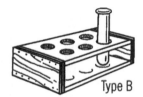

Type A   Type B

Can You Work Like a Scientist?

Can you make a test tube holder for light bulb test tubes? Remember the rack should be able to hold both empty and full light bulb test tubes.

Retort and Liebig Condenser

Purpose: A retort is used for distilling water and other liquids. "Distilling" is heating a liquid until it boils and changes into vapor. Then the vapor condenses (cools and turns back into liquid) and is collected. This process is a means of separating (decomposing) substances by heat.

Materials: Two-hole size #1 rubber stopper, burned-out 150-watt light bulb, and a piece of glass tubing about 16 to 24 inches long.

What to Do: Take the inside out of the light bulb (see "Light Bulb Chemistry Flask" on page

---

28 *Glass* baby bottles. See Note 5 in Appendix E.

5). Bend one end of the glass tubing with your alcohol burner. Cut the bend off by using the edge of your file (see "Cutting Glass Tubing" on page 6). This piece needs to be only two or three inches long. Bend the rest of the glass tubing as shown. Insert both pieces into the light bulb with the short piece pointing up when the light bulb is lying on its side. Slip a small piece of rubber tubing over the open end. Clamp the tubing with a clothespin or a plug made from glass tubing.

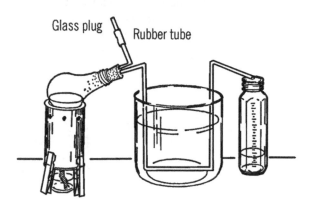

Glass plug   Rubber tube

Operation of Equipment: Place the size 303 tripod over the alcohol burner. Put some water or other liquid in the light bulb. Insert the rubber stopper. Set the light bulb on the tripod so that it leans to the side as shown. The glass tubing should rest inside a large bowl or gallon jug aquarium (see "Aquarium" on page 230). The bowl or aquarium is filled with cold water. A collecting graduated cylinder (baby bottle) can be used to measure the amount collected. If you want to fill the retort while heating the liquid, remove the clothespin clamp or plug and add the liquid either by a funnel or a siphon connected onto the hose by a piece of glass tubing.

Safety Tips

1. Steam burns like hot water. You are using both when you distill water.

Modern Safety Practice

Read Note 2, Note 3, and Note 4 in Appendix E for safety practice when cutting glass, working with open flames, and working with hot glass, respectively.

Can You Work Like a Scientist?

1. Why do you run the glass tubing through cold water? Why does the vapor turn back to a liquid? What happens to the water in the aquarium after you have used the retort for a while? Can you change the water with a siphon?

2. Can you distill muddy water? How about salt water? What effect does placing ice cubes in the aquarium water have on the rate of collecting?

3. Place food coloring or ink in the water in the bowl. Does the color go with the vapor?

4. Try distilling other liquids, such as milk, sugar water, or a mixture of salad oil and water. Which will distill first? Can you separate these two liquids by regulating the temperature?

5. Mix different chemicals in the water. Can you regain the chemicals by distilling? Do any of the chemicals evaporate before the water boils?

6. Can you think of any way to take the temperature of the water in the light bulb? If you could, you could tell the temperature at which the different substances boil and evaporate.

Distillation Condenser

Purpose: A distillation condenser is used in the process of heating a liquid, changing it into a vapor, cooling the vapor, and changing the vapor back into a liquid.

Materials: Two cans as shown, two rubber stoppers (one-hole, size #1), glass tubing, rubber tubing, 6″ piece of copper tubing, alcohol burner, bottle.

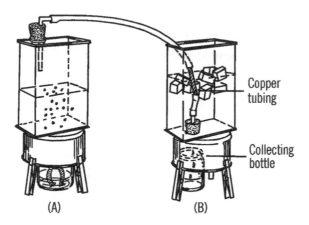

(A)   (B)

What to Do: Bend a short piece of glass tubing and insert into a rubber stopper (A). Attach to the glass tubing a piece of rubber tubing. Cut the bottom out of can (B). Turn the can over and insert a rubber stopper from the inside of the can. Attached to the rubber stopper should be a piece of glass tubing on the inside of the can and a short piece of glass tubing pulled into a nozzle and inserted into the rubber stopper from the outside of the can.

A short piece of copper tubing is used in the can to condense the vapor. The copper tubing is connected to can (A) by the long piece of rubber tubing. It is connected to the stopper in can (B) by a short piece of rubber tubing.

Operation of Equipment: The liquid to be distilled is placed in can (A). The rubber stopper is inserted in the can. As the liquid is heated by an alcohol burner, the vapor rises and goes through the rubber tubing into can (B). The liquid in can (B) plus the ice cubes cool the copper tubing. As the vapor passes through the copper tubing it condenses and drips through the glass nozzle into a collecting bottle.

Modern Safety Practice

1. Follow the safety guidance given previously when bending the glass tubing ("Bending Glass Tubing" on page 6).

2. In operation, this project uses an open flame. See Note 3 in Appendix E for basic safety practice.

3. If distilling flammable liquids, consider using a heat source other than an open flame. Do you have access to a hot plate that could do the job?

Can You Work Like a Scientist?

A mixture of dry ice and alcohol produces an extremely low temperature.[29] What gases could you condense at this temperature? What changes might you have to make in your setup?

Clothes Hanger Chemistry Stands—Filter Paper

Purpose: Chemistry stands are used to support various pieces of laboratory equipment during chemistry and other experiments.

Materials: Clothes hangers,[30] electrician side-cutting pliers, and a second pair of pliers.

What to Do: Cut the hooks off the clothes hangers. Shape the hangers to form the stands shown below.

---

29 See Note 14 in Appendix E about working with dry ice.

30 Steel music wire (3/32″ in diameter) or 12 gauge solid-core bare copper wire may be substituted.

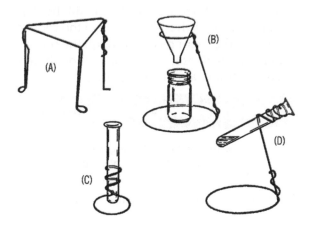

Operation of Equipment: A screen can be placed over the tripod stand (A). An object can be heated by placing it on the screen with an alcohol burner under the screen.

Fold   Cut

The support stand (B) is used primarily to support funnels and similarly shaped pieces of glassware. The funnel is used in a process known as filtration. A paper towel makes a good substitute for filter paper.[31] Fold a piece of towel as shown. Place the towel into the funnel and pour the liquid into the funnel. The

Modern Safety Practice

When using side cutting pliers to cut wire, small bits of wire can go flying off at high speeds. Wear safety glasses with side protection, and make sure that everyone else in the vicinity is wearing a set as well.

clear liquid can be collected in a baby bottle. This cleared liquid is called "filtrate."

Graduated Beaker

Purpose: A beaker is used for mixing chemicals. It has a pouring lip and should be made of material that can be heated.

Material: A beaker can be made by cutting a bottle. A simple beaker is a wide-mouth one-pint mason jar. A better substitute for the beaker is the one- or two-cup measuring cup made out of pyrex with a pouring lip. Measuring cups are marked in ounces and in fractions of a cup.

Stirring Rod

Purpose: This is used to stir chemicals in beakers or flasks. The rod is usually made of glass so that it is not affected by acids. The rod can also be used for static electricity experiments.

Material: Piece of glass tubing about 12 inches long.

What to Do: Heat one end of the glass tube in the flame of an alcohol burner. Turn the tubing slowly until the glass flows and seals the hole. Allow the tubing to cool and then seal the other end.

---

31 Coffee filters are also available at any grocery store.

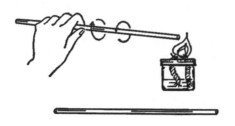

Modern Safety Practice

This project involves cutting (cold) glass, open flames, and hot glass. Follow the guidelines in Note 2, Note 3, and Note 4 of Appendix E, respectively.

Petri Dish

Purpose: This is a small shallow dish used for growing cultures of bacteria or simple one-celled animals such as paramecia.[32]

Materials: Small round bottles, and glass bottle cutter already described.

What to Do: Etch (scratch) the bottle with your glass cutter. Cut the bottom of the bottle off with the nichrome wire and the salt water rheostat. Smooth the glass with wet or dry emery paper.

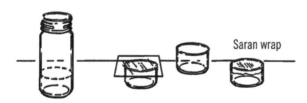

Saran wrap

Operation of Equipment: Place a few kernels of wheat in your Petri dish. Add distilled water. You may either distill your own water with your light bulb retort and glass condenser, or you may collect rain water. The important thing is that the water does not contain chlorine.[33] Cover the dish and let it stand in a warm place for about one or two weeks. A good cover can be made by cutting a piece of glass with your glass cutter or by using Saran Wrap. A regular saucer or jar lid is also satisfactory. After two weeks there should be a large number of bacteria and some small micro-organisms. If you put a few drops of any culture in this dish of bacteria, they will feed and reproduce. Another method is to use bits of hay or lettuce in place of the wheat kernels. Try all methods and compare your results.

Measuring Spoon

Purpose: The spoon is used to measure out chemicals. A plastic spoon set from the variety store is a good substitute for chemical measuring spoons, and much less expensive.

Acid Bottle

Purpose: Acids can only be kept in glass containers.

---

32 Today, the term "animal" is carefully defined to mean only *multicellular* organisms that have certain characteristics. One might properly describe a paramecium as being an *animal-like* single cell organism. Paramecia (the plural of paramecium) will appear again in Part III.

33 The ultraviolet light in sunlight will typically break down the chlorine in tap water. As an alternative to collecting rainwater, you can leave a (clear) container of water for several days in a sunny place such as a windowsill.

Materials: Ideal acid bottles are vitamin or nose drop bottles.[34] They have a medicine dropper built in the lid. *Be sure to label bottles.*

Wash Bottle

Purpose: The wash bottle is valuable as a ready source of water. It is very necessary in a room or laboratory that does not have running water.

Materials: Gallon jug, 2-hole #7 rubber stopper, glass tubing, and rubber tubing.

What to Do: Bend a piece of glass tubing to the shape shown. The tubing should be long enough to reach down to the bottom of the jug. Rubber tubing should be connected to a very short piece of glass tubing.

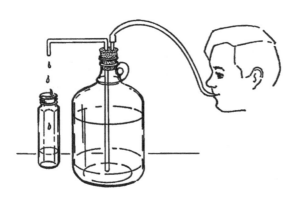

Operation of Equipment: Fill the jug almost full of water. Insert the rubber stopper. The glass tubing is the faucet. It can be directly over a bowl or plastic dishpan. In order to get water, blow into the rubber tubing. The air is compressed and pushes on the water. The water is pushed up the glass tubing and out.[35]

Can You Work Like a Scientist?

1. Can you design a wash bottle that works on a siphon and uses a clothespin as a pinch clamp to shut off the water?

2. Would a small hand pump work for your air supply? Can you design a better air supply?

Asbestos Board

Purpose: Asbestos will not burn. Hot bottles or pans can be placed on an asbestos board to prevent burning of a desk or counter top.

Material: An asbestos shingle can be obtained from almost any lumber store. Usually a student can get one without cost.

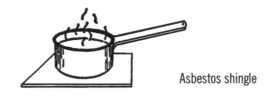

Asbestos shingle

Drilling Glass

Purpose: Many pieces of equipment may require the drilling of a hole in a glass bottle. Uses include cloud chambers, wash bottles, electrical wires in bottles, etc.

Materials: Triangular file, turpentine, bottle, and rat-tail file.

---

34 Vitamins and nose drops now more commonly come in plastic containers. See "Bottle, Dropper" in Appendix A for sources.

35 Modern laboratory wash bottles have the same function, but they are usually made of squeezable plastic with an air valve. Can you design a version that works this way?

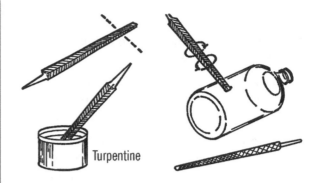

Turpentine

What to Do: Break off the tip end of a file (about one inch). The jagged end of the file serves as a cutting tool. One of the triangular points is placed against the glass, and the tool is rotated as pressure is applied with the thumb. After starting the hole, dip the point of the file in turpentine. Then place the tip of the file in the hole and work in a circular motion. Keep dipping the cutting edge of the file in turpentine. The turpentine turns the glass powder into a cutting paste and also cools the glass. If the file stops cutting, use the second edge of the triangular file. When the point comes through the glass, use a thin rat-tail (round) file to increase the size of the hole to the desired size. Be sure to dip the file in turpentine before using.

Operation of Equipment: Seal the hole with sealing wax if you are just going to insert an electrical wire. If you want to insert glass tubing, slip rubber tubing over the glass tubing and force the rubber tubing into the hole. A vacuum cement can be used if the container is to be used under a vacuum.

Can You Work Like a Scientist?

Can you make a gravity wash bottle from a gallon jug?

Modern Safety Practice

1. Read Note 2 in Appendix E about safety when working with glass.

2. When using turpentine, avoid breathing its fumes, getting it on your skin, or getting it in your eyes. Work in an area with good ventilation. Wear eye protection (also because you are working with glass). Nitrile rubber gloves can protect your skin from turpentine.

Litmus Paper

Purpose: Litmus paper is used to tell if a liquid is an acid, base, or salt.

Materials: Red flower petals (for example, red rose, hydrangea, or hibiscus petals), purple cabbage leaves, water, paper,[36] and a pan.

What to Do: In order to make a red litmus liquid, boil flower petals in water until most of the water evaporates and the color is very strong. Dip pieces of paper into the liquid and dry. These colored strips of paper serve as red litmus paper.[37]

Blue litmus paper is made in the same way except you use purple cabbage leaves instead of flower petals. The dry colored paper is called an indicator.[38]

Can You Work Like a Scientist?

1. Does vinegar turn blue litmus paper red or red litmus paper blue? Vinegar is an acid.

2. What effect does a base like baking soda water have on litmus paper?

Sensitive Gram Scale

Purpose: A gram scale is an accurate weighing device. This one can measure weight changes as small as 1/100 of a gram.

Materials: Yardstick (free[39] at many lumber yards and hardware stores), two glasses or jars, broom straw, small piece of tin[40] or cardboard for slider, three thumbtacks, a finishing nail, and two pieces of paper 4" square (you may use two cone-shaped paper cups instead).

What to Do: Drive the nail through the middle of the yardstick. The hole should be about one-fourth of an inch from the top edge as shown. Make two paper cones and tack one at each end of the yardstick. Place the two glasses about one inch apart and balance the yardstick by letting the nail ends rest on the edge of the glasses. Glue or Scotch tape the straw perpendicular to the ruler, as shown. Now balance the scale by using the third tack. Move the tack along the lighter end until you find the place the scale just balances. Cut a narrow piece of tin or cardboard and bend it so it slips over the yardstick and will slide easily. Slide this to the middle of the yardstick and balance again with the tack.

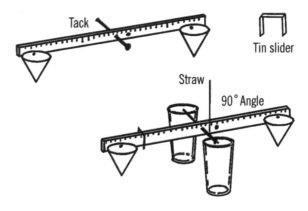

Operation of Equipment: Place the object to be weighed in the paper cone at the opposite end of the yardstick from the slider. In order to

---

36 Various versions of this project call for paper towels, filter paper, or acid-free paper from an art or office supply store. Can you perform an experiment to find out if there is a difference?

37 An easier approach is to obtain hibiscus herbal tea bags (or dried hibiscus flowers) from a grocery store, lightly wet them, and dip the paper into the resulting red liquid.

38 Other natural indicating substances include rhubarb stems, blueberries, and turmeric. What properties do these indicators have?

39 Or nearly free.

40 E.g., a strip cut from the lid of a tin can. Use tin snips or aviation snips (heavy-duty scissor-like tools available at most hardware stores). Use caution because the cut edges of the tin can be very sharp.

be exact, weigh one square inch of newspaper first. Fold the paper and drop it in the cone. One square inch of newspaper weighs .033 of a gram. Move your slider along the other end of the yardstick until the yardstick balances. You can tell because your indicator, the broom straw, will be straight up and down. Now mark on the yardstick the spot of the slider as .033 of a gram. Take three square inches of paper and do the same. This time mark the spot for 1/10 of a gram. Continue to do this until the slide is all the way out to the end of the yardstick. To weigh objects heavier than this, use a standard weight in the cone on the slider side. Place the unknown object in the other cone. Add standard weights until the scale almost balances. Then move the slide until it exactly balances. To find the total weight, add your standard weights together, plus the mark shown by the slide. Some objects are suggested for use as standard weights. Using these, and remembering that each square inch of newspaper weighs .033 of a gram, you should be able to make your own set.

Metric Weights

Purpose: A set of weights will enable you to weigh accurately objects of almost any weight on your balance or beam scale.

Materials: Newspaper, pins, paper clips, dime, penny, quarter, half dollar, and a nickel. Aluminum foil paper is also helpful.

What to Do: A square inch of newspaper weighs about .033 of a gram. A common pin made of steel weighs .075 of a gram. From these, see if you can find the weights of the other objects listed by using your gram balance scale. A piece of newspaper ten inches by ten inches weighs 3.3 grams. By trimming your newspaper, you can make any desired weight.

Fold your paper weight when you place it in the paper cone.

List of Known Weights: In order that you may check for accuracy, the following are the weights of objects as measured on a druggist's delicate gram scale:[41]

Newspaper (1 sq. inch)	0.033 of a gram
Common pin	0.075 g
Paper clip	0.75 or ¾ of a gram
Dime	2.268 g
Penny	2.500 g
Nickel	5.000 g
Quarter	5.670 g
Half Dollar	11.340 g
Dollar (gold tone)	8.1 g

From this list you can make your own weights by balancing the above against rocks or aluminum foil folded into a ball.

Can You Work Like a Scientist?

1. Do Canadian and American nickels weigh the same? Are they made of the same material? Test with a magnet. Remember, a magnet attracts nickel.

2. Do rocks of the same size have the same weight?

3. What happens if you place the nail through the yardstick on the nine-inch mark? How much more weight must be placed in the cup to balance the scale? If the cup is too small, could you hang things from a tack if you bal-

---

41 This table has been updated with the weights of modern US coins. See Note 15 in Appendix E for more information.

anced it with another tack on the other end of the scale?

4. Do things weigh as much in water as in air? Hang an object in the water from string and tack attached to the end of the yardstick.

5. Why do you use a cone to place the weights in-why not just a flat cup? Why do you just fasten the cone from one point with the tack?

6. Do all liquids weigh the same? Could you take an equal amount of each and weigh them? Could you compare how many times heavier one is than another? What effect has the weight on how high a piece of wood or cork floats in the liquid?

7. Can you weigh the amount of air in a balloon? Does air really have weight?

8. If you could fill a balloon with helium or hydrogen gas, could you measure the upward pull with your gram scale?

9. How much water will a sponge hold? Can you weigh this? Can you weigh moist and dry bread? What's the difference?

10. Can you determine the amount and percentage of water in different vegetables and fruits?

Soda Straw Chemical Balance

Purpose: The chemical balance is used to weigh out accurately small amounts of chemicals for experimental purposes.

Materials: Soda straw, screw, common pin, two glass slides, block of wood as shown, and material for the scale stand as shown.

What to Do: Make the balance stand by fastening the glass microscope slides to a block of wood with a rubber band. Stick the pin through the straw. Screw the wood screw into the end of the straw. Attach a pin hook and a paper container to hold the chemicals being weighed.

Balance the straw and pin on the microscope slides. Move the pin in the straw until the straw almost balances. Then your fine adjustment is made by screwing the wood screw in or out of the end of the straw.

Make a stand to hold a scale. Mark the scale off in grams by using sample known weights in your paper container. A nickel weighs five grams. Cut a piece of tin so that it will balance a nickel on a balance scale. Then cut the strip of tin as shown to make sample weights.

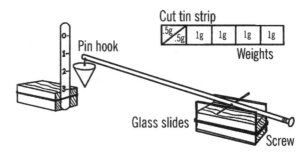

Operation of Equipment: Balance the straw so that when the container is empty, the tip of the straw points to zero on your scale. The one-gram mark on the scale is the place where the tip of the straw points when a one-gram weight is added to the paper container. Make the rest of your scale in a similar manner.

Can You Work Like a Scientist?

See the end of the chapter for suggested experiments.

Bridge for Pneumatic Trough

Purpose: A pneumatic trough is used to collect gases generated in a chemistry experiment. The bridge is used to support the collecting bottle.

Materials: Plastic water tray or gallon jug aquarium (see "Aquarium" on page 230), coffee can, tin snips, baby bottle, and rubber tubing.

What to Do: Cut the bottom off the coffee can with a can opener. Cut the side of the can and flatten the tin into a rectangular sheet as shown. Cut a strip about 2½" wide and bend it into the shape shown in the illustration. Drill or punch a hole in the center of the "bridge." Insert the end of the rubber tubing in the hole. Set the "bridge" into the water tray or gallon jug aquarium jar as shown. Fill the container until the "bridge" is covered with water.

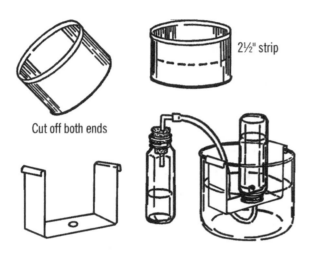

Cut off both ends

2½" strip

Operation of Equipment: Fill the collecting bottle with water. Slide a piece of cardboard over the top of the bottle and then turn the bottle upside down and lower it over the hole in the bridge. When the opening of the bottle is just under the surface of the water, remove the cardboard. The water should stay in the baby bottle container.

Connect the rubber tubing to any gas generating bottle. As the gas is formed, it travels through the rubber tubing and into the baby bottle. The gas pushes the water out of the baby bottle. After all the water has been pushed out of the collecting bottle, carefully slip a cardboard under the opening and remove the bottle from the pneumatic trough (water tray).

Modern Safety Practice

Use care with the tin snips. The edges on the cut tin can be very sharp. File or sand them down.

Can You Work Like a Scientist?

Fill a baby bottle one-fourth full of hydrogen peroxide (3 % solution). Add a pinch of manganese dioxide (the black powder from the inside of a flashlight battery[42]). Insert the stopper as shown and collect the gas in the trough. Is the gas oxygen or hydrogen? Can you test the gas with a glowing splint?[43]

Thistle Tube

Purpose: The thistle tube has many uses. One use is to pour liquids into glass tubing, such as is shown in the hydrogen generator below.

Materials: Plastic funnel, glass tubing that fits tightly into bottom opening of funnel. (The 3" plastic funnel exactly fits over 6 mm glass tubing.)

What to Do: Slip the opening of the plastic funnel over the glass tubing as shown. You can seal the tubing in the hole by melting wax around the glass inside the funnel, but this usually isn't necessary.

---

42 Inside a specific type of battery. Please see Note 12 in Appendix E.

43 A glowing splint is a long, thin strip of wood that is used in the laboratory for testing how a sample reacts to a weak flame. A good example of a glowing splint is a long fireplace match, after the head finishes, leaving a long strip of wood with a weak flame on the end. *Always wear safety glasses when testing a gas for flammability.*

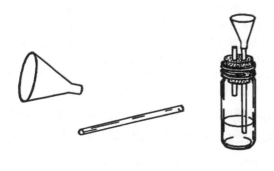

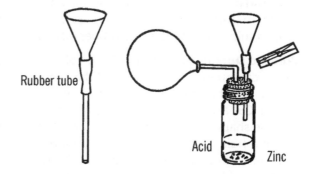

Can You Work Like a Scientist?

1. Hold your finger over the end of the glass tubing. Fill the funnel with water. Does the water come out when you remove your finger?

2. Place a cork in a large jar of water. Place the funnel over the cork. Hold your finger over the end of the glass tubing. Push the funnel down over the cork. Where does the cork float?

3. Could you make a thistle tube by cutting off the top of a bottle, and using a rubber stopper and glass tubing?

Hydrogen Generator

Purpose: This generator produces gases by means of acids working on bits of metal. The gas can be collected in bottles or in balloons as shown.

Materials: Flask (light bulb with large opening or baby bottle with #6½ stopper), thistle tube (as above), two-hole rubber stopper, short piece of rubber tubing, and a clothespin or metal clip for a pinchcock.

What to Do: Set up your generator as shown. The thistle tube can be supported by your curtain rod ring stand. The metal case around old flashlight batteries furnishes a good source of zinc.[44] Most acids will work on the zinc metal. Acids such as sulfuric can be purchased from the drugstore.[45] When you need to dilute (thin) the acid, pour the acid slowly into the water. Never pour water into acid. The acid will spatter. Wash your hands immediately if you get acid on them.

Operation of Equipment: Drip the acid slowly on the zinc or aluminum foil. Control the amount of gas given off by pinching the rubber tube with a clothespin.

Oxygen Generator

Purpose: This is a simple way to produce oxygen for use in experiments to discover the properties of this gas.

Materials: Hydrogen peroxide (called peroxide) from dime or drug store, manganese dioxide, and a jar with a lid. The manganese dioxide may be gotten from an old flashlight battery. The black paste is part manganese dioxide.

What to Do: Cut open a flashlight battery.[46] Add about a spoonful of the black paste to

---

44 See Note 12 in Appendix E about batteries. See Appendix A for additional sources.

45 No longer; see Note 16 in Appendix E about acids.

46 Zinc-carbon battery. Note 12, Appendix E.

Modern Safety Practice

1. Always wear safety glasses while working with chemicals such as acids. Ideally, also wear nitrile rubber gloves to protect your skin from contact with the acid. Read Note 17 in Appendix E for additional discussion about safety around chemicals.

2. Hydrogen is a flammable gas that can potentially cause explosions. Keep away from open flames and ensure that you have excellent ventilation.

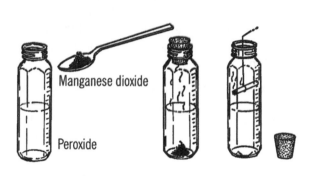

Manganese dioxide

Peroxide

about an inch of peroxide in the jar. Tighten the lid and wait about fifteen minutes.

Operation of Equipment: Remove the lid and place a glowing splint or a burning cigarette into the bottle. The cigarette can be held by a wire.[47]

Can You Work Like a Scientist?

1. Does the oxygen come from the manganese dioxide or the peroxide?

2. What happens to the splint when you place it in the bottle?

3. Do you get more gas if you shake the bottle?

4. Why do you keep the lid on the bottle?

5. Ask your mother what hydrogen peroxide is used for.[48] Try dripping hydrogen peroxide on colored paper.

Modern Safety Practice

Note 17 in Appendix E discusses safety practice around chemicals. Hydrogen is a flammable gas that can potentially cause explosions. As you will be working with flame (possibly, a small explosion!), also read through the Note 3 in Appendix E about fire safety.

Chemical Source of Hydrogen

Purpose: A quick source of hydrogen for experiments.

Materials: Bits of aluminum foil or zinc, a few drops of sulfuric acid (from a storage battery), test tube, and a cork.

What to Do: Draw a little dilute acid out of the storage battery in your car[49] with an eye dropper. You should have about an inch of liquid in a test tube. Add a few small pieces of zinc and place the cork in the test tube. Wait about five minutes and then remove the cork and quickly place a lighted match near the end of the tube.

Wrap a handkerchief or towel around the test tube for safety. Don't point the test tube at anyone.

---

47 A long fireplace match ("glowing splint") will reach without a wire. (Not even *this* is a good reason to use a cigarette!)

48 What kind of a question is *that*? Ask *any* parent, or even better, do the research on your own!

49 It is better to start with clean acid, rather than acid from a battery. Hydrochloric acid can be used as a substitute for sulfuric in this experiment. See Note 16 in Appendix E for more about acids.

Can You Work Like a Scientist?

1. What effect would heat have on the production of the gas? Watch out for the cork and some liquid flying out of the test tube. Remember, the liquid is an acid.

2. Try the same experiments with other metals and acids. How about vinegar?

Safety Gas Generator

Purpose: This generator is used to generate various gases. Since the acid is kept in one bottle and the metal strips in a second bottle, the action can be controlled by adding small amounts of acid to the metal as desired.

Materials: Two short pieces of rubber tubing, glass tubing as shown in "Bending Glass Tubing" on page 6, two rubber stoppers (two-hole, size #6½), and two baby bottles.[50]

What to Do: Bend the short pieces of glass tubing as shown in the drawing. Use the double wick alcohol burner and follow the directions for bending glass tubing (see "Bending Glass Tubing" on page 6). Rub soap on the outside of the pieces of glass tubing before you try to insert the glass tubing into the rubber stopper.

Attach the short pieces of rubber hose as shown. See the directions for making a clamp out of a clothespin ("Adjustable Clamp" on page 37).

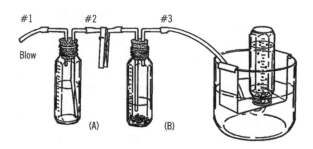

Operation of Equipment: Place small pieces of gravel in the bottom of the reaction bottle (B). Add strips of metal (zinc, aluminum foil) to this bottle. Be sure to add enough of the metal strips. You cannot add more metal during the reaction.

Fill the acid bottle (A) about half full of an equal mixture of water and acid (hydrochloric or sulfuric). In making the acid mixture, add the acid slowly to the water. *Do not pour water into acid.*[51]

Connect up the reaction bottle with a collecting bottle in a water tray. Blow into tube #1. The air forces the acid up the second glass tube and into bottle (B). Gas is formed and goes up and out tube #3. Be sure to clamp tube #2 as soon as you add acid to the metal strips. This keeps the gas from going back into bottle (A).

Add small amounts of acid at a time so you can control the reaction. When you have finished

---

50 See Note 5 in Appendix E about baby bottles.

51 It is a general rule to always pour acids into water, rather than water into acid. There are several reasons for this, both for safety and for good mixing. Can you think of what the reasons are?

collecting all the gas you wish, place the clamp on rubber tube #3. The gas that builds up in bottle (B) will force the excess acid back into the acid in bottle (A). When the acid no longer touches the metal strips, the reaction stops.

Can You Work Like a Scientist?

Can you use the gas generator to make carbon dioxide gas? Put baking soda in the reaction bottle (B) and vinegar and water in the acid bottle (A).

X Connector

Purpose: The X connector is used in experiments requiring two connections to two outlets.

Materials: Cork, short pieces of glass tubing, end of a file or a cork borer.

What to Do: Bore holes through the cork as shown. Insert four short pieces of glass tubing.

T Tube

Purpose: The T tube or T connector can be used in the chest cavity and the water faucet vacuum pump, plus many chemistry experiments requiring two lines to join in one outlet.

Materials: Alcohol burner, two pieces of glass tubing.

What to Do: Seal one end of the glass tubing. Heat the middle of the tubing and when hot, blow on the other end and pop a hole in the

tubing. Heat around both this hole and the end of the glass tubing that will be the side arm. Bring these two heated spots together and join firmly. Seal the other end as shown, and then blow gently into the one open end to smooth the joint. Heat the joint and blow gently until the joint is strong and smooth. Cut off the ends that are sealed.

A second way to make a T tube is to drill only one side hole in the cork as mentioned above.

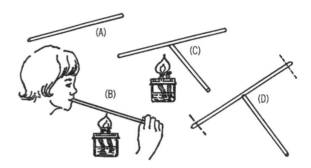

Modern Safety Practice

Follow the safety guidance given previously for working with cold glass, hot glass, and open flames, as given in "Bending Glass Tubing" on page 6, as well as Note 2, Note 3, and Note 4 in Appendix E.

Adjustable Clamp

Purpose: This clamp is used to seal rubber tubing completely when working with high pressure.

Materials: Clothespin, small bolt, and two nuts.[52]

What to Do: Drill two small holes in the legs of the clothespin. Insert the bolt. Place the nuts on as shown. In order to tighten the pinchcock,

---

52 There are many other possible ways to build an adjustable clamp of this nature. Can you design one that could be 3D printed?

turn the adjustable nut to force the legs farther apart.

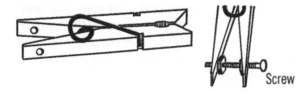

Screw

Crystal Coal Garden

Purpose: Nearly all solids are made from crystals. By growing and experimenting with crystals you will get a basic understanding of the world of solids.

Materials: Several pieces of coal or charcoal briquettes, ¼ cup table salt,[53] ¼ cup water, ¼ cup laundry bluing, and a tablespoon of ammonia.

What to Do: Place the coal or charcoal in a bowl. Mix the water, ammonia, and table salt into a solution and then pour this solution over the coal or charcoal. You might try putting different coloring materials at various places on top of the coal. Colored ink, food coloring, stains, and Mercurochrome[54] are among the coloring liquids you might use. Place the bowl where it will remain undisturbed.

Food coloring stains

Part of charcoal above surface

Operation of Equipment: Look at a lump of coal or a briquette with a magnifying glass. Do you notice the many small holes in the surface of the coal? Water is drawn into these holes, and along with the water is the salt you mix in the solution. The water is constantly evaporating, and as it evaporates, salt left behind by the coal or briquette crystallizes on something solid, such as the lump or the side of the bowl. These crystals are not solid, but contain many tiny spaces in them. The water is drawn up through these crystals. Again the water evaporates, depositing new crystals on the old. In this manner the crystals seem to grow and soon fill the entire dish. Your coal garden is made up of salts of sodium chloride (table salt), ammonia, and bluing.

Modern Safety Practice

Read Note 17 in Appendix E about working with chemicals.

Can You Work Like a Scientist?

1. What effect does the temperature of the room have on the rate of growth? Try growing a garden in the refrigerator.

2. What effect does the humidity of the air have on the rate of growth of the crystals? Use your wet and dry bulb thermometer to determine the relative humidity.

3. If the garden is made up of salts of different materials, could you experiment with salts from other materials instead of those you just used in your coal garden?

4. Why does your crystal garden collapse if you attempt to move it?

---

53 Some recipes for this "garden" call for using non-iodized salt. Others say to use either iodized or non-iodized salt. Can you do an experiment to learn whether and how iodine affects the growth of the garden?

54 See Note 18 in Appendix E about mercurochrome.

5. Why should part of your lump of coal or charcoal be above the water? Try a coal garden in which the lump is completely covered.

6. Observe some of the crystals under the microscope or microprojector. Do you notice any particular shape to the salt crystals?

Growing Crystal Candy

Materials: Sugar, water, baby bottle, pencil, string, and paper clips.

What to Do: Pour two cups of water into a small kettle.[55] Add as much sugar as will dissolve in the water. Heat the water. As it becomes warmer, you should find that you can dissolve more and more sugar into the water. When the water is boiling, you should be able to dissolve three to three and a half cups of sugar in the two cups of water. Let the solution cool slowly. While it cools, hang a string down into the baby bottle as shown. Use a paper clip for a weight and suspend the string from a toothpick or pencil. Be sure the string and paper clip are very clean.

When the sugar solution has cooled, pour it slowly into the baby bottle. Crystals should start to form on the string in a few hours. If you don't disturb the bottle, you may get sugar crystals up to one inch on a side.

Can You Work Like a Scientist?

1. Why does hot water dissolve more sugar than cold water? Can you keep a graph of the temperature of the water and compare this with the amount of

---

Modern Safety Practice

Note 17 in Appendix E talks about working with chemicals, and (among other things) reminds you to keep food out of your chemical work area, and to keep chemicals out of food prep areas.

Thus, you need to make a choice. To make *edible* sugar crystal rock candy, do not take it to your chemical work area, but instead perform all of these steps in your kitchen, treating this as a cooking project. To make sugar crystals to study, take the sugar to your lab and treat it as you would any other potentially hazardous chemical.

---

sugar that will dissolve at this temperature?

2. When you dissolve sugar in water, does the water level rise? If you dissolve two cups of sugar into two cups of water, do you get four cups of solution?

3. Why do the crystals form on the string?

4. What effect does the rate of evaporation have on the formation of the crystals? Cover the jar completely so that no evaporation takes place. Will the crystals still form? You can slow down

---

55 You may find it easier to use a bowl that can be put in the microwave. One hazard to watch out for is that water in a microwave can become *superheated* and boil instantly (and sometimes violently) when you add something to it. You may have seen this phenomenon when adding pasta or a tea bag to water or when stirring hot water with a spoon. Related question: Why do chemists use boiling chips?

the evaporation rate by covering the jar with a damp cloth.

5. Can you grow salt crystals the same way you grew sugar crystals?

6. Does the temperature have to be constant in order to grow crystals? Try growing crystals in a temperature that varies or changes quite often.

7. Can you grow both sugar and salt crystals together on the same string?

8. Is water really a solution? Heat water and notice bubbles rising to the surface of the water. Where do these gas bubbles come from?

9. Will sugar dissolve in alcohol? Can you grow sugar crystals in alcohol?[56]

10. Will water dissolve in alcohol? Will alcohol dissolve in water? What effect has temperature on the amount of one liquid that will dissolve in another liquid?

11. Solubility is the amount of a material that will dissolve in a liquid at a given temperature. Can you plot a curve on a graph for the solubility of various chemicals as the temperature increases from freezing to boiling? Is the solubility curve the same for all chemicals?

Growing Gem Crystals

There are two general ways to grow large single crystals–the sealed jar method and the evaporation method. In both methods, a seed crystal is suspended in a jar containing a solution of a particular salt.

The Sealed Jar Method

The first way to grow crystals is the sealed jar method. Again you hang a seed crystal in the jar by a thread. The solution is made supersaturated by heating and dissolving as much salt material as possible in the liquid. The seed crystal is then placed in this supersaturated solution, and the jar is sealed to keep the water from evaporating. The excess salt will slowly crystallize on the seed and cause the seed to grow. This method is usually the quicker way to grow various crystals.

The growth of large crystals requires a constant temperature and a place where the crystals will not be disturbed.

To prepare a supersaturated solution, heat the liquid and dissolve all of a particular salt possible. Pour the solution into a mason jar. As the solution cools, crystals are deposited at the bottom of the jar. The crystals at the bottom take the solid out of the solution and leave the surrounding liquid less dense. This less dense solution rises and is replaced by the part of the solution containing more solids. These solids are again deposited as the solution seems to stir itself by this action of less dense liquid rising and being replaced by denser material.

It is wise to seed the supersaturated solution after it cools to start this forming of excess crystals on the bottom of the mason jar. You can seed the solution by placing a pinch of the salt used in the jar. The jar should then be sealed, shaken well, and then allowed to stand at the temperature at which you wish to grow the crystal, at least for two days. You should shake the jar twice each day in order for the solution to be well mixed.

When crystals stop being deposited in the bottom of the jar, you know that the solution has reached its saturation point and contains all the crystal it normally can at that particular temperature. Pour off the clear solution into another mason jar and seal the jar. Try to avoid getting any of the crystallized salt from the bottom of

---

56 As alcohol is flammable, it isn't safe to just fill the kettle with alcohol and put it on the stove. How could you heat it safely, without risk of fire?

the jar into this clear solution. Remove the deposited crystals from the bottom of the jar, dry, and place them back into your supply bottle. The mason jar should then be washed out and dried.

Preparing a Seed Crystal: Pour about an ounce of your saturated solution into a small container and place it where it will be undisturbed. As the liquid slowly evaporates, the solution becomes supersaturated, and crystals form on the bottom of the container. If your solution does not begin depositing crystals, you can seed your solution by adding a small amount of the crystal salt you are using. Examine your container several times a day. When the crystals have grown large enough for easy handling but not so large that they touch each other and interfere with normal growth, remove the best crystals with tweezers. Place the seed crystals on a piece of paper toweling and allow them to dry.

Cut out a cardboard disc just large enough to fit inside the lid of the jar. Punch a hole in the cardboard and fasten the thread to the cardboard. Make a slip knot at the other end of the thread and carefully slip the knot over the largest seed crystal you have grown. The thread should be just long enough to suspend the seed crystal about an inch from the bottom of the jar. You are now ready to grow your "gem crystals."

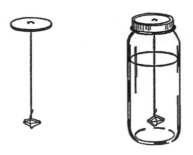

Preparing the Growing Solution: The growing solution is made up of the saturated solution you have saved in the jar and some extra salt crystals you have added to the saturated solution. In each of the formulas given, the A part is the proportion or ratio necessary to make the original supersaturated solution. This is, of course, cooled and poured off as a saturated solution.

The B part of the formula is the amount of salt necessary to add to the saturated solution in order to make the growing solution. A sample formula is given below:

1. To form the supersaturated solution for Rochelle salt, dissolve in the ratio of 130 grams of salt to 100 cc of water.

2. In order to make the growing solution, add 9 grams of Rochelle salt to each original 100 cc of water you used.

For each 100 cc of saturated solution you have saved in the jar, you will use 9 grams of Rochelle salt. If you used 500 cc of water, you would use 45 grams of Rochelle salt to make the growing solution. You add this salt to the top of a double boiler and then pour the saturated solution over it and allow the solution to heat gently until all the crystals are dissolved. This growing solution is then poured into a clean mason jar, sealed, and allowed to cool.

When this growing solution is about 5 °F above the room temperature, suspend the seed crystal with a thread attached to a cardboard disc. The jar should then be sealed by screwing the lid over the cardboard disc.

The seed crystal will first start to dissolve because of the warmer temperature. As the temperature cools, the solution becomes supersaturated, and the crystal grows. You can observe this by watching the currents in the water. If the current is descending, the crystal is dissolving. If the current is rising, salt crystals are being deposited on the seed crystal, and the crystal is growing.

Harvesting Your Crystal Garden: Most crystals will be "grown" in about a week. Remove the crystal and dry it with a soft cloth. Don't handle the crystal directly as the perspiration from your skin will dissolve some of the crystal. Store the crystal by wrapping it in a cloth and placing it in a sealed jar.

You can use the growing solution over again to grow another crystal. Just add salt equal to the weight of the crystal you have just grown plus the weight of the crystals collected on the bottom of the jar, warm and stir and you have a new growing solution.

Evaporation Method

Purpose: Large crystals can be grown easily by the evaporation method. The crystals are usually not as perfectly formed as in the sealed jar method, but the greater variety of possible crystals and the ease of setting up a crystal-growing tank make this method a favorite for beginners.

Materials: Saucepan to dissolve the chemical, baby bottle, clean string, paper clip, and a piece of cardboard. Any of the materials listed in the next section can be used for crystal growing. This experiment will describe the use of borax in growing borax crystals.

What to Do: Pour one-half cup of water into a baby bottle. Set the bottle into a saucepan containing water. Heat the pan until the water boils. Slowly add borax to the baby bottle until the water will not dissolve any more of the borax. The heated liquid will hold more of the chemical in solution at the boiling point of water than it will at normal room temperatures. We say the liquid is supersaturated.

Tie a paper clip or button on the end of a piece of string and hang it in the baby bottle from a cardboard disc. Several holes should be punched in the disc. Cover the holes in the cardboard with Scotch tape. Do not disturb the crystals by bumping or moving the bottle, which must remain perfectly still in a place where the temperature does not change. Most crystals grow well around room temperature.

After one day, the string should be covered with many small crystals.

Operation of Equipment: The slower the liquid cools and the slower the water evaporates, the larger the crystals will form on the string. If you want just one large crystal, remove the largest and best formed crystal from the string to use for a seed. Next, reheat the liquid in the baby bottle. Then tie the seed crystal to the string and hang it in the baby bottle or carefully drop the crystal into the bottle. The dissolved borax will form around the seed crystal instead of forming many small crystals.

You can control the rate of evaporation by covering and uncovering the holes in the cardboard disc. If the crystal seems to have stopped growing, uncover more holes. If the crystal is growing too rapidly, many small crystals will start to appear, and the large crystal will be poorly formed. In this case, you should slow down the evaporation rate by covering the holes.

Modern Safety Practice

Strictly follow standard procedures for working with chemicals, Note 17 in Appendix E.

Can You Work Like a Scientist?

1. You can use many chemicals in crystal growing. Below are listed a few common chemicals that can be used to grow beautiful, colorful crystals.

Alum	Magnesium sulfate (Epsom salt)
Boric acid	Ammonium sulfate
Washing soda	Potassium chromate
Baking soda	Potassium dichromate
Salt	Potassium permanganate
Sugar	Sodium thiosulfate (used by photographers as hypo)
Zinc sulfate	Nickel sulfate
Cobalt chloride	Manganese sulfate
Copper sulfate	Chromium potassium sulfate
Ferrous sulfate	

2. Can you determine the amount of each chemical it takes to saturate a certain amount of boiling water?

3. Can you dissolve more of the chemical if you place the chemical in water and then heat the water under pressure in a pressure cooker?

4. Can you keep a graph of the amount of a chemical that will dissolve at different temperatures up to the boiling point of water?

5. Is this graph a straight line or does it suddenly curve upward at a certain temperature for each chemical?

6. If you can slow down the rate of cooling of the solution, larger crystals will form. Can you devise a way in which the solution will cool very gradually?

7. Crystals usually have certain shapes. Can you make up a classification system for crystals based on the shape of the crystals?

8. You might be able to control the rate of evaporation of the solution by using a wick and running the wick through a jar lid that covers the jar. The wick can be a piece of clothesline rope.

9. Can you use a crystal of one type as a seed for another type of crystal?

10. Place small seed crystals under the microscope or microprojector. Try using the polarizer filter and examining the crystals under polarized light. Some crystals are left-handed and some are right-handed. To determine this, turn the polarizer filters so that light is almost completely shut out (cross the two filters). Place the crystal on the stage. You should notice that the crystal now transmits light. Turn one polarizer until you again shut out or darken the light going through the crystal. Measure how many degrees you had to turn the polarizer. This is your angle of polarization for the crystal. Check to see if you turned the filter to the right (clockwise) or to the left (counterclockwise). This will tell you if the crystal is right- or left-handed.

11. Is each type of crystal either right- or left-handed, or can a crystal of a certain chemical be both? What seems to determine this?

12. What crystals exhibit the piezo-electrical effect? This effect is an electrical current given off by the crystal when struck with a hammer and jarred.

13. Can you experiment with splitting your crystals along certain faces or planes?

Recipes for Growing Crystals

Purpose: These recipes will help you grow a varied crystal collection. With the many colored crystals you can try various experiments as to electrical, mechanical, optical, and geometric properties of these crystals.

Materials: Each recipe will include the amount of material necessary to make the supersaturated solution as described under growing gem crystals (A). The second part of the formula is the material you add to the saturated solution to make the growing solution (B).

The proportion mentioned in the recipe is quite exact and should be followed closely. Any major variation from the formula will result in poorly grown crystals. However, the formula cannot take into consideration the many variables that may exist from one growing session to another. Therefore, the experimenter is encouraged to alter the formula carefully to suit his own conditions. If the crystal grows too quickly, it will have many "veils" and look milky. If this occurs, reduce the amount of the crystal material in solution (B). A second problem is caused if solution (A) is too strong. In this case the veiled crystal plus the bits of crystals in the bottom of the jar weigh more than the crystal material used in solution (B). This means that solution (A) was more than saturated when you combined it with the growing solution (B). Reduce the amount of crystal material in solution (A).

You can make any amount of solution using the proportion mentioned. If you wish to grow very large crystals, you should double or triple the suggested amount given in the formula.

Potassium Aluminum Sulfate
(Potassium alum—a colorless cube type)

1. Supersaturated solution from which you make a saturated solution:

   a. Proportion of 20 grams of potassium aluminum sulfate to each 100 cc of water. Minimum amount—4 ounces of alum.

2. Add to the saturated solution to make a growing solution:

   a. 4 grams of alum for each of the original 100 cc of water. Minimum amount—22 grams of alum.

Potassium Chromium Sulfate
(Chrome alum—purple cube type)

1. Supersaturated solution from which you make a saturated solution:

   a. Proportion of 60 grams for each 100 cc of water. Minimum amount of potassium chromium sulfate—120 grams.

2. Add to the saturated solution to make a growing solution:

   a. 5 grams of potassium chromium sulfate for each of the original 100 cc of water. Minimum amount—10 grams.

This crystal is usually quite dark and hard to observe in the growing solution. You can lighten the final color by substituting some potassium aluminum sulfate for the potassium chromium sulfate. In this case the molecules of aluminum sulfate will substitute for the chromium sulfate when combining with molecules of potassium sulfate. Since aluminum alum is colorless, the mixture will dilute the final color . You can grow an aluminum alum crystal for a while and then place the seed in chromium alum. The dark purple will form over the light aluminum alum. You can then place the seed crystal back into aluminum alum. Thus you can make a layered crystal of unusual size.

Potassium Sodium Tartrate
(Rochelle salt—colorless orthorhombic type)

1. Supersaturated solution from which you make a saturated solution:

   a. Proportion of 130 grams of Rochelle salt to each 100 cc of water. Minimum amount— one pound of Rochelle salt.

2. Add to the saturated solution to make a growing solution:

   a. 9 grams of Rochelle salt (potassium sodium tartrate) for each of the original 100 cc of water. Minimum amount—one ounce or 31 grams of Rochelle salt.

Rochelle salt crystals are somewhat harder to grow than alum crystals. It is sometimes difficult to get the seed crystals to grow in your supersaturated solution. In that case add extra Rochelle salt to the supersaturated solution. The seeds grow rapidly once they are started.

The solubility, or dissolving of Rochelle salt in water, depends on the temperature of the water. Slight changes in temperature affect the solubility greatly. If you are not very careful, you will dissolve your seed crystal off the string when you place it in the growing solution.

The final crystal is orthorhombic or block-shaped. The faces of the crystal vary in size, with the length being greater than the width. Rochelle salts are difficult to keep since the crystals lose water easily. In order to prevent this, wrap the crystals in cloth or cotton and store in a closed mason jar.

You can experiment with different forms of Rochelle salt. Add a solution of copper acetate (one gram per 10 cc of water) to each 100 cc of growing solution of Rochelle salt. Very long thin crystals should form. If you add one large piece of sodium hydroxide to the copper acetate solution, the crystal form will change again.

Sodium Chlorate

(NaClO$_3$— similar to table salt but containing three atoms of oxygen in each molecule. It is a colorless cube type of crystal.)

1. Supersaturated solution from which you make a saturated solution:

   a. Proportion of 113 grams (¼ Lb.) to 100 cc of water. Minimum amount—one pound of sodium chlorate.

2. Add to the saturated solution to make the growing solution:

   a. 4 grams of sodium chlorate per 100 cc of water originally used. This crystal is quite easy to grow. You can experiment with this crystal's shape by adding borax to the solution. You should add 6 grams for each 100 grams of sodium chlorate in the growing solution.

Sodium Bromate

(This is a colorless cube type of crystal)

1. Supersaturated solution from which you make a saturated solution:

   a. Proportion of 50 grams of sodium bromate with each 100 cc of water. Minimum amount —8 ounces or 240 grams of sodium bromate.

2. Add to the saturated solution to make the growing solution:

   a. 2 grams of sodium bromate for each original 100 cc of water.

Sodium bromate crystals must be grown very slowly to prevent veils from forming on the faces of the crystal.

Sodium Nitrate

(This is a colorless hexagonal-type crystal)

1. Supersaturated solution from which you make a saturated solution:

    a. Proportion of 110 grams of sodium nitrate for each 100 cc of water. Minimum amount—1 pound of sodium nitrate.

2. Add to the saturated solution to make the growing solution:

    a. 3 grams of sodium nitrate for each of the original 100 cc of water.

Growth is very sensitive to temperature changes. If your seed dissolves when it is placed in the growing solution, reseed the solution at a slightly lower temperature.

Sodium nitrate is an ideal crystal to study since it can be used to study double refraction (gives two images), cleavage (property of breaking cleanly along a certain face), and glide (part of the crystal shifting position or moving apart as the result of pressure such as the blade of a knife on the corner of a crystal.

Potassium Ferricyanide
(Red Prussiate of potash—red monoclinic type)

1. Supersaturated solution from which you make a saturated solution:

    a. Proportion of 46 grams of potassium ferricyanide to each 100 cc of water. Minimum amount—92 grams of potassium ferricyanide.

2. Carry on experimentation to determine the amount of potassium ferricyanide you should add to the saturated solution to make the growing solution. If you add too much, the crystal will grow too rapidly. If the amount is too small,

the crystal will grow only to a very small size.

> ## Modern Safety Practice
>
> Several of these chemicals that happen to grow nice crystals are also toxic or irritants. It is standard practice to consider all chemicals that you encounter to be hazardous, unless you have firm evidence to the contrary. Crystals that you have grown may look harmless (and even beautiful), but they are still chemicals. Review safety procedures for working with chemicals, Note 17 in Appendix E.

Polarimeter

Purpose: A polarimeter is used to polarize or screen out certain light rays so that the effect of particular rays on crystals can be measured.

Materials: Cardboard tube from a toilet paper roll, piece of thin cardboard such as a file card, Polaroid filter[57] or lens from a pair of Polaroid sunglasses.

What to Do: Wrap the file card around the cardboard tubing and fasten the file card with Scotch tape. Slide the file card tubing off the toilet roll tubing. Cut two round cardboard discs just large enough to cover the end of each of the two pieces of tubing. Cut a one-inch diameter hole in the center of each disc and glue or Scotch tape a Polaroid filter over each disc. Slip the smaller disc in the end of the toilet paper roll tubing and glue in place. Slip the other disc over the end of the roll tubing you made and glue in place. Slip the two pieces of tubing together. Look through the smaller tube end (analyzer) while you slowly turn the file card tubing (polarizer). You should reach a position where the light is completely blacked out. Mark the outside of both tubes for this position and label it 0°. Now divide the outside of

---

57 Most often called "Linear polarizing film."

the file card tubing into 360°. The mark just opposite the 0° mark should be labeled 180°. In this position you should not be able to see light through your polarimeter.

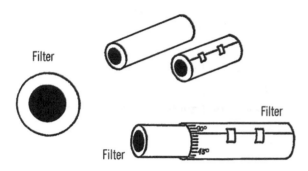

Operation of Equipment: In order to use the polarimeter, slide the two pieces of tubing apart and insert the crystal or material to be viewed into the end of the smaller tube. Slide the two pieces of tubing together and point the polarizer end toward a strong light source, such as a light bulb. Turn the polarizer end slowly until the light through the crystal is at its darkest, probably a dark blue. Look at the scale on the analyzer end of the tube and see how many degrees you had to turn the polarizer from zero and what direction you had to turn it. Each crystal has an angle of polarization. Some crystals require you to turn the polarizer clockwise while others require a counterclockwise direction. This is determined by the way the crystal is grown.

Can You Work Like a Scientist?

1. Twist a piece of Scotch tape and fasten it into the opening of the small tube. When the polarizer is set on zero, do you see any light coming from the Scotch tape? Which way do you have to turn the polarizer in order to darken the light rays to dark blue?

2. Try several thicknesses of Scotch tape. Does the thickness of the material have anything to do with the number of degrees the polarizer is turned?

3. Place crystals in the tube. What direction and how many degrees must you turn the file card tube in order to turn the light rays to dark blue?

Hydrometer

Purpose: A hydrometer is used to measure the specific gravity[58] or weight of molecules in different liquids, as compared with water.

Materials: Test tube, cork or stopper, and some BBs or small shot.[59]

What to Do: Put some shot in the bottom of the test tube. Seal the test tube with the cork. Place the test tube in a glass of water. The test tube should float and be upright in the water. Half the test tube should be under the surface of the water. Add or subtract shot until the test tube floats in the correct way.

Place a strip of adhesive vertically up the side of the test tube, as shown.

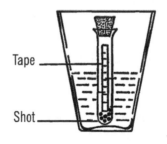

Specific Gravity Table	
Water	1.00
Ice	.92
Gasoline	.75
Alcohol	.8
Milk	1.03

Can You Work Like a Scientist?

1. Place your test tube hydrometer in a glass or jar full of water. Note the level of the water on the side of the test

---

58 See Note 19 in Appendix E about specific gravity vs. density.

59 BBs and other types of metal shot can be found at sporting good stores.

tube. Place a mark on the tape at this level and label the mark 1.00. Most hydrometers use the density of water as a standard and call this standard 1.00.

2. If the molecules of water were closer together, would the hydrometer float higher or lower in the water? What if the molecules were heavier?

3. If the molecules were farther apart (less dense), would the hydrometer float higher or lower in the water?

4. Make a mark on the glass or jar to show the level of the water. Be sure your test tube hydrometer is floating in the water.

5. Add salt to the water. Does the salt raise the level of the water in the glass or jar?

6. How much salt must be added before the water level starts to rise?

7. What happens to the salt you add to the water? Does all of the salt sink to the bottom of the jar?

8. If some of the salt dissolves in the water, is the water more or less dense? Remember you added salt molecules (NaCl).

9. Does the hydrometer float higher or lower in the water? Try adding sugar instead of salt.

10. Try other liquids. Make a mark on your test tube hydrometer for each of these liquids. Above is a chart of some liquids and their specific gravity as compared with water. Place these numbers on your test tube. Specific gravity is the weight of a sample of a liquid as compared with an equal amount (volume) of water. Specific gravity depends on the weight of the molecules and the number of molecules in a certain space.

11. Can you find the specific gravity of cream? Of milk? Which is the highest?

12. Can you find the specific gravity of different weights of motor oil?

13. What effect has heat on the specific gravity of liquids? (Be careful not to heat liquids that will burn, such as gas or alcohol.)

14. What effect has cooling on the specific gravity of liquids? (No danger here. Remember to use your refrigerator.)

15. Could you separate different liquids by their specific gravity? Try ice, alcohol, water, gasoline. What are your problems?

Cartesian Diver

Purpose: A cartesian diver is used in the study of the buoyancy of objects in water and other liquids.

Materials: *Type A*: a gallon jug, balloon or sheet rubber, and an eyedropper. *Type B*: a flat-sided bottle with a screw lid and an eyedropper.

What to Do: *Type A*: Fill the gallon jug with water. Let the water stand in the jug until it is room temperature. Fill the eyedropper partly full of water. Try to float the eyedropper in a glass of water. If the dropper sinks, remove a drop or two of water until the amount of water in the dropper is just enough to allow the dropper to barely float. Place the dropper in the gallon jug and then cover the opening to the jug with rubber from a balloon or sheet rubber.

The rubber can be fastened in place with tape or rubber bands.[60]

Type B: Fill the flat-sided bottle completely full of water. Add water to the eyedropper until it barely floats in a glass of water. Carefully place the eyedropper in the bottle and screw the lid down tightly.

Operation of Equipment: The eyedropper is normally less dense than water and will float. The air inside the eyedropper prevents any water from entering. When you add water to the eyedropper, you make the dropper heavier, and yet you don't increase its volume or size. When you reach a point where the eyedropper plus the water it holds exactly balances the weight of the water the eyedropper pushes aside because of its size, the dropper will barely float on the surface. If any water is added to the dropper, the weight of the dropper will become greater than the water it pushes aside, and it will sink. In type A, when you push on the rubber, you compress the air in the neck of the bottle. This air pushes on the surface of the water. Since water can't be compressed, the added pressure pushes against the air in the opening to the eyedropper. As more pressure is applied, the air in the eyedropper is compressed and more water enters it. This makes the eyedropper heavier, and it sinks. When you release the rubber, the pressure is reduced, and the air inside the dropper pushes the excess water out. The eyedropper becomes lighter and floats.[61]

Can You Work Like a Scientist?

1. Notice the level of the water in the eyedropper as you press on the rubber. Can you see the level of the water change?

2. When you press on the sides of the flat bottle, why does the eyedropper sink?[62]

3. Can you stop the eyedropper halfway down the jar?

Problems to Investigate in the Study of Chemistry

(P)-Primary

(I)-Intermediate

(U)-Upper

1. Can you list the materials around you which are solids? Liquids? Gases? *(P)*

2. Can you change materials (matter) from one state to another (solid to liquid)? *(P)*

3. What does the flame of a candle consist of? Blow out a candle and then bring a lighted match into the gas given off.[63] *(P)*

---

60 As an alternative, you can use an old 2-liter soda bottle for type A. Since a 2-liter bottle is squeezable (unlike the glass gallon jug), you can directly screw the cap on and omit the rubber sheet. Can you understand why this is equivalent?

61 Apparently, you can use a (sealed) ketchup packet instead of an eyedropper in this experiment. How and why does that work?

62 Related questions: What else could make the eyedropper float or sink? How does a Galilean thermometer work? Can you now build one?

63 A long fireplace match is a good type of match to use for this experiment.

4. Can you collect the vapors from a candle and change them into a solid? Conduct the vapors into a cold bottle by using bent glass tubing. *(I)*

5. Is hydrogen gas given off during the burning of a candle? Hold a cold glass over a candle flame. Test the inside of the glass with your finger. *(P)*

6. Does a candle contain carbon? Hold a dish over a candle flame. *(P)*

7. What part of a candle flame is the hottest? *(I)*

8. Does a candle flame give off carbon dioxide? Collect gas from a candle flame. Pour Lime water into the bottle containing the gas. Lime water turns cloudy if carbon dioxide is present. *(P)*

9. Can you measure the amount of heat given off at various parts of a candle flame? *(U)*

10. What is the difference between a mixture and a compound? Mix iron filings and sulfur together. Is this a mixture or a compound? Can you separate the iron and sulfur? *(I)*

11. Mix two grams of sulfur and 3.5 grams of iron filings in a test tube. Heat with an alcohol burner. Can you separate the iron and sulfur? Is this a compound or a mixture? *(I)*

12. What materials will dissolve in water? What materials will not dissolve in water? *(P)*

13. What effect does heat have on the ability of water to dissolve materials? Try dissolving different materials in water of varying temperatures. *(P)*

14. What effect does the amount of surface area have on the rate at which a substance dissolves in a liquid? Try dissolving a large piece of material. Then try breaking the material into small pieces. *(I)*

15. In what reactions is water a catalyst? *(U)*

16. What is the water cycle? Can you make an artificial water cycle? *(P)*

17. Can you determine the amount of minerals in different water samples by the process of distillation? Do you lose any of the water during the process? Be sure to measure and keep track of your findings. *(I)*

18. How can you separate water electrically by electrolysis? If you use washing soda as an electrolyte (substance in water to help conduct electricity), what effect does the amount of electrolyte in solution have on the rate of gas production of hydrogen? Of oxygen? *(I)*

19. In breaking water apart by electrolysis, what does increasing the distance between the electrodes have on the production of hydrogen and oxygen? *(I)*

20. What is the ratio of hydrogen to oxygen in various types of water? Is the ratio the same in sea water? *(I)*

21. What is the effect of increasing the amount of current on the rate at which water breaks up during electrolysis? Will electrolysis work with both AC and DC current? *(I)*

22. What pole gives off hydrogen? Oxygen? Is this always true? *(I)*

23. What part of different materials consists of oxygen? Can you think of a way to determine this? *(U)*

24. Does soil contain oxygen? Does water contain oxygen? Do rocks contain oxygen? Place soil in water. Observe closely. Watch for gas bubbles in water. *(P)*

25. Oxygen forms what part of air? Pour water around a burning candle attached to the bottom of a bowl or pie plate. Place a baby bottle over the candle. How high did the water rise in the bottle? This distance indicates the amount of oxygen used up in the burning. *(P)*

26. What effect does oxygen have on burning wood? Get a small bottle of oxygen from a shop that has a cutting torch. Light a splinter of wood and blow out the flame. Quickly lower the wood into the bottle of oxygen.[64] *(P)*

27. What effect does oxygen have on burning? Heat some steel wool in the flame of an alcohol burner. Place the red hot steel wool into a bottle of oxygen. Light some sulfur with a match. Lower the sulfur into a bottle of oxygen. *(I)*

28. What materials will burn? What materials will not burn? *(P)*

29. What temperature is necessary for different materials to start to burn? This temperature is called the kindling temperature. *(I)*

30. Will all wood start to burn at the same temperature? *(P)*

31. Is oxygen heavier or lighter than air? Can you determine this experimentally? *(I)*

32. Is water given off when hydrogen is burned? Hold a cold glass over a hydrogen flame.[65] *(I)*

33. What materials can be used to generate hydrogen? *(I)*

34. Is hydrogen lighter than air? *(P)*

35. To what altitude will a homemade hydrogen balloon rise? *(I)*

36. What is the effect of temperature and pressure on the rate at which a balloon filled with hydrogen will rise? *(U)*

37. Can you set up a carbon cycle in a closed aquarium? *(P)*

38. How does a carbon dioxide fire extinguisher work? Can you make a homemade extinguisher?[66] If you use baking soda and vinegar, what effect does the temperature of the vinegar have on the thrust of the water out of the nozzle? *(I)*

39. Can you pour invisible carbon dioxide? Is carbon dioxide heavier than air? *(P)*

40. What effect does carbon dioxide have on burning? *(P)*

41. Why does lime water turn milky when exposed to carbon dioxide? Make lime water by mixing a teaspoon of hydrated lime with one pint of water. After the lime sinks to the bottom of the bottle, filter the liquid and then screw the lid down tightly. *(I)*

42. Does your breath contain carbon dioxide? Blow through a straw into a test tube containing lime water.[67] *(P)*

43. Does the air around us contain carbon dioxide? Place a dish of lime water on a

---

64 A "bottle" in this context could be (for example) a single *mason jar* filled with oxygen at atmospheric pressure. Colloquially, a "bottle" of oxygen usually refers to a container of highly compressed oxygen—which you *do not want*. Compressed oxygen is extremely dangerous.

65 Ask a glassblowing shop to help you with this; they have special hydrogen torches.

66 Even though the answer to this is "yes," do not consider a homemade extinguisher to be a substitute for having a "proper" one on hand for safety!

67 Take care not to ingest any of the lime water. A better procedure is to blow into a long rubber tube attached to the straw.

table. Observe later. If carbon dioxide is present, there should be a scum on the water. *(P)*

44. What effect does exercise have on the production of carbon dioxide? Run a distance and then test your breath by blowing into a test tube containing lime water. *(I)*

45. What rocks when combined with acids will produce carbon dioxide? Try marble, limestone, granite, and other common rocks. *(I)*

46. What part of the atmosphere is nitrogen? Burn a candle in a bowl of water. Cover the candle with a baby bottle. The gas remaining in the bottle after the water rises is almost all nitrogen. *(P)*

47. Can you determine the solubility of various gases? *(U)*

48. What effect does temperature have on the solubility of gases in water? *(U)*

49. What gases are soluble in liquids other than water? How does the solubility in other liquids compare with that of water? *(U)*

50. How can you produce ammonia gas? Heat household ammonia gently and collect gas. What are the properties of ammonia gas? Is it heavier or lighter than air?[68] *(I)*

51. Is ammonia gas soluble in water? Place a test tube of the gas upside down over a bowl of water. Remove your finger covering the opening. The solubility of the gas is shown by the amount of water that rushes up the test tube. *(I)*

52. What effect does ammonia have on litmus paper? Check both the liquid and the gas forms. *(I)*

53. What causes the white smoke when the fumes of hydrochloric acid come in contact with the fumes of ammonia? Dampen the inside of a jar with hydrochloric acid. Fill a second jar with ammonia gas. Place a piece of cardboard over the jar with hydrochloric acid. Turn the jar containing ammonia gas over and place it on the cardboard. Remove the cardboard. *(I)*

54. What effect does chlorine have on living things? Place a drop of Clorox or other bleach in a drop of water containing protozoa and other microscopic plants and animals. Examine the life through a microscope or microprojector. *(P)*

55. What liquids and other materials contain chlorine? Mix a half teaspoonful of starch with about 60 ml of water. Bring the water to a boil. Dissolve a very small amount of potassium iodide (about as much as 4 grains of rice) in the mixture. Dip strips of filter paper or paper towels in the mixture and then dry them. A strip will turn blue in the presence of chlorine. *(I)*

56. Can chlorine be made from hydrochloric acid? Put a gram of manganese dioxide into a test tube. Add 6 ml (about ¼ of the test tube) of hydrochloric acid and heat gently. Test with chlorine test paper. Be careful not to breathe into the gas directly. Move your hand back and forth over the test

---

68 Ammonia gas is toxic. Use extreme caution and excellent ventilation. *If you can smell it, you are risking excess exposure.*

tube and sniff the air as the gas moves toward you. *(I)*

57. How can chlorine gas be made in large amounts?[69] Pour about an inch of Clorox or other bleach into the bottom of a baby bottle. Add about half a teaspoonful of sodium bisulfate. Collect the gas by using a stopper with a piece of glass tubing leading to a second bottle. A third bottle containing lye mixed with water is connected to the second bottle by glass or rubber tubing. The lye water then absorbs the excess chlorine gas formed. Be sure chlorine gas does not escape into the room. *(U)*

58. Is chlorine water soluble? Add some water to a bottle containing chlorine. Cover the mouth of the bottle with the bottom of your hand. What should happen if some of the chlorine dissolves in the water? *(I)*

59. Will chlorine react with hydrogen and hydrogen compounds? Lower a burning candle into a bottle of chlorine. The candle is made of hydrogen and carbon. If the chlorine combines with the hydrogen, what should be given off? *(I)*

60. Will chlorine combine with different metals? Twist a piece of wire around some steel wool. Heat the steel wool with a match. Lower the steel wool into a bottle containing chlorine. If the chlorine mixes with the iron in the steel wool, iron chloride should be formed. Iron chloride is a brownish gas. *(I)*

61. Does chlorine bleach cotton and linen? Hang a colored strip of cotton or linen cloth in a bottle of chlorine. Cover the bottle. Try a second bottle of gas, but this time moisten the cloth. If it is the chlorine that does the bleaching, the dry cloth should turn white. *(I)*

62. What effect does chlorine have on living things? Place an insect in a bottle containing chlorine gas. Can you devise a way of testing the effect of chlorine gas on micro-organisms such as paramecium? *(I)*

63. Is chlorine heavier than air? Test to see if chlorine will rise out of a bottle. Be careful not to breath the gas. *(I)*

64. How can you recognize an acid? Dilute acid, such as hydrochloric, in a ratio of one part acid to three parts water. Dip the tip of one finger in the acid water mixture. How does acid taste?[70] How does an acid feel to the touch? Wash your hand immediately in cold, soapy water. *(P)*

65. How can you detect an acid? Place a drop of acid on a strip of blue litmus paper. Does the litmus paper change color? Try other acids. Are the results the same? *(P)*

66. What effect do acids have on metals? Place small bits of a metal in a test tube. Add a strong acid, such as hydrochloric. What happens to the metal? Is a gas given off? *(P)*

67. Do acids have the same effect on all metals? *(P)*

---

69 We recommend reading about and *not experimenting with* chlorine gas. Chlorine gas is toxic, and producing large amounts of it carries significant safety risks. If you have a compelling reason to generate chlorine gas, consult with a professional chemist or chemistry professor about proper safety practice first.

70 Only ever taste or touch very well diluted *food-grade* acids that are sold as food. Other acids, or concentrated versions, can be extremely harmful to your body. Food-grade hydrochloric, citric, acetic, phosphoric, and malic acids are potentially good choices if you wish to taste one.

68. Do all acids react the same on a certain kind of metal? *(I)*

69. How can you recognize a base? Dissolve a teaspoonful of lye[71] in a half glass of water. Place about ten drops of this solution into a glass of water. Dip your finger into this very diluted solution and then taste. *(I)*

70. Do all bases have the same characteristic taste? Try other bases.[72] Be sure to dilute the base with a large amount of water. *(I)*

71. How can you detect a base? Place a drop of lye on blue litmus paper. Try placing another drop on red litmus paper. *(P)*

72. What liquids in your household are bases? *(P)*

73. How can you make litmus paper? Slice leaves of red cabbage into strips. Boil these cabbage strips in hot water and let stand for about a half hour. The liquid can then be used as an indicator. You can soak: paper towel strips in the colored water and then let dry. Try making other indicators by using blueberries, cherries, different flowers, and other plants and vegetables. *(P)*

74. What effect do acids have on bases? Make a lye solution. Place a drop of phenolphthalein solution into a small solution of lye in a test tube. The phenolphthalein should color the solution. Now add an acid such as hydrochloric. Is the solution of hydrochloric acid and lye base an acid or a base? *(I)*

75. What effect does a base have on fat material? Drop a lump of fat into a test tube containing a lye solution. Heat gently. What happens to the fat? What does the solution feel like? *(I)*

76. What effect do bases have on acids? Put a drop of phenolphthalein solution into a small amount of diluted hydrochloric acid. Pour this solution into a lye solution. The ratio of hydrochloric acid to lye solution should be about two parts acid to five parts base. Try other acids and bases and see if all bases act on acids in the same manner. *(I)*

77. How is phenolphthalein solution made? Get some phenolphthalein powder from a drugstore.[73] Mix a pinch of the powder in a one-ounce bottle of denatured alcohol. Try out your mixture on different bases. *(U)*

78. What solutions conduct electricity? Wire up a flashlight bulb in series with a container for liquids. Attach carbon rods to the two ends of the wires going into the liquid solution. When current passes through a liquid which conducts electricity, the bulb will light up. If the liquid will not conduct electricity, the bulb will not light. *(I)*

79. What effect does temperature of the water have on the amount of a material that will dissolve in it? Try dissolving a measured amount of salt in cold water. Try dissolving the same amount in warm water. Can you keep a graph of the amount of salt that will dissolve at different temperatures? *(I)*

---

71 Use *food-grade* lye.

72 Tread carefully: some bases are outright poisonous. Don't taste anything if you're not certain that it is safe to eat. A good one to try is hydrated lime (calcium hydroxide, not the fruit), which is used in food preparation. Another to consider is baking soda, which is technically *amphoteric*, meaning that it can act as either an acid or base in different circumstances.

73 Likely *not* available at your local drug store. See Appendix A.

80. What is a saturated solution? A super-saturated solution? Add a chemical to water. Stir until you find that some of the chemical won't dissolve with any amount of stirring. The solution is saturated. Heat and add more of the chemical. The solution is now supersaturated—it holds more of the chemical than it normally can at room temperature. Cool the liquid. What should happen? *(I)*

81. Will all chemicals dissolve in the same amounts in water at given temperatures? Can you keep a chart of the amount of different chemicals that will dissolve at a given temperature? How does the amount that will dissolve compare with the atomic weight of the substance? *(U)*

82. What effect does temperature have on the formation of crystals? Make a supersaturated solution of alum or Epsom salt. Pour some of the solution on a warm piece of glass or microscope slide. Pour another sample of the solution on a cold pane of glass. *(I)*

83. Can you determine the temperature at which different chemicals in solution will crystallize? *(U)*

84. If you add materials to a liquid in order to make a solution such as salt to water, does this raise or lower the freezing and boiling point of the liquid? Try various chemicals and liquids. *(I)*

85. What effect has the rate of evaporation on the formation of crystals? Make a supersaturated solution of sugar. Pour some of this into several jars. Suspend a string into the center of the sugar solution in each jar. Control the rate of evaporation by using jars with different size openings. *(I)*

86. What foods and materials found around the house contain acid? Which ones contain bases? Try ammonia, tea, soda pop, lye, aspirin, grapefruits, oranges, milk of magnesia, lime water, cleanser, tomatoes, vinegar, milk, cream, and similar materials. *(I)*

87. Can you determine the pH (acidity or alkalinity) of the soil in various areas around your home or city? Use litmus or pHydrion paper. *(U)*

88. What effect does the pH of the soil have on the type of native plants found growing in it? *(U)*

89. Can you determine the normal pH of the cells of various plants and animals? *(U)*

90. Can you determine the normal pH condition in the mouths of various animals and humans? *(I)*

91. Can you determine the strength of various bases by titration? Make up a weak solution of ammonia and water. Add a drop of phenolphthalein to color the solution pink. Add drops of dilute hydrochloric acid until the color is gone. Compare the amount of acid needed in order to neutralize the base to the amount of ammonia solution. This is the relative strength of the base. *(U)*

92. Can you experiment with different acids and bases in order to determine what kind of salts are formed when the acids and bases neutralize each other? *(U)*

93. Can you make a salt from a metal and an acid? Try a few drops of hydrochloric acid on zinc or aluminum strips. Test with blue and red litmus paper. A salt should not affect either colored litmus paper. *(I)*

94. How can iodine crystals be collected? Mix two parts of sodium bisulfate with one part of potassium iodide and one part of manganese dioxide. Heat the mixture gently. Collect the violet gas fumes on the bottom of a pan containing ice and water. The gas vapor cools rapidly and crystallizes on the bottom of the pan. Scrape the crystals off the bottom of the pan and store in a tightly closed bottle. Be careful not to breathe the gas fumes. *(U)*

95. How do iodine crystals appear under polarized light? Use the polarizing filters on either a microprojector or a microscope. Examine a crystal on a microscope slide. *(U)*

96. In what liquids is iodine soluble? Place a few crystals in a test tube containing water. Try the crystals in water containing potassium iodide. Try dissolving the crystals in alcohol, fingernail polish remover, and other liquids that evaporate quickly. *(U)*

97. How is iodine used as a test for starch? Mix starch with water, and then bring to a boil. Mix a drop of the starch solution with 10 cc of water. Place a drop of iodine into the water and starch solution. *(P)*

98. What foods contain starch? Use the iodine test. *(P)*

99. Is iodine freed when chlorine is added to a solution of potassium iodide crystals? Add a few drops of Clorox or other bleach to a solution of potassium iodide crystals. *(I)*

100. How can iodine stains be removed? Stain a cloth with iodine. Mix a few hy-po (sodium thiosulfate) crystals in water. Place drops of hypo on the stain. *(I)*

101. What do you observe when you melt sulfur? Does the sulfur seem to go through stages? *(I)*

102. In what form does sulfur exist as crystals? Examine flowers of sulfur under the microscope. Heat sulfur in a test tube. Filter the sulfur through a paper towel or filter paper. Examine crystals formed on filter paper under the microscope. Examine under polarized light. *(I)*

103. What effect do sulfur fumes have on the color in different materials? Heat a small amount of sulfur powder in metal lid. Hold the lid by wrapping a stiff wire around the lid for a handle. When the sulfur in the lid starts to burn well, lower the burning sulfur into a mason jar. Sulfur dioxide fumes will be given off in the jar. Place differently colored materials into the jar and then cover the jar so that the fumes cannot escape. *(I)*

104. How is sulfuric acid made? Lower burning sulfur into a jar. After the fumes fill the jar, remove the burning sulfur and add a few cc of water. Shake the bottle and then test with litmus paper. *(I)*

105. Is sulfur dioxide soluble in water? Fill a baby bottle with sulfur dioxide fumes. Insert a one-hole stopper containing a piece of glass tubing. Invert the gas-filled bottle over a container of water. If the gas is water-soluble, the water should run up the glass tubing and into the bottle containing the gas. *(I)*

106. Will sulfur dioxide support burning? *(I)*

107. How can you make hydrogen sulfide in your home laboratory? Fill a test tube one-eighth full of powdered sulfur. Add a small lump of candle wax. Heat the test tube. Hydrogen sulfide has the smell of rotten eggs and is quite unpleasant.[74] *(I)*

108. How is hydrogen sulfide used in analyzing the type of metal found in an unknown salt? Generate hydrogen sulfide. By the means of rubber and glass tubing, bubble the gas through the unknown metal salt. The color of the solution indicates the metal present. Try this method in making an analysis of unknown salt solutions. *(U)*

109. Will hydrogen sulfide burn? *(I)*

110. How do you grow a silicon garden? Place a layer of sand on the bottom of a wide-mouth mason jar. Fill the jar with an equal mixture of water and water glass (sodium silicate). Add crystals of different salts such as copper sulfate, alum, Epsom salt, zinc sulfate, and sodium sulfate. Let the jar stand undisturbed. *(P)*

111. How is borax used in chemical analysis? Make a small loop by wrapping the end of a piece of nichrome wire around the end of a pencil. Insert the other end of the wire into a piece of heated glass tubing or a cork. Either the cork or the glass tubing will serve as a handle. Heat the wire loop and dip the loop into melted borax to form a bead. Touch the bead to the chemical to be tested and then heat the bead again in a very hot flame. You may use a blowpipe with an alcohol lamp. The color of

the bead when cold compared to the color when the bead is hot is used to determine the metal. *(I)*

112. What are the properties of boric acid? Make boric acid by heating a solution of two parts borax. to five parts water. Add one part hydrochloric acid to the boiling solution. The boric acid crystallizes out as the solution gradually cools. Filter the liquid and then wash with cold water to remove salt formed with boric acid crystals on the filter paper. Let the crystals dry on the filter paper. Examine under polarized light. *(U)*

113. How can you test for boric acid? Make an indicator paper by dipping strips of paper toweling in mustard.[75] Wash the mustard off and allow the strips to dry. The strip turns brown when exposed to boric acid. This is because of the coloring matter (turmeric) in the mustard. *(I)*

114. Can you use the flame test in order to identify sodium and potassium compounds? Clean your nichrome wire loop with hydrochloric acid and then heat the loop. Dip the loop into the unknown compound and then hold the loop in the flame. Sodium gives off a bright yellow-red color. Potassium gives off a violet color and can be viewed best by observing through blue glass or cellophane. Try table salt and potassium nitrate (saltpeter).[76] *(U)*

115. Can you test the hardness of water around the area in which you live? Make a test solution by dissolving about a gram of soap flakes in about twenty cc of denatured alcohol or duplicator fluid. Filter the solution. Test

---

74 Hydrogen sulfide is not just unpleasant, it is toxic. *Never* directly smell chemicals by placing them close to your face. Allow the smell to diffuse towards you naturally, or waft the air towards you with your hand.

75 Classic American-style yellow mustard, not Dijon!

76 What other chemicals and elements will give interesting flame tests? How can you find out?

the unknown sample by filling a baby bottle half full of the water. Add about ten drops of your soapy test solution to the water. Cover and shake the baby bottle. The amount of foam indicates the degree of hardness, with very hard water making little foam. Check the amount of foam formed by using rain water and distilled water. (I)

116. Can you distill hard water and remove impurities? Test your distilled water with a hardness test. (I)

117. What are the properties of magnesium? Hold a piece of magnesium ribbon with a pair of tweezers. Light the end.[77] Test the magnesium also for reaction with acids such as vinegar. Place a piece of magnesium ribbon in boiling hot water. (U)

118. Can you determine the density of different metals? Weigh the sample. Determine the volume by the amount of water displaced when you submerge the metal in water. Figure the weight per unit volume. (I)

119. Do metals give off a characteristic color when they burn? Sprinkle bits of different metals into the flame of an alcohol burner. Record the colors given off. Aluminum is easily available in the form of pie tins or foil. (I)

120. What is the reaction of aluminum with a base? With an acid? Drop strips of aluminum into HCl. What gas is formed? Make up a weak-base solution of sodium hydroxide (lye). Drop the aluminum strips into the lye. What gas

is given off? What is left in the bottle? (I)

121. Can crystals be grown with double salts? Make a supersaturated solution of potassium aluminum sulfate or ammonium aluminum sulfate. These double salts are called alums. Heat your solution and add your salt until no more can be dissolved. Strain off the liquid and allow the undissolved alum to cool. Save the liquid. Pick out the largest of the crystals and discard the rest. Replace the discarded alum crystals with an equal amount of fresh alum salt. Heat to dissolve the new crystals in the liquid you have saved. Allow this solution to cool and then pour the solution into a baby bottle. Tie a thread around the largest crystal you have saved. This crystal is then suspended in the solution in the baby bottle. The solution should be allowed to evaporate slowly and remain undisturbed. (P)

122. How can alum be used to clear water? Add a spoonful of dirt to two jars of water. Stir to mix the dirt throughout the water. In one of the jars add about a half teaspoonful of alum and two teaspoonsful of ammonia. Why does the dirt seem to settle out? What effect does temperature have on this? (I)

123. How small is a molecule? Dissolve a gram of potassium permanganate in 100 cc of water. This gives a solution of 1/100 or 1 to 100. The color is due to the $KMnO_4$ molecules moving around in the water. Remove 10 cc of this solution and add to 90 cc of fresh water.

---

77 Once you light it, magnesium burns rapidly at extreme temperatures and emits a blinding light. Use a #14 welding shade or equivalent to protect your eyes. See Note 13 in Appendix E for more information. burning metal presents a special hazard, in that fire extinguishers will only make a magnesium flame *bigger*. Have a plan for extinguishing the flame if you need to. A metal flame can usually be extinguished under a pile of *fully dry* (not moist) sand. However, it is usually best if you can set the piece of burning magnesium down on a completely fireproof surface and just allow it to burn out on its own.

You now have a solution of 1 to 1000. Can you still see a color? Repeat this with several additional bottles of water. Be sure always to take your colored solution from the bottle containing the weakest solution. Can you still see the molecules after you have diluted the solution to 1 to a million parts? *(P)*

124. Will all iron rust? *(P)*

125. What effect does humidity have on rusting of iron? Wedge a piece of steel wool into the bottom of a glass. Invert the glass over a pie tin containing water. For your control, repeat the experiment but don't use any water. Compare the results after several days. If water rises in the glass, something must have been used up out of the air in the glass. *(P)*

126. How can rust be prevented? If rust is iron reacting with oxygen in very slow burning, could you coat iron nails with different materials to prevent the oxygen from reaching the iron? Will rust occur without moisture? *(P)*

127. How is copper sulfate used in chemical analysis? Crush some copper sulfate crystals and pour them into a test tube. Heat and stir these crystals until they have formed a white powder. You have removed all the water from the copper sulfate. If you add a drop of a liquid that does not contain water, the crystals will not change. If you add a drop of water, blue crystals form. You can test many liquids for the presence of water. Try rubbing alcohol, gasoline, vinegar, and others. *(I)*

128. What is the replacement series for metals? Metals vary in their amount of activity. The replacement series is a list starting with the most active metal, potassium, and going down to the least active metal, gold. You can discover the correct order of the metals in this series by simple experiments. If a metal such as iron (a nail) is placed in a solution of a salt of a less active metal such as copper sulfate, the more active iron will replace the less active copper. The copper then will form around the nail and plate the nail. Iron then is more active than copper. If we place a piece of copper in a solution of silver nitrate, we find the copper replaces the silver, and the silver is plated on the copper metal. Therefore, copper is more active than silver. All of these activity experiments can be performed under the microscope or microprojector. Place a strand of wire of the metal being tested on a blank microscope slide. Add a drop of the metal salt solution to the wire. *(I)*

129. How many different kinds of plating can you do? Remember to use one of the metals listed in the replacement series and a salt of another metal that is listed below the first metal (one that is less active). *(I)*

130. What causes silver to tarnish? *(I)*

131. Can you make your own photographic paper and make photographs with it? Mix silver bromide with gelatin and spread on a heavy paper. Fasten the paper to a piece of plywood and place it in the sunlight. Place some object such as a leaf on the paper and then cover with a piece of glass or Saran Wrap. In order to fix the print after the paper has turned a dark violet, soak the paper in a solution of hypo for about ten minutes. *(U)*

132. What forms of carbon will conduct electricity? Try different types of coal, graphite (pencil lead), diamonds, charcoal, and others. *(I)*

133. What chemicals compose coal? Crush a lump of bituminous coal into a powder. Fill a test tube about one-quarter full of this powdered coal. Place a wad of cotton near the mouth of the test tube to act as a filter. Insert a rubber stopper containing an L-shaped piece of glass tubing. The opening to the tubing should be drawn to a jet point. Heat the coal in the test tube over an alcohol burner. Fumes will form and escape through the jet point. Will these fumes burn? Test the gas by inserting litmus paper into the test tube. Ammonia will turn red litmus paper blue. Acetic acid will turn blue litmus paper red. *(I)*

134. What chemicals compose wood? Try the same experiment as above. *(I)*

135. What foods contain carbon? Heat small bits of such foods as bread, potatoes, cheese, and sugar. What is the final product formed after you have heated the food material until it "burns?" *(I)*

136. Will sugar burn? Can the vapors given off by heated sugar be ignited? *(I)*

137. Does methane gas come from coal? Break lumps of bituminous coal into a powder. Fill a funnel with the coal powder and place a bottle over the funnel. Turn the bottle over so that the funnel containing the coal powder is resting on the bottom of the bottle. Fill the bottle with water and then place a test tube containing water over the opening to the funnel. If a gas is given off, the gas will rise in the test tube and slowly force the water out. You may have to wait several days. *(I)*

138. What are the properties of methane? First make sodium acetate by adding washing soda (sodium carbonate) to a half cup of white vinegar until all the carbon dioxide possible is given off. Evaporate the liquid slowly at a low heat. The white powder that remains is sodium acetate. Now mix equal amounts of sodium acetate, calcium oxide, and sodium hydroxide in a test tube and heat slowly. Collect the methane gas given off by bubbling through water.[78] *(I)*

139. Can a gas be turned directly into a solid? Crush moth balls and then heat gently.[79] The gas given off is naphthalene. Place a jar containing ice over the gas vapor being given off by the heated moth balls. If the gas can be turned directly into a solid (sublimation), crystals will form on the bottom of the jar. *(I)*

140. Is turpentine a hydrocarbon? Pour a little turpentine into a jar lid. Place a short piece of clothesline rope or heavy string in the lid to serve as a wick. Light the wick. Hold a jar over the flame. If the turpentine is a hydrocarbon, it should give off carbon when it burns and a black soot should form inside the jar.[80] *(I)*

141. What sweet-tasting foods contain glucose sugar? Make your test solutions "A" and "B." Solution A is made by dissolving 5 grams of copper sulfate into 70 cc of water. Solution B is made by dissolving 7 grams of lye (sodium hydroxide) into 70 cc of water. Then add 25 grams of Rochelle salt (sodium potassium tartrate) to this solution. In or-

---

78 Use caution; methane is a highly flammable gas. It is the same "natural gas" that is used in most gas stoves and heaters.

79 See Note 20 in Appendix E about mothballs.

80 Remember proper safety practice when working with flames. See Note 3 in Appendix E.

der to use the test solutions, heat a mixture of 3 cc of solution A and 3 cc of solution B. Add a few drops of the material to be tested to the mixture. If glucose is present, a red precipitate (solid) of cuprous oxide will be formed. Test fruits, honey, molasses, corm syrup, cane sugar, maple syrup, and beet sugar. *(I)*

142. Can one type of sugar be changed into another? Dissolve 2 grams of cane sugar (sucrose) into 20 cc of water. Add about 15 drops of hydrochloric acid. Heat gently and then test with solutions A and B. If the sucrose turns to glucose a red precipitate should be formed. *(I)*

143. How do you make a test solution for starch? Dilute one part of tincture of iodine with nine parts of water. Iodine gives a blue color to materials composed of starch. *(I)*

144. Does a potato contain starch? Grate up a potato and then place gratings into a cheesecloth. Dip the cheesecloth into a bowl of water and squeeze the gratings in the cheesecloth. Repeat many times until the juice is in the bowl. Let the material in the bowl settle and then pour off excess water. Let liquid in the bowl evaporate. Test the dried material left in the bowl with an iodine test solution. *(U)*

145. What effect does saliva from your mouth have on starch materials? *(U)*

146. Can you make and test the properties of different kinds of alcohol? *(U)*

147. What foods contain fats? Crush food material and drop some in the bottom of a test tube. Cover the food material with a few drops of carbon tetrachloride.[81] Let the material stand for about ten minutes and then pour a few drops on a piece of white paper. After the carbon tetrachloride has evaporated, examine the paper. If the food contains fat, there should be a transparent grease spot on the paper. Remember, carbon tet vapors are dangerous to breathe. *Handle with care! (I)*

148. What kind of soap or detergent gives the most suds? Fill test tubes with different kinds of detergents and soaps. Add oil drops. Which detergents and soaps mix with the oil? Add one part lime water to two parts solution. Shake the test tube and note the amount of foam compared with other soap products. *(I)*

149. What is the composition of egg white? Mix a half-and-half mixture of egg white and water. Add an equal amount of denatured alcohol. If albumin is present, it will coagulate into white flecks. *(I)*

150. What is the composition of albumin? Heat coagulated egg white in a jar lid. Test for ammonia by smell and litmus paper.[82] Continue to heat the albumin. If the albumin turns black, it also contains carbon. *(I)*

151. What does egg yolk contain? *(I)*

152. What liquids are colloids? Test by shining a pen light through the test liquid. If the liquid is a colloid, the large particles reflect light and the light beam can be seen. Try shampoos, hair oil, gasoline, and other liquids. *(U)*

---

81 Do not use carbon tetrachloride. See Note 21 in Appendix E.

82 Again: allow the smell to diffuse towards you naturally, or waft the air around with your hand.

153. Can you devise a burning test for different kinds of fabrics such as wool, silk, nylon, linen, cotton, Orlon, cellulose acetate, rayon, and others? Burn small pieces of the material in the flame of an alcohol burner. Note and record characteristics of the flame, the smell, and the remaining ash after the burning. *(U)*

154. Can you devise a chemical test for different fabrics? Try fabrics in sodium hydroxide solution and then in hydrochloric acid. *(U)*

155. Can you make rayon? Compare the rayon you make with commercially made rayon strands. *(U)*

156. Can you test different plastics to determine if they are thermoplastic (molded by heat and pressure and can be remelted and remolded many times), or thermosetting (cannot be reheated and remolded). The hot tip of a glass rod makes a dent in thermoplastic but not in thermosetting plastic. Thermoplastics burn while in flame while thermosetting plastics give off a very strong smell but do not burn. *(U)*

Physics

Astronomy and Light

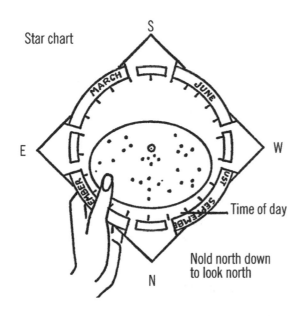

Star Chart

Star chart

S

MARCH

JUNE

E

W

DECEMBER

SEPTEMBER

Time of day

Nold north down
to look north

N

Purpose: A star chart is used as a map of the sky. It can be set for any hour of any day in the year. With this chart you can predict the position of any star, planet, constellation, comet, or satellite.

Materials: A star chart, either store-bought or you can make one yourself.[1]

You may make a large-size star chart, if you have one for a pattern, by using the following material: one large piece of black cardboard, a second piece of cardboard of a lighter color, silver or white paint, and the use of an opaque projector.[2]

What to Do: You can make a direct copy of a star chart by removing the inside black star sheet and tracing with tracing paper. The dots for stars can then be transferred to your black cardboard. You then trace the outer cover. Mount your star chart on a piece of plywood and use a thumbtack in the center of the black star dial so the dial will turn. Your stars will stand out better if you use silver or white paint.

A second way to make a star chart is to use an opaque projector. Place the sample black dial in the projector and project the larger image on the wall. Fasten your cardboard on the wall and mark the spots or stars on your cardboard with a pencil. You can paint them later. Copy the outside cover in the same manner. This enlarged star chart can be mounted on a bulletin

---

1 See Note 22 in Appendix E about getting a star chart like this. Download-and-print is a good method!

2 E.g., a document camera plus digital projector.

board with a single tack through the center of the dial. The chart could also be mounted on plywood.

Operation of Equipment: Your star chart has four corners. Hold the corner marked NORTH up. Notice the time of day right under the word NORTH. To the right of MIDNIGHT is 11 PM, 10 PM, etc. Eight o'clock is a good time to watch the stars, so turn the black dial so that the month and day is exactly over the correct hour. You have now set the chart to show you the position of the heavens at 8 PM this evening

Now turn the whole chart, depending on the direction you wish to look. If you are looking north, place the corner marked NORTH *down.* The stars you see on the lower edge of your chart are the stars just above the horizon. As you look toward the center of the chart, the stars you see are those higher in the sky. The ones in the center of the chart are those directly overhead. Can you find the Big Dipper (Ursa Major)?

If you wish to locate stars in the southern sky, place the corner marked SOUTH *down.* Again, the stars near the lower edge of the circle are those just above the horizon. As you look toward the center of the chart, the stars are found higher in the sky.

In the same manner, if you wish to look east, you would place EAST *down,* and if you wanted to study the western sky, you would place the corner marked WEST *down.*

The larger dots show you the brighter stars. By using these bright stars as signposts, you can locate all the constellations in the sky. Notice that some of these larger dots have a Greek letter by them.[3] If you look on the back of the chart, you will find the name of the star in that constellation and also the brightness or magnitude of the star.

In order to locate planets, look at the bottom on the back of the chart.[4] You will see the four planets mentioned: Venus, Mars, Jupiter, and Saturn. Use the chart for the correct year. By using the month of the year and the name of the planet, you will find the name of the constellation that the planet is either in or near. The abbreviation of the constellation is usually given. To find out the full name, look at the list of abbreviations.

In order to locate the planet, look around the outer edge of the circle for the constellation. All planets are found very close to the horizon. If the planet is visible for the hour you set your chart, the constellation will be within the black circle. Notice the direction of the constellation on your star chart. This is the direction in which you would look for the planet.

To help you recognize the planets, here are some tips: Venus is exceptionally bright, the brightest object in the sky except for the sun and the moon. Jupiter is very bright and white. Saturn is dull, greenish-yellow, and found near Jupiter.[5] Mars is bright and red.

Can You Work Like a Scientist?

1. Hold the chart over your head. Turn the dial around once. This is what happens to the sky each complete day. Do the stars seem to move? What is the real cause of this movement?

2. Locate the Big Dipper. Hold your chart with NORTH down. Rotate your star chart for one complete day. What happens to the Big Dipper? Does the Big Dipper ever disappear from sight? Can

---

3 Notations vary by star chart.

4 See Note 23 in Appendix E if yours does not have an up-to-date planet chart.

5 This was the case at the time that the original book was written but is not true in general. We can expect Jupiter and Saturn to be close to each other again in the years 2020 and 2040.

you find other stars that are always in view? Where are these stars located?

3. Where is Polaris, the North Pole star, located on the chart? Why does it seem not to move when all other stars do move?

4. Would your chart be accurate for every place on Earth?

5. Have other stars been our North Pole star during the history of the Earth?

6. Is there a South Pole star for people living in the Southern Hemisphere?

7. Are all stars in the Big Dipper single stars? Look carefully at the stars in the handle of the Big Dipper. You will need either a telescope or a pair of binoculars.

Refracting Telescope

Purpose: A telescope can be used for many purposes. A simple telescope can be used for viewing craters on the moon, the four largest moons of Jupiter, and the rings of Saturn. It is also excellent for observing sunspots. *Never look at the sun directly. Never look directly through a telescope at the sun. The light can damage your eyes and perhaps blind you!* (See "Sunspot Viewer" on page 76.) A telescope can also be used for observing objects in nature, such as birds and animals.

Materials: Two convex lenses,[6] a mailing tube or toilet paper roll, cardboard, rubber cement, and paper.[7]

What to Do: Find the focal length of your lenses. This can be done by focusing the rays of the sun or the light from a ceiling light on a piece of paper. The distance from the lens to

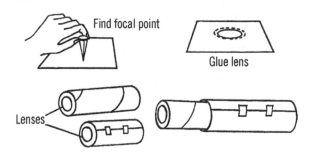

the paper is the focal length. This will help you know how long you must make your telescope. The length of each tube should be longer than the focal length of the lens. First, mount the lenses. Cut a round hole in a piece of cardboard. Glue the lens to the cardboard with the cement. You will want to mount the smallest lens[8] in the mailing tube. Place the end of the tube on the cardboard. Center the lens and then trace around the end of the tubing. Cut the cardboard just inside the line so the cardboard will just fit in the end of the tubing. Glue the cardboard lens holder in with cement. Now to make the slide tube, wrap one sheet of paper around the tubing, fastening with Scotch tape. Wrap a second sheet over the first sheet and fasten with Scotch tape. Slide both sheets off the tubing and remove the inside sheet. The outside sheet can be used for your slide tube and will just fit. Mount the holder of the second lens in the end of the paper tubing with cement. Slide the two pieces of tubing together, and your telescope is completed. You can shorten the slide tube by cutting the paper with scissors.

Operation of Equipment: The eyepiece of your telescope is the end with the mailing tube. Look through this end and sight on a book or sign at the other end of the room. Move the

---

6 See Note 24 in Appendix E for some sources of lenses.

7 Rather than starting with paper and paper tubes, can you design a version of this that could be 3D printed?

8 The one with the shorter focal length.

slide tube back and forth until the object comes in focus.

Can You Work Like a Scientist?

1. Is the object that you see upright or upside down?

2. Turn the telescope around and view from the other end. Is there any change?

3. Refracting means bending. When light goes through glass, the glass is thicker than air, and part of the light is slowed down. Part of the light from the object then is traveling faster than the other part. The object seems bent. Can you test this theory by using water for your lens? Fill a glass with water. Place a pencil in the water. Can you find a position where the pencil seems to be bent? Does the water seem to slow down or speed up light? Can you try other liquids and see if light bends more or less than in water?

4. Light bends more going through some things than others. Does the angle the light enters the object have anything to do with the amount of bend?

5. Could you measure the amount of bend? Could you compare the amount of bend? This is called the refractive index. Can you find out more about different materials bending light?

6. Galileo made a telescope similar to yours and discovered many things. Find out more about the work of Galileo. Perhaps his work will suggest other uses for your telescope.

7. Sir Isaac Newton bent light rays with a prism to find out what the rays of light were made of. Can you break light up by using a mirror and a pan of water?

8. Can you measure the power of your simple telescope? Try to think of a way to tell how many times your telescope enlarges what you see. Galileo's telescope was only 25 power.

9. Fill a sink or bathtub to a depth of several inches. Make waves by dripping water into the sink or tub. Can you arrange the light so you can see the shadows of the waves? Can you make the waves bend? These rays are very similar to light waves.

10. From the drawing, can you see why a convex lens bends light?

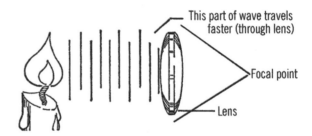

Spectroscope

Purpose: This instrument is used to examine light. The light from a star can give us a clue as to what the star is made of. This instrument can also be used to examine light from burning objects on Earth. The colors given off through the spectroscope give a clue as to the make-up (composition) of the object.

Materials: Cardboard mailing tube or roll from toilet tissue, diffraction grating.[9]

What to Do: Cut a small slit in a piece of cardboard. Glue this piece to the end of the mailing tube. Trim the edges of the cardboard so that

---

9 See Note 25 in Appendix E about gratings.

the piece just fits. Glue your diffraction grating to the other end. Before you fix the diffraction grating in position, be sure it is turned the way you want it. Read how to operate the spectroscope below. Then you will know if the grating is turned correctly.

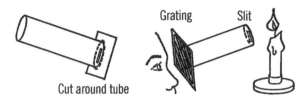

Cut around tube · Grating · Slit

Operation of Equipment: Point the end with the slit at a bright light. Look through the end with the grating. Don't look directly at the slit, but off to the side. If the diffraction grating is turned right, you should see the colors that make up the light.

Can You Work Like a Scientist?

1. Would the spectroscope work if you didn't have the slit at one end? Why does the slit help? What would happen if you made the slit in the form of an X?

2. What happens to the colors as you turn the grating?

3. What does the grating say about the lines on it? How many are there? How does this help to break up the light?

4. What are the colors of the spectrum of sunlight? Can you see these in your spectroscope? Are the colors always in the same order? Is this the same as in a rainbow? What in a rainbow works as a diffraction grating?

5. Use your spectroscope on a candle, regular light bulb, and on a fluorescent light. Are the spectrums the same? What colors are different?[10]

6. On a bright moonlight night, find the spectrum of the moon. Is this the same as the sun? Why doesn't the moon give off a brighter spectrum?

7. What are the problems of examining the spectrums of distant stars?

8. What is the spectrum (colors) of different chemicals when they are burned in the flame of an alcohol burner?

9. You can use the diffraction grating to project the spectrum on a wall or screen. Punch a hole in a piece of 2" × 2" cardboard and then place this cardboard slide in a slide projector. Hold the grating at a 45° angle in front of the beam of light.[11]

Astrolabe

Purpose: The astrolabe is used in navigation and astronomy to sight and record positions of stars. It can be used to find the latitude and the bearing of a star.

Materials: Wooden base, 2" × 2" wood upright, small piece of tin or cardboard for pointer, a protractor, and a soda straw or piece of glass tubing for the sighting tube.

What to Do: Glue the pointer to the bottom of the upright. Use the protractor and draw on the cardboard a circle of 360°. Glue this cardboard on the wood base. Fasten the upright to the base with a screw. Glue the straw or glass tubing to the protractor. Fasten the protractor to the wood upright by means of a screw. Tie a piece of string to the screw and hang a weight on it. This string will help you level the astrol-

---

10 Also interesting to look at: a white LED, colored LEDs, light from a laser pointer, and different types of streetlights.

11 The 2" × 2" size is that of a traditional photographic slide, and the hole is to make a small beam of light with a traditional slide projector. How can you make a similar narrow beam of light with a modern computer projector?

abe and also serve as the pointer to tell you the number of degrees of elevation.

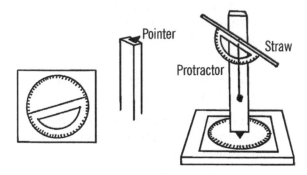

Operation of Equipment: Sight the North Star through the tube. Turn the base so that the base says 0°. Now sight any other star or object in the sky. The protractor and string line show the angle of elevation (how high the object is in the sky). The number of degrees you turn the sighting tube from true north is shown by the number of degrees the pointer moves from 0°. This shows how many degrees east or west the object is from true north.

Can You Work Like a Scientist?

1. Take readings of the North Star at different times and on different days. Are the readings always the same?

2. Take readings of a star in the Big Dipper. Are these readings always the same? Why?

3. Take readings on the moon at a certain time each evening for a week. What can you say about the path of the moon?

4. Can you plot the path of the moon across the sky by your readings? Will it follow this same path one month later?

5. Sight on the North Star. Is this true north as shown on your compass?

6. Take the readings on the sun during a day. Can you plot the path of the sun across the sky? Is it the same as the moon? (*Don't look directly at the bright sun.* See "Sunspot Viewer" on page 76.)

7. Are the readings of the sun exactly the same at the same times day after day? Why would this be?

8. From some spot in the room take a reading on the clock. Have a classmate who was not in the room try to figure the spot from which you took the reading. Try this on the playground. (Use your compass to find north.) How would this help you to navigate a ship?

Dip Circle Meter

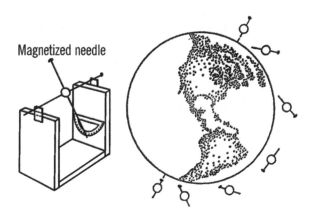

Purpose: This instrument is used to measure the angle of dip or angle of inclination at different places around the Earth. Since the needle points toward the magnetic pole, this instrument can be used to determine latitude. The greater the dip from the horizontal position the farther north a person is located. At the North Pole the needle would dip straight downward.

Materials: Wood for frame as shown, two glass slides, two rubber bands, protractor, two steel knitting needles, and a cork.

What to Do: Build the U-stand as shown. Glue the protractor on one end of the stand. Glue the microscope slides to the wood supports.

Magnetize one of the knitting needles either by rubbing the needle with a magnet or by placing the needle in a solenoid coil (see "Solenoid" on page 118).

Insert the non-magnetized needle through the cork to serve as an axle. Insert the magnetized needle as shown. Place the axle needle on the ends of the glass slides and let the needle turn freely.

Operation of Equipment: The needle is attracted to the poles of the Earth. In the Northern Hemisphere, the north-seeking end of the needle will point toward the North Magnetic Pole. If you are at the equator (0° latitude), the needle will point horizontally or parallel with the Earth. The angle of dip is zero. As you move farther north, the needle points more and more vertically. When you cross the 45th parallel, the dip circle meter is pointing at a 45° angle as compared with the protractor. Therefore, for every degree north you move, the angle of dip increases one degree.

Can You Work Like a Scientist?

1. If you were over true north, would the dip circle needle point straight down?

2. If the needle did not point straight down, what would you conclude?

3. Where on Earth would the needle point straight down?

4. The Earth's magnetic poles are constantly moving. At one time the North Magnetic Pole was over the state of Oregon. Scientists can tell this by examining the remains of adobe fireplaces built by Indians[12] long ago. The bricks contained bits of iron. When the bricks were fired, the magnetic material became fixed as a permanent magnet pointing toward the Magnetic Pole.

How can you account for the fact that these fine compasses don't point at the Magnetic Poles today?

Universal Sundial

Purpose: This sundial shows the season of the year, sunrise and sunset, and the hour of the day when the sun is shining. You can use this sundial to show the effect of the sun's rays on any spot on the Earth.

Materials: Large ball or globe of the Earth, a base to hold the ball or globe.

What to Do: If you are using a large ball, trace on the ball the outline of the continents and other features of the Earth. If you are using a globe, use the type that sets in a round circular base. In order to make such a base, cut the top and bottom off a one-pound coffee can. Set your globe sundial outdoors where the sun will shine directly on it. The globe is an exact copy of our Earth. When lined up properly, the spot where you live will be on the top side of the globe.

In order to line the globe up properly, turn the North Pole of the globe in the direction of North. At twelve noon, notice the angle of the shadow cast by a stick that is straight up and down. The axis of your globe should be set at that same angle.

You can check the position of your globe by setting the axis to point directly at the North Star on a clear night.

After you have set your globe at the correct angle, rotate the globe (east or west) so that the spot on Earth where you live is on the top side of the Earth. The sun should fall directly on the longitude lines that run through your town (or close to it).

---

12 *Native Americans.*

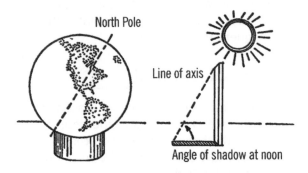

North Pole

Line of axis

Angle of shadow at noon

Operation of Equipment: After you have set your globe, fix it so that it cannot turn. The globe can be covered at night. If the globe is not disturbed, you can observe the seasons of the year, the changing angle of the sun throughout the year, difference in sun time at any time on the Earth, the speed of rotation of the Earth, and even the light as cast by the different phases of the moon on our model Earth. In order to observe the changes for a complete year, you must make regular observations for that length of time.

Can You Work Like a Scientist?

1. How many degrees does the Earth turn each hour?

2. Could you make a time clock and attach it to the top of your model Earth?

3. What part of the Earth has the longest day? The shortest day? Can you predict sunrise and sunset at different locations around the Earth?

4. Are the seasons of the year the same in the Southern Hemisphere as in the Northern Hemisphere?

5. What days of the year do we have an equal amount of daylight and darkness? Can you predict this? Is this date the same throughout the world?

6. When the sun is shining brightly, use a thermometer and take the temperature at various places on your model Earth. Can you see why the angle of the sun produces the seasons on Earth?

7. Can you tell at what spot the sun is directly overhead on your model Earth? This would be the spot where there would be the shortest shadow.

8. Can you use this idea to plot the path of the sun as it seems to travel around the Earth? Does the sun really move around the Earth?

9. Could you use a flashlight or a light bulb inside a room to demonstrate how the universal sundial works?

10. Make a tripod as shown. You can use this tripod to show the exact spot on Earth where the sun is shining directly overhead. The spot where the pin does not cast a shadow is the part of the Earth that directly faces the sun.

11. Could you use a sundial of this type for navigating a ship?

Cardboard

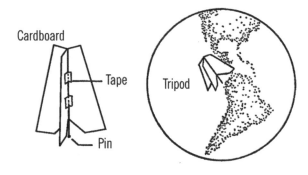

Tape

Tripod

Pin

Planetarium Model of the Solar System

Purpose: This planetarium produces the motion of the planets in their orbits around the sun. The motion of the moons around their host planet, as well as the rotation of the planets, can be shown with this model Solar System.[13]

Materials: Large wooden rod about two inches in diameter and two feet in length, old playground ball (rubber) about 6" in diameter, wood for the base, a long nail, coat hangers, and clay or styrofoam balls for the planets.

What to Do: Drill a hole into a piece of 2" × 6" wood. The hole should be the same size as the wooden rod. Cut off the top four inches of the wooden rod. Glue the remainder of the rod into the hole. Be sure the rod is straight up and down. Nail the 2 × 6 to the larger wooden base. Drill a small hole into the short piece of rod about an inch deep. Carefully drive the nail into this hole. Cut off the head of the nail.

Drill a hole into the top of the larger upright rod. The hole should be deep enough and large enough so the nail can turn easily in the hole. Next cut a small hole in the top and bottom of the rubber ball and slip the ball over the wooden rod. This represents the sun.

Drill holes in the short top wooden rod and insert coat hanger wires. These wires will be of different lengths depending on a planet's distance from the sun. These wires should balance each other.

Paint your sun with two or more coats of yellow enamel paint. Make your planets of papier-mâché, styrofoam, or clay. Since Jupiter would be quite heavy, use a ping-pong ball for such a large planet and coat with clay. You may insert dressmaker pins (pins with large colored heads)

for the moons of the planets. Hang the planets from the coat hanger with black thread.

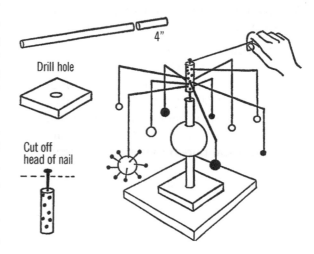

Operation of Equipment: Place a tack in the top of the short wooden rod and tie a piece of thread to it. Wrap the thread around the short rod several times. Pull on the thread, and your model Solar System should revolve around your sun. To make a planet rotate, twist the thread on the planet. The planet then will rotate as the planets orbit the sun.

Can You Work Like a Scientist?

1. Do all of the planets normally move in the same direction and on the same plane?

2. How many times would you wrap the thread around the rod to represent your lifetime? What would happen if all of the planets speeded up in their path?

Umbrella Planetarium

Purpose: A planetarium projector is made by shining a light through small holes in a round cover that goes over the light. The spots are

---

13 While "planetarium" is correct in this context, the term "Orrery" is more common.

then shown on a ceiling or dome and resemble the stars at night.

Materials: An old umbrella and a star chart or a book about stars showing star patterns.

What to Do: The cloth of the umbrella represents the dome-like sky. Punch small holes with a pin or nail to represent the stars. Use your star chart so the holes will be in the proper positions to represent various stars and constellations. Perhaps it would be best to layout your sky before you punch the holes by using chalk and placing dots on the umbrella for stars. The center of your umbrella should be the North Star. Small holes should represent faint stars. Larger holes should represent the brighter stars. If possible, paint the inside of the umbrella with flat black paint.

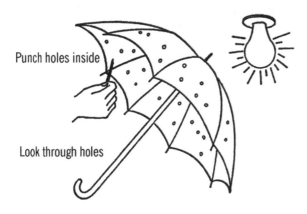

Punch holes inside

Look through holes

Operation of Equipment: Hold the umbrella over your head. The lights should be out and the shades drawn so the room is dark. There should be a light over the umbrella. As you look up through the umbrella, you should see the stars as they appear overhead. Slowly turn the umbrella. The stars will seem to move across the sky.

Can You Work Like a Scientist?

1. Which way do you turn the umbrella, left to right, or right to left in order to imitate the movement of the stars across the sky?

2. Does the sky really move?

3. If you were turning the umbrella with the movement of the Earth, how long should you take to make one complete turn of the umbrella?

4. Which stars seem to move the most? What is the location of stars that seem to move the least?

5. If you hold the umbrella directly overhead, from what position on Earth would you be viewing the stars?

6. How would you hold the umbrella to show the stars at your latitude?

7. How would you hold the umbrella to show the stars as you would view them from the equator?

8. What does the handle of the umbrella represent?

Planetarium Projector

Purpose: A planetarium projector is used to project spots of light representing stars on a dome or ceiling.

Materials: Large rubber ball, flashlight bulb and socket. As a substitute for the ball, use a large balloon and papier-mâché materials.

What to Do: Cut out a small hole in the bottom of the ball. Drill holes in the ball to represent the stars. Place the ball over the flashlight bulb and socket. Connect the flashlight bulb up to the power supply or battery.

A second type of projector can be made by using a balloon as a form. Inflate the balloon and then cover the balloon with several layers of newspaper strips soaked in wheat paste. After the papier-mâché strips dry on the balloon, remove the balloon by deflating it. Cut out part of the bottom of the papier-mâché ball and insert the bulb. Drill or punch small holes in the ball to represent the major stars and constellations. The bulb is fastened in the bottom of the ball. The North Pole Star should be punched directly over the bulb at the top of the ball.

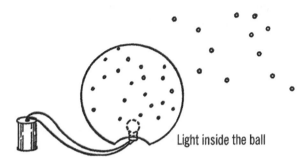

Light inside the ball

Operation of Equipment: As you turn the ball, the "stars" seem to move or rotate on the ceiling and side walls. Since you rotate the ball around the North Pole Star, this star seems to stand still while the stars farther down on the ball (south) seem to move very rapidly. When these stars seem to make one complete trip around the North Pole Star, one day and one night is completed.

Can You Work Like a Scientist?

1. Can you turn the ball so as to show the North Pole Star and constellations in the same position as they normally appear at your latitude?

2. As you move farther south, how would the position of the North Star seem to change?

3. Can you find out how the position of the stars can be used for navigating ships and planes?

4. At your latitude, what stars never seem to set, but can be seen throughout the year?

5. Can you use your planetarium to show the stars at the different seasons of the year?

Foucault Pendulum

Purpose: The Foucault pendulum is used as a proof that the Earth rotates, or turns on its axis. This model pendulum shows the principle that the French scientist, Foucault, used back in 1851 to prove that the Earth does rotate.

Materials: Three clothes hangers, cork, button, marble, thread, cake tin or metal lid to a round candy tin, and a phonograph.[14]

What to Do: Cut and straighten the coat hangers so that you have three pieces 18 inches in length. Force a hole through the cork with a nail. Insert a thread through the hole. Fasten a button on the top side of the cork and glue a marble to the other end of the thread. Insert the top ends of the coat hanger wires in the bottom of the cork as shown. Fasten the other end of the coat hanger wires to the edge of the cake tin with Scotch tape. Drill a hole in the center of the cake tin so that the tin will fit on a phonograph turntable in much the same way as a record.

Operation of Equipment: Place the cake tin and pendulum apparatus on a phonograph turntable. Start the pendulum swinging back

---

14 Steel music wire may be used as a substitute for the coat hangers. You can find a used phonograph (record player) at a thrift store or substitute a hardware store lazy susan, rotated by hand.

and forth. Then turn on the phonograph so that the turntable rotates. The rotation of the turntable is similar to the rotation of the Earth except that it is greatly speeded up.

Can You Work Like a Scientist?

1. Start the pendulum moving back and forth before you turn on the phonograph. Does the pendulum seem to follow the same path or does it change directions?

2. Time how long it takes for the pendulum to make one full swing. Does it take longer to make a long swing of the pendulum than it does to make a short swing? Why is this constant time period of a pendulum swing helpful in designing a clock?

3. If you increase the length of the cord or thread, do you increase or decrease the time for one complete swing?

4. If you shorten the thread or cord, do you increase or decrease the time for one complete swing of the pendulum?

5. Can you experiment and determine the length of thread necessary in order for the swing to take exactly one second?

6. Would the time it takes for the pendulum to make one swing be affected by changes in altitude or different positions on Earth?

7. Start the pendulum moving and then turn on the phonograph. Does the pendulum change directions when the platform beneath the ball turns?

8. If the platform represents the turning Earth, how does the Foucault Pendulum prove that the Earth turns? What would happen if the Earth did not turn?

9. Will the pendulum gradually slow up? Why? Could you design an electromagnet that could keep the pendulum moving?

Sunspot Viewer

Purpose: The sunspot viewer is used to protect the eyes when looking at the sun. *Never look directly at the sun with either a telescope or your eyes alone.* The sun is so bright that it can blind you. The effects from looking directly at the sun sometimes show up years later.

Materials: A dark exposed negative (one that was overexposed and is almost completely black) or a smoked piece of glass[15] and a shoe box.

What to Do: Cut a square hole in the end of the box. The hole should be smaller than the negative. In the event you are using smoked glass instead of the negative, hold the clear glass over a burning candle. The incomplete burning will deposit carbon on the glass. Tape the negative or smoked glass to the inside of the box over the square hole. Cut a small viewing hole in the other end of the box.

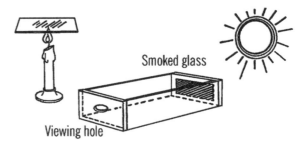

Smoked glass

Viewing hole

Operation of Equipment: Again, *don't look directly at the sun.* Be sure when you view the sun to look through the shoe box filter. The nega-

---

15 This is actually not quite good enough for the job. See "**Modern Safety Practice**" on page 77.

tive or smoked glass should dull the appearance of the sun and enable you to view the sun without harm to your eyes. Caution: If the sun appears bright through your viewer, look away quickly. Your negative or smoked glass is not dark enough. You should be able to see dark spots on the sun. These are called sunspots.

Modern Safety Practice

Protecting your eyes is a serious concern. For directly viewing the sun, you need to use a special type of dark filter—not just smoked glass or sunglasses. Use a #14 (or darker) welding shade or purpose-built solar viewing film. See Note 13 in Appendix E for extended discussion.

Can You Work Like a Scientist?

1. What are these sunspots? What causes them?

2. Are these spots always in the same place from day to day?

3. Could you keep a record of sunspots over a period of time? What relation do the sunspots have to weather conditions?

4. Are sunspots really storms on the sun? Read about what was found out during IGY (International Geophysical Year).

5. Could you use your sunspot viewer to take a picture of the sun with your camera?

6. Does the size of the sunspots change from day to day?

7. Does the sun turn (rotate), or do the sunspots move around the sun?

8. In which direction do the sunspots move? Do all the spots seem to move in the same direction? Why?

9. Are the sunspots related to the "northern lights" and magnetic disturbances that upset radio communication?

Moon Range Finder

Purpose: The moon range finder is used to determine the relative distance of far-off objects, such as the moon, distant buildings, mountain peaks, and ships far out at sea.

Materials: Yardstick, 3″ × 5″ file card or tin.

What to Do: Cut a slit in the piece of tin or file card large enough so that the tin or cardboard can slide easily back and forth. Cut a second slit about half an inch high and a quarter inch wide.

Operation of Equipment: In order to understand how your moon range finder works, try this simple experiment. Hold out a pencil at arm's length. Sight a distant object and compare the apparent height of the object with the pencil. Now move the pencil closer to your eye. Notice that the object seems to get smaller as compared with the size of the pencil. The size of the object didn't change, but its apparent size as compared with movable object (the pencil) did. Your moon range finder works on this principle. The distant object to measure can be the moon. The movable object or point of reference is the hole in the file card.

Hold the yardstick up until it touches the cheekbone under your eye. Sight through the slit in the file card on some distant object such as the moon. Move the file card back and forth until the size of the moon just fills the vertical slit in the card. Notice on the yardstick how far the card has to be held from your eye.

Try the experiment again. Does the card have to be held at the same distance from your eye in order to fill the hole in the file card? If the moon were larger, would you have to move the file card toward your eye or farther away? What if the moon were smaller?

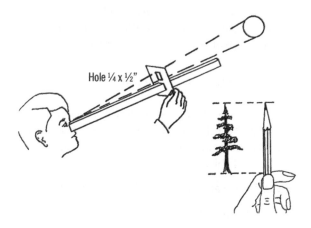

Hole ¼ x ½"

You can make a scale of distances on your yard-stick by using objects of a known or easily measured distance as a comparison. For objects close by, have a person hold an object a known distance from your eye. Move the file card back and forth until the object just fills the hole in the card. Mark on your moon range finder the distance in feet. Have the person move away. Repeat the operation for the new distance.

Can You Work Like a Scientist?

1. The moon appears to be larger when it first rises and gets smaller as it moves overhead. Does the moon change in size during the evening? Remember, if the distance the card is held away from your eye doesn't change, the size remains the same. If the moon increases in size, you must move the card closer to your eye in order to fill the hole.

2. Does the moon change in size at different periods of the year such as during the harvest moon?

3. Does the moon change in size during the phases of the moon?

4. Is the moon always the same distance from the viewer? Assume in this case that the moon does not change in size.

Water Faucet Vacuum Pump

Purpose: This vacuum pump is available any place there is a water faucet. With this simple pump you can conduct many experiments requiring a low air pressure.

Materials: Cork glass T tube (see "T Tube" on page 37), one-hole rubber stopper to fit opening in faucet, rubber tubing, and a bell jar.[16] For the bell jar you can use either the wide-mouth gallon jar or a regular gallon jug.

What to Do: Connect the T tubing to the rubber stopper as shown. Slip the stopper tightly into the faucet. Connect up the vacuum jar to the T tubing with rubber tubing.

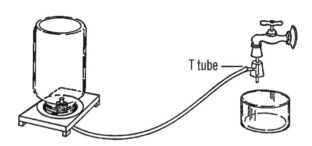

T tube

Operation of Equipment: Turn the faucet on slowly. Then open the faucet up. Be sure to hold the stopper in place. The pressure of the water going through the T tube will suck the air from the bell jar. In order to close off the bell jar, pinch the rubber tubing.

---

16 A *bell jar* is a glass vessel designed to hold a vacuum.

Vacuum Jar Pressure Gauge

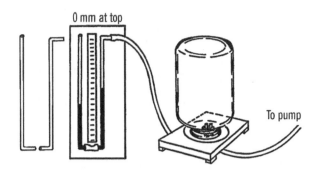

0 mm at top

To pump

Purpose: This mercury pressure gauge enables you to measure the amount of air pressure remaining in the jar. Since air pressure decreases with altitude, the pressure reading can be compared with altitude by using the chart.[17]

Materials: Two 3' pieces of glass tubing, short piece of rubber hose, meter stick or ruler that is marked off in both inches and millimeters, wood for support, and about 2 oz. of mercury.

What to Do: Bend the glass tubing as shown. Connect the bottom pieces together with rubber tubing. Fasten a meter stick to the wood support. The 1000-millimeter mark should be down. If you do not have a meter stick, use a piece of tape for a scale. Mark a scale on the tape using a plastic ruler which is marked in millimeters. Call the weatherman and ask for the barometric pressure reading in millimeters.[18] Pour the mercury into the tube until the mercury reaches this mark on the scale. The height probably will be somewhere between 650 and 750 millimeters. *In pouring the mercury, be sure that you don't touch it.* It can be a deadly poison if it gets into your body through a scratch or a cut. Use a plastic funnel thistle tube for pouring.

Modern Safety Practice

For health reasons, mercury is no longer considered safe to work with. Please see Note 26 in Appendix E for discussion.

Operation of Equipment: Connect the bent end of the glass tubing to a vacuum jar as shown. As the pump draws the air out of the jar, the normal air pressure pushes down the mercury in the open glass tubing. The mercury rises in the other tube as the air is withdrawn. See how high the mercury climbs and then note the number of millimeters on the scale. Compare with your chart for an altitude reading. (These are approximate readings.)

Millimeters of mercury	Altitude in feet	Millimeters	Altitude
760	0	255	30,000
716	1600	215	31,000
675	3200	205	32,000
632	5000	195	33,000
598	6300	186	34,000
562	7500	179	35,000
523	10,000	175	36,000
466	13,000	161	37,500
429	15,000	140	40,000
350	20,000	110	45,000
315	23,000	85	50,000

---

17 Using mercury—and building this project as described—is not advised. Please read Note 26 in Appendix E for alternatives and discussion.

18 Rather, visit weather.gov and look up the current barometer reading for your zip code!

282	25,000	70	55,000
274	26,500	56	60,000
247	28,000	43	65,000
236	29,000	33	70,000

Can You Work Like a Scientist?

Can you use the principle of the vacuum jar pressure gauge in making a mercury barometer? See the section on meteorology (Chapter 7).

Vacuum Pump

Purpose: The vacuum pump is used to withdraw air from vacuum jars and other containers in order to conduct various experiments concerning air pressure, altitude effects, and various biological research projects.

Materials: Tire pump,[19] rubber tape, 5" piece of 7/16" plastic garden hose, 2" piece of glass tubing.

What to Do: Unscrew the cap off the top of the pump. Remove the inside of the pump. Unfasten the screw and turn the rubber plunger over. Tighten the screw and insert the handle and plunger back into the outside case of the pump. Unscrew the hose and valve. You will reverse this valve later.[20]

Cut off the screw cap at the end of the tire hose with a razor blade. Then cut off about five more inches so you have a short piece of tire hose. Insert a short piece of glass tubing into the longer hose. Wrap rubber tape around the other end of the glass tubing and force the tubing into the hole in the base of the tire pump. Slide a short piece of 7/16" plastic garden hose over

the valve and over the rubber tire pump hoses. Insert a short piece of tire hose into the other end of the plastic hose. You can then slip this hose over glass tubing to make connections on a bottle.

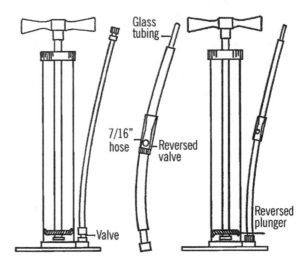

Operation of Equipment: When you pull up on the pump, air is drawn into the pump. When you push down, the valve closes in the pump, and the air is pushed out around the sides of the gasket and out the top of the pump. Remember, seal up any leak that shows up in your hose, etc., or air will leak in from the outside.

Modern Safety Practice

Wear safety glasses when working with a vacuum pump. Not every vessel is designed to withstand a vacuum. What happens if you pull a vacuum on an empty 2-liter plastic soda bottle? Can a glass bottle respond the same way?

---

19 Preferably, an old bicycle pump that you don't mind destroying

20 Your tire pump may have a different type of valve mechanism. If so, can you figure out how to reverse it?

Can You Work Like a Scientist?

1. How much of a vacuum can you produce? Use a pressure gauge. Be sure you have tested your bell jar with a regular vacuum pump before you try this. Can you improve the efficiency of your pump?

2. What effect on plant growth does altitude have? What plants will grow at higher altitudes? (Reduce the air pressure with your pump.)

3. If the amount of air on Mars is about equal to that present on the top of Mount Everest (altitude about 30,000 feet), what plants might grow on Mars?

4. At what altitudes can clouds form? See your cloud jar ("Cloud Jar" on page 190).

5. What animal would be the best space traveler and withstand a drop in air pressure? Try frogs, fish, worms, rats, etc.[21]

6. What happens to fish in a fishbowl or jar with reduced air pressure?

7. Will your alcohol burner continue to burn in your vacuum jar as air is removed? At what altitude will the burner go out?

Vacuum Jar—Bell Jar

Purpose: A vacuum jar is used for carrying out experiments that require a different atmospheric pressure or much less air than is normally available.

Materials: Block of ¾" wood, rubber sink stopper, wide-mouth gallon jar,[22] a one-hole rubber stopper, and glass tubing.

What to Do: The block of wood should be larger than the opening of the jar. Drill a hole in the center of the block to fit the size of your rubber stopper. Glue the rubber stopper to the wood block. Cut a hole in the stopper to match the hole in the wood. Insert the rubber stopper and a short piece of glass tubing as shown.

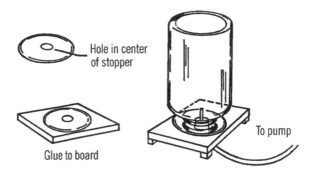

Hole in center of stopper

Glue to board

To pump

Operation of Equipment: The bell jar can either be used upright or tipped upside down. Grease the rubber sink stopper with heavy lubrication grease (from a service station). You can try vaseline, but it is sometimes too thin.[23] Set the jar on the board base. Connect the glass tubing to a vacuum pump with rubber hose. In order to close off the jar and remove the pump, you can either install a valve in the rubber hose or bend the rubber hose and clamp the end. The clothespin adjustable clamp will also serve as a clamp to seal the hose.

---

21 Important: Read Note 1 in Appendix E about working with animals.

22 Some people suggest using a solid Nalgene container as a bell jar. The same safety precautions should be used with plastic as with glass.

23 See *Vacuum Grease* in Appendix A.

Safety Tips

1. Some jars are thicker in one place than another. Therefore, thoroughly test your wide-mouth gallon jar by connecting the jar to the vacuum pump and then completely covering the jar with a metal waste paper basket, or a strong wooden box, etc. The jug might implode if it is weak in one spot. Draw as high a vacuum as your pump will permit in your test. The safest method would be to have the pump (and all people) outside a closet, the jar inside, and the tube running under the closet door.

2. When in use, don't draw a full vacuum unless you set up adequate safety provisions. Use of a pressure gauge is very helpful (see "Vacuum Jar Pressure Gauge" on page 79).

3. Allow the air to return into the bottle before you attempt to open the jar. Otherwise, the air will rush in with tremendous force and this sudden force might break the jar.

Modern Safety Practice

Recall that it only takes a small scratch and some pressure to break glass ("Cutting Glass Tubing" on page 6): *Don't scratch the glass of your bell jar.*

See Note 2 in Appendix E for additional notes about safety practice around glass that may break.

Solar Furnace

Purpose: A solar furnace is used to focus the sun's rays and thus produce a high degree of heat upon a small area.

Materials: A headlight reflector or an old umbrella.[24]

What to Do: If you use an umbrella for your solar furnace, the inside surface of the umbrella should be covered with aluminum foil. Care should be taken that the foil is spread on as smoothly as possible. The foil can be glued or stitched to the cloth surface of the Umbrella.

An old headlight reflector can also be used to make the solar cooker or furnace. The shiny surface should be polished.

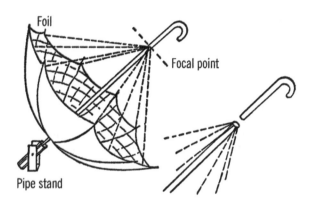

Operation of Equipment: You will need to build some kind of stand to support your furnace. The furnace should be pointed toward the sun or some other source of light. The rays of light strike the curved surface and are reflected to one spot. At this spot all the rays converge or come together. This is the focal point.

If you are using an umbrella, the handle can be cut off just before this focal point. Objects to be heated can then be placed near or attached to the remaining part of the handle.

Can You Work Like a Scientist?

1. Can you determine the focal point of your umbrella furnace? Move a thermometer back and forth until you find the point where the most heat is gen-

---

24 An old satellite dish may also be used like an umbrella.

Modern Safety Practice

1. Keep your eyes (and other body parts) away from the focal point at all times.

2. A solar furnace can potentially start a fire. Just in case, keep a fire extinguisher handy, and don't leave the furnace unattended in sunlight.

erated. You may have to use a candy thermometer if the heat is greater than that which can be measured with a regular household thermometer. Be careful that the heat doesn't cause the liquid in the thermometer to break it.

2. You can use your solar furnace to determine the degree of heat given off by the light from the sun. How does this heat vary throughout the year?

3. Can you boil liquids and melt certain solids with your furnace?

4. Can you design a furnace in which you use a lens to focus the rays on a very small area? How hot can you get the point?

5. If you had a larger surface from which you could reflect the sun's rays, could you increase the temperature at the focal point?

Solar Distillation Apparatus

Purpose: One of the greatest problems scientists face is that of securing enough fresh water for drinking and irrigation purposes. The solar distiller uses the energy of the sun to remove the salt from sea water and thus provide fresh water.

Materials: Cake tin, one pane of glass slightly longer than the cake tin, a second pane of glass several inches longer than the first (you can determine this length when you start to build your distiller), black paint, and a tray to collect the distilled water.

What to Do: Paint the inside of the cake tin with black paint. Also paint the shorter of the two pieces of glass black on one side. Support the glass as shown in the drawing by cementing the glass to a block of wood with rubber cement. The longer piece of glass is fastened to the top of the short piece with masking tape so that the two pieces of glass will hinge and fold together. When the apparatus is set up, the long piece of glass is raised slightly above the cake tin by gluing matchsticks[25] to the glass. A tray is placed under the edge of glass that projects beyond the cake tin. This tray is used to collect the distilled water.

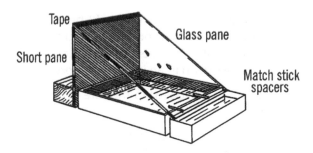

Tape
Glass pane
Short pane
Match stick spacers

Operation of Equipment: Salt water (mix salt and water in a quart jar) is added to the cake tin. As the sun's rays pass through the glass pane, the water is heated and begins to evaporate. The vapor strikes the bottom of the slanted pane and condenses. The drops of water flow down the inside of the glass into the tray.

---

25 or toothpicks

Modern Safety Practice

1. You can cut the glass to shape for this project; see "Glass Cutter" on page 17. Whether or not you cut your own glass, remember that glass edges are very sharp. As described in that project, you can sand the edges off glass under water.

Can You Work Like a Scientist?

1. Why should you paint the inside of the cake tin and the vertical glass pane with black paint?

2. Can you determine the rate (pints per hour) at which your solar distiller produces fresh water?

3. Is the water collected really free of salt?

4. What areas of the Earth would be helped if a cheap method of distilling sea water could be developed? Could crops be grown in an irrigated desert?

5. How can you speed up the rate of distillation?

6. If you closed in the sides of the still, would this increase the rate of distillation?

7. Could you design a portable distillation apparatus that could be used by people stranded on life rafts in the ocean?

8. Can you distill sea water? Is sea-water salt the same as the salt we buy in stores?

9. Is the salt in your tears the same as the salt from the ocean?

Constellarium

Purpose: A constellarium is used for projecting the image of constellations so that the shape of the constellation can be studied.

Materials: Wood for box, base for light bulb, 60-watt bulb,[26] cord, and cardboard blanks.

What to Do: Build a long box with slots so that pieces of cardboard can be inserted in the slots. At the far end of the box, a base for a light bulb should be fastened. A small hole should be cut in the near end of the box so that the viewer can look toward the card and the light on the other side of the card.

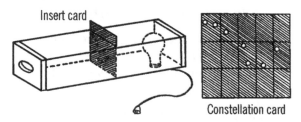
Insert card — Constellation card

Operation of Equipment: Black art paper is glued on one side of the cards. Holes are then punched with various sized nails, pins, etc. to represent the stars in a particular constellation.

The card is placed in the slot in the box. The viewer looks through one end of the box and sees the pinpoints of light coming through the holes punched in the card. These pinpoints of light represent the stars in a particular constellation.

Can You Work Like a Scientist?

1. Stars are rated according to brightness or magnitude. The lower the number of magnitude the brighter the star. Can you select a nail to represent each order of magnitude?

---

26 Could you design a version of this that uses LEDs (Light-Emitting Diodes) instead?

2. If you rule off each card into small squares, these squares could represent distances in the sky. As an example, one square could represent the distance in the sky covered by the width of one thumb held at an arm's length away from you. Using such a scale, could you accurately place the stars in a constellation on a constellation card? You will have to record the distances at night by using your thumb and measuring the apparent distances between the stars in the constellation. Perhaps your thumb might be too large as a measuring unit. Could you use something smaller?[27]

3. Ancient people imagined the constellations formed pictures in the sky. Can you trace around the holes in the card and form the imaginary figures for each constellation?

Problems to Investigate in the Study of Astronomy and Light

(P)-Primary

(I)-Intermediate

(U)-Upper

Space

1. Can you determine the time it takes for each of the moons of Jupiter to make one complete revolution around it? Observe the moons through a telescope.[28] Make drawings of the positions of the moons each night. *(I)*

2. Can you classify the major stars in the sky as to their color? How can you tell a red star from a white star? *(U)*

3. How do the position of the stars vary from night to night? Can you measure the change in position of certain stars over a period of a month? *(I)*

4. Do sunspots all move at the same rate of speed across the sun? WARNING-Be sure you do not look directly into the sun. See "Sunspot Viewer" on page 76. *(I)*

5. Can you take pictures of falling meteorites? Set your camera on time lapse.[29] You should be able to catch the streaks of the meteorites on your camera. *(U)*

6. If a nut were loose on the outside of your spaceship, could you leave the ship in a spacesuit and tighten the nut with a normal wrench? *(I)*

7. How close to the different planets could you get and still remain weightless? *(U)*

8. What would be the weights of different objects around you if these objects were sent to the various planets and other solar bodies? *(I)*

9. Can you measure how far an object falls in one second? *(I)*

10. Can you measure how long it takes an object to fall different distances from a tall building? How does the time seem to be related to the distance an object falls? *(U)*

11. Is there any such thing as a micrometeorite? Can you find one? *(U)*

12. Can you measure the speed sound travels? *(U)*

---

27 The moon range finder ("Moon Range Finder" on page 77) comes to mind!

28 See Note 27 in Appendix E for telescope suggestions.

29 Really, you want a *long exposure*, not time lapse. What is the difference?

13. How does temperature affect the speed of sound? *(U)*

14. Can you take a picture of the moon through a pinhole camera? *(I)*

15. What is Kepler's Law? Can you draw to scale the orbit of various planets and the time it takes each planet to reach particular spots in the orbit? *(U)*

16. From your drawings can you predict when the Earth will be closest to each of the other planets in the Solar System? *(U)*

17. How does the amount of humidity in the air affect the rate at which a person will sunburn? You might focus the sun's rays a certain distance from a piece of paper and see how long it takes before the paper is set on fire. Does this depend on the humidity?[30] *(I)*

18. Does land cool off more rapidly than water? How would this affect the climate for cities on the coast and inland? Can you compare the winter temperatures for Portland, Oregon, and Minneapolis, Minnesota? *(I)*

19. How do the temperatures vary around the world? Get a pen pal in the Southern Hemisphere who is about the same distance from the equator. Compare temperatures and other weather conditions.[31] *(I)*

20. What planets experience winter and summer? Are there any planets that have more than four seasons? *(U)*

21. What are some ways we could communicate to one another if we were on the different members of the Solar System? *(I)*

22. What causes a sonic "boom?" Can you experiment with such "booms?" Try snapping a towel or cracking a whip. What causes the pop? *(I)*

23. As an object increases its speed when moving through a fluid such as water or air, it encounters a greater resistance or drag. If a measured amount of air (say in a balloon) blows against an object, the force it takes to hold the object from moving is the *drag* on the object. Can you devise a way to measure the drag on differently shaped objects? *(U)*

24. What effect does the weight of the object have on the drag? *(U)*

25. What effect does the surface of an object have on the drag? *(U)*

26. What is the effect of different shaped nose cones? *(U)*

27. Will a liquid evaporate faster in a vacuum? *(I)*

28. If you observe and measure the angle to a star at the same time as a friend does who is several miles away, can you figure the distance you and your friend are apart? *(U)*

29. You can make exact constellation slides. Use very fast film.[32] Attach your camera to a tripod and aim the camera at the constellation you wish. Focus the camera at infinity and then expose it about fifteen to twenty seconds. After the negative is developed, punch

---

30 Related question: What is sunburn, and how does it relate to paper burning?

31 Modern interpretation: Check worldwide weather at worldweather.org or forecast.io, and compare weather conditions at similar lattitudes to your own, both *north and south*.

32 That is, a high ISO number on your digital camera!

---

holes with a pin for each star.[33] The size of the hole should match the apparent brightness of the star. Use the slide in a constellarium. *(I)*

30. What is the effect of the shock of "landing" on various animals? How fast can one safely approach Earth? *(U)*

31. How large should the parachutes be in order to safely lower objects of different weight? *(I)*

32. Can you plan a "moon" vacation resort? *(I)*

33. Can you disprove by experimentation Ptolemy's theory of the universe? *(I)*

34. Can you measure the distance to the moon? *(U)*

35. Can you weigh the Earth? You might want to try the method used by the German scientist, von Jolly, in 1881. *(U)*

36. How long would it take a Foucault pendulum to make one complete turn at the equator? (See "Foucault Pendulum" on page 75.) *(I)*

37. How is a gyrocompass used in navigation? Can you build a gyrocompass? *(U)*

The Moon

1. How often do full moons occur? (Keep a chart of your observations from month to month.) *(P)*

2. Does the moon always travel the same path through the sky? Use an astrolabe to take readings as to elevation (how high up in the sky) and declination (how many degrees from true north). *(I)*

3. Does the size of the moon change? (Use a moon stick and "measure" the size of the moon at various heights. See "Moon Range Finder" on page 77.) *(P)*

4. Does the size of the moon change from season to season? *(I)*

5. What are the dark spots on the moon? (Use a telescope.) *(P)*

6. Are there mountains on the moon? (Use a telescope.) *(P)*

7. What causes an eclipse of the moon? (Experiment with models of the sun, Earth, and moon to see how the light given off by the moon can be shut off.) *(I)*

8. What causes the phases or different shapes of the moon? (Darken a room and then use a projector and a large rubber ball or basketball.) *(P)*

9. What effect does the shape of the moon have on tides? (Keep a record of the phases of the moon and compare with tide charts.) *(I)*

10. Will plants grow by moonlight? (Try bean plants.) *(U)*

11. What kind of light is given off by the moon? (Use a diffraction grating spectroscope.) *(U)*

12. What is the intensity of moonlight? (Use a light meter or radiometer.) *(U)*

13. Can you use photographic paper to measure the intensity of the light of the moon? *(U)*

14. What animals are more active at night? Does this activity depend on moonlight? *(I)*

15. Can you make a map of the moon through your observations? (telescope) *(U)*

---

[33] With a digital camera, print your image and use it the same way.

16. What causes a harvest moon? *(I)*

17. What causes a hunter's moon? *(I)*

18. What would be the weight of various objects in your room if these objects were up on the moon? *(I)*

19. What is the difference between weight and mass? Would your weight change if measured on the moon? Would your mass change? *(U)*

20. Has the moon ever been to the same spot in the universe more than once? *(U)*

21. Could plants grow in soil found on the moon? *(I)*

22. Can you determine the size of various physical features on the moon? *(U)*

23. Does the moon revolve? (Note the positions of various features on the moon.) *(U)*

24. What would happen to the orbit of the moon if the moon should speed up in its path around the Earth? (Whirl a ball on a rubber band.) *(I)*

25. Can you tell time by the moon? *(U)*

26. Is the moon always the same distance from the Earth? Can you determine this experimentally? *(U)*

27. What is the average number of craters per surface area on the moon? Determine this with a telescope. Can you determine the depth by the shadows? *(U)*

28. What is the best time of the month to study the appearance of the moon? *(I)*

29. What effect does the shape of the moon have upon success in fishing? *(I)*

30. Can you make a material with about the same density as the moon? Will this material float in water? *(U)*

31. Do people all over the world see the same moon as we do? *(I)*

32. How was the moon formed? (theories of the moon's origin) *(U)*

33. Does a full moon affect your ability to think and reason? *(I)*

34. Are there more babies born during a full moon? (Check hospital records.[34]) *(U)*

35. Does the moon always seem to rise from the same place? Does this position vary with the season? *(P)*

36. How good a mirror is the moon? What part of light does it reflect? (Measure the light from the sun and then from the moon.) *(U)*

37. How does the surface of the moon compare with that of the Earth? *(P)*

38. How long would various insects, plants, and animals continue to live if exposed to an atmosphere similar to that of the moon? *(I)*

39. What plants and animals could live in the temperature extremes of the moon? *(U)*

40. Are there different kinds of tides? What causes these? Compare a tide chart with a calendar showing phases of the moon. Be sure to figure in the position of the sun at the times shown on the tide chart. *(I)*

41. Does the moon always look the same? (observations) *(P)*

42. What happens to the moon the nights we don't see it? *(P)*

---

34 You may be able to look up *Vital records statistics* online.

43. Is the moon larger than the sun? Why does it appear larger? *(P)*

44. Can you see the stars as well during a full moon as when there is no moon? *(P)*

45. What are some early legends about the moon? *(P)*

46. What do we know about the make-up of the moon? *(I)*

47. Why is the moon yellow? *(P)*

48. What are the problems in taking a trip to the moon? *(I)*

49. Could you eat on the moon? *(I)*

50. What kind of weather is there on the moon? *(I)*

51. Could man live on the moon? *(I)*

52. What makes the moon move across the sky? *(P)*

53. Is there only one moon in our Solar System? *(P)*

54. If you approached the moon, would the light be brighter? *(I)*

55. Could bacteria live under conditions similar to those found on the moon? *(U)*

56. Could we mine different minerals from the moon? *(U)*

57. Could you plan a summer resort on the moon? What are the problems? *(U)*

58. Are there volcanoes on the moon? Any moonquakes? *(I)*

59. What is the man-in-the-moon? Has man ever reached the moon? *(I)*

60. What experiments could scientists do on the moon? *(U)*

61. How could you talk on the moon? (Can you experiment with talking through solids, radio in a vacuum, etc.?) *(I)*

62. What kind of telescope did Galileo use to first view the moon? Can you make one of the same power? *(I)*

63. How can we find out what the back side of the moon looks like? *(I)*

64. How can you measure the height of the mountains on the moon? *(U)*

65. Is the moon gradually moving toward or away from the Earth? *(U)*

66. Are all the craters on the moon alike? *(I)*

67. Why do the craters on the moon seem to run in a straight line? (What is your theory?) *(U)*

68. What is the difference between the volcano and the meteor theory as to the cause of the craters on the moon? What is the bubble theory? *(U)*

69. What causes the "seas" on the moon? Can you measure some of these "seas?" *(U)*

70. Why are the mountains on the moon taller and more jagged than those on Earth? (Try some experiments with erosion and weathering.) *(I)*

71. Can you locate domes on the moon? *(U)*

72. Can you find rays on the moon? (Use binoculars or a telescope. Look for white craters.) *(U)*

73. Can you find clefts or deep cracks in the moon? *(U)*

74. Would the temperature change very much inside the caves on the moon? (Experiment with caves found on the earth.) *(U)*

75. Can lichens live on the moon? Experiment by growing lichens under different conditions. *(I)*

76. Could we get oxygen from the rocks on the moon? Can you get oxygen from rocks found on Earth? *(U)*

77. Would you see more stars from the moon than from the Earth? *(I)*

78. How would the Earth appear from the moon? *(I)*

79. Is the moon round? What causes its strange shape? *(I)*

80. What causes the strange moving shadows on the moon? *(U)*

81. What causes a "mist" in the crater Plato? *(U)*

82. Can weather predictions be made accurately from the appearance of the moon? *(I)*

Heat, Light, and Color

1. Is heat matter? Weigh different objects. Then heat the objects. Do the objects gain weight? *(P)*

2. Can you disprove the theory that heat is a fluid that flows in and out of objects? *(I)*

3. Is heat produced when you bore a hole with a drill? What effect does the sharpness of the drill have on the amount of heat produced? Can you measure this heat with a thermometer? *(I)*

4. What objects become hot when rubbed together? Can you rub various objects and measure the amount of heat given off? Try rubbing a coin on a piece of wood. *(P)*

5. Is heat kinetic or potential energy? Put a few eyedroppers of ink[35] in cold wa-

ter. Try the same experiment with hot water. What do you observe? *(I)*

6. Can you determine the amount of energy in different liquids at various temperatures? *(U)*

7. Can you determine the amount of energy possessed by different gases at various temperatures? *(U)*

8. What is the effect of heat energy on the movement of molecules? Heat water containing a few drops of food coloring. *(P)*

9. Can you detect the movement of molecules? Observe the Brownian movement through a microscope. *(U)*

10. Is the Brownian movement the same in all liquids and gases? *(U)*

11. What effect does added heat energy have upon the Brownian movement? *(U)*

12. Is temperature the same as heat? Place a pan of boiling water in the refrigerator. Place alongside the water a pan of sand heated to the same temperature. Place in the same refrigerator a kettle of lukewarm water. Which one reaches 45 °F first? *(I)*

13. What factors affect the heat content of an object? What effect does temperature, mass, and kind of substance have on the heat content? *(U)*

14. Can heat exist as potential energy? *(U)*

15. Does the temperature of ice change while the ice is melting? *(P)*

16. How hot can water become in the liquid state? Can water become hotter as vapor? *(I)*

---

35 Food coloring.

17. Do all solids melt at the same temperature? Try ice, butter, grease, various crystals, and other solids. *(P)*

18. How accurate are your senses in determining temperature? Place a finger of one hand in ice water. Place your other hand in hot water. Then place both fingers in lukewarm water. *(P)*

19. What parts of your body are the most sensitive to changes in temperature? *(I)*

20. Are girls or boys better detectors of temperature? Try some experiments to see which students are best at detecting changes of temperature. *(U)*

21. How do various forms of matter change when heat is given off or absorbed? Can you detect and measure these changes? *(I)*

22. How did Galileo make the first thermometer? Can you make one using the same principle? How accurate is this type of thermometer? *(I)*

23. Can you make a thermometer using a glass tube and water? How accurate is your thermometer? *(U)*

24. Can you determine the freezing point of water? Is it the same as the melting point of ice? *(P)*

25. Does salt water freeze at the same temperature as fresh water? What effect does the amount of salt have on the temperature at which the liquid freezes? *(I)*

26. Can you determine the freezing point of various liquids? *(U)*

27. Can you determine the boiling point of water? *(I)*

28. What effect does adding salt to water have on the temperature at which the salt water boils? *(I)*

29. How was the Fahrenheit thermometer developed? Can you make a thermometer and scale using Fahrenheit's methods? Why is the freezing point of water 32° on a Fahrenheit scale? *(U)*

30. Can you make a mercury thermometer?[36] What is the freezing and boiling point of mercury? How can you determine and measure this? *(U)*

31. How can you change (convert) the readings of a Fahrenheit scale to those of a Celsius scale? Can you make a thermometer of each type and make the conversion by an experimental method? *(U)*

32. What are the advantages and disadvantages of mercury and alcohol thermometers? *(I)*

33. How does a clinical thermometer work? Is the mouth temperature of all animals the same? *(I)*

34. What is the difference in temperature between mouth and rectal temperature of various animals? *(U)*

35. How are thermometers calibrated by using hydrogen and helium gas? Can you devise a means to do this? *(U)*

36. Can you determine experimentally the accuracy of various thermometers sold in stores? *(I)*

37. Can you set up an experiment to change mechanical energy into heat energy? Can you measure the amount of mechanical energy necessary to pro-

---

36 Working with mercury is not recommended, due to its toxicity. Here is a real challenge: Can you make a mercury-free liquid-metal thermometer?

duce a certain amount of heat energy? *(I)*

38. Does all matter expand when heated? *(I)*

39. Can you detect and measure the expansion of different metals? *(I)*

40. In which directions do different kinds of matter expand? Is this expansion uniform in all directions? *(U)*

41. What effect does temperature have on the expansion of different metals? *(P)*

42. Can you determine the coefficient of linear expansion of various metals? *(U)*

43. Can you determine the distance certain metals should be separated at various temperatures in order to allow for expansion? Why is the maximum and minimum temperature of an area important in laying metal lengths such as railroad rails? *(U)*

44. Can you identify metals by determining their coefficient of linear expansion? *(U)*

45. What is the total seasonal change in the length of steel cables used to support bridges in your area? Can you determine this? *(U)*

46. Can you determine the expansion rate of various non-metals such as glass, ice, cement, plastic, etc.? *(I)*

47. What effect does heat have on a bimetallic bar? Can you make a bimetallic bar by riveting strips of two different metals together? *(I)*

48. What effect does cooling have on different bimetallic bars? *(I)*

49. How does a thermostat work? Can you make a bimetallic bar by using alumi-num foil and paper fastened together for the two strips? *(P)*

50. Can you make an operating thermostat? How accurate is it compared to a regular thermostat? *(I)*

51. Can you devise a thermostat to control the temperature inside an incubator so you can hatch eggs? *(U)*

52. How does expansion of metals affect the accuracy of a pendulum clock? How can you compensate for this expansion? *(U)*

53. What effect does temperature change have on the expansion of water? Place a thermometer (preferably Celsius) in a bottle of water. Insert a long piece of glass tubing in a one-hole stopper. Have the glass tubing stick up about two feet above the top of the jar. Push the stopper into the bottle until the water is forced quite high up the glass tube. Place the jar in packed ice. Note the height of the water in the tubing and compare this with the temperature reading. Keep a graph if you can. *(U)*

54. Does salt water cooled below 32 °F expand or contract as the temperature decreases?[37] *(U)*

55. Why does ice float on the top of the water? *(P)*

56. What is the temperature of water at different depths in ponds and lakes? *(I)*

57. What effect does the temperature of air have on the temperature of the water at different depths? *(U)*

58. What is the temperature of salt water (ocean or bay) at different depths? Does this vary with the season of the year? *(I)*

---

37 Is this the same as *supercooled* water? If not, how is it different?

59. Why does a pond freeze over only on top and not on the bottom? *(I)*

60. What effect does the freezing over of a pond have on the life underneath the surface of the ice? *(U)*

61. What is the effect on different gases if the volume is the same but the temperature changes? Can you use the size of a balloon as a measure of gas pressure? Now vary the temperature by using a refrigerator, outside temperatures, and heating. *(I)*

62. Can you determine the pressure coefficient of different gases? *(U)*

63. Can you devise and convert the different temperature scales to a standard scale? You must find out about Celsius, absolute or Kelvin, and perhaps other scales. *(U)*

64. What is the relation between the absolute (Kelvin) temperature and the pressure of a gas? Try changing temperatures and recording the pressure. Change your findings to the absolute scale. *(I)*

65. What is the relationship between the volume of a gas and its temperature? If you hold the pressure the same (by using a standard weight to push on the gas in a container), what happens to the volume as you increase the temperature? Keep a graph of the change and the temperature. *(U)*

66. Why might the surfaces of large rocks chip during hot summer days? *(I)*

67. Why are concrete sidewalks made with spaces between the sections? *(P)*

68. What materials would be better than steel to use for surveyors' tapes? *(U)*

69. Why do fountain pens which are nearly empty tend to leak while being used? *(U)*

70. Why is a thick glass less likely to crack than a thin glass when hot water is poured into it? *(P)*

71. Can Pyrex glass stand the shock of a sudden temperature change? Why? *(I)*

72. Why does bread dough rise? What effect does the amount of kneading have on the volume of bread dough expansion? Check for carbon dioxide. *(P)*

73. Will balloons rise if filled with hot air? Why? *(P)*

74. Why should the moving engine parts be made of the same material? *(I)*

75. What effect does the size of the bulb and the diameter of the tube have on the amount of liquid that rises inside a thermometer? How can you make a thermometer more accurate? Experiment with glass tubing inserted in a bottle of water and sealed with a rubber stopper. *(P)*

76. Why should wires sealed in glass have the same coefficient of expansion as the glass?[38] *(I)*

77. Why does the level of liquid in a thermometer suddenly dip when the bulb is placed in cold water? *(I)*

78. Can you use a hydrometer and a thermometer to determine the temperature at which water has its greatest density? *(U)*

79. Can you determine experimentally the heat lost or gained by water? What effect does changing the amount of water (mass) have on the heat loss or gain at a given temperature change? *(I)*

---

38 What is Kovar?

80. Can you experiment with mixing varying amounts of water at different temperatures? What is the resultant temperature of the mixture? An example might be mixing 50 cc of water at 20 °C with 30 cc of water at 40 °C. Can you predict the resulting temperature? *(I)*

81. Can you determine the specific heat of various metals, air, steam, ice, and water? *(U)*

82. Does metal gain and lose heat at the same rate as water? Measure the temperature of a piece of metal. Measure the temperature of very hot water, and drop the metal into an equal volume of water. Remove the metal and record the temperature of the metal and the water. *(I)*

83. Does it take more heat, weight for weight, to heat water or earth? Place samples of each in containers. Heat for the same period of time and take the temperatures. *(P)*

84. Why does a shallow lake warm up quicker in the spring than a deep lake? *(I)*

85. What effect does a large body of water in the area have on the growing season? *(I)*

86. If water contains ice, does the water start to heat immediately when heat is applied? What happens to the heat energy? *(P)*

87. Can you determine the heat of fusion (extra heat required to change a substance from a solid to a liquid) of ice and other solids? Mix equal amounts of ice (0 °C) and hot water (90 °C). The temperature should be 45 °C. The actual temperature is much less. The difference changed to calories is the heat of fusion. *(I)*

88. Can you determine the fusing point (temperature at which substance turns from a solid to a liquid) of different solids such as ice, frozen salt water, beeswax, mercury,[39] butter, and tin? *(I)*

89. Why is ice a good refrigerant? *(P)*

90. Does wrapping ice in paper or burlap help or prevent refrigeration? *(I)*

91. What is the temperature of dry ice? What is the heat of fusion of dry ice?[40] *(P)*

92. What does dry ice consist of? Test the gas given off with a burning match. Test the gas with lime water. *(P)*

93. What effect on the melting rate of ice does the exertion of pressure on the ice have? Attach two weights to a wire. Hang the weights over the block of ice so the wire presses against the ice. Try different weights and kinds of wire. *(I)*

94. Why does a snowball pack? Why does some snow pack easier than other snow? *(I)*

95. If you stand on ice, why do the skates freeze to the ice? *(U)*

96. Do liquids evaporate more rapidly when hot or cold? Can you keep a chart on the rate of evaporation of different liquids and varying temperatures? *(P)*

97. Do wet clothes dry more quickly on a calm or windy day? *(P)*

---

39 Is it safe to use a sealed mercury switch to do this experiment?

40 Dry ice can cause severe frostbite injuries. Handle only briefly, using oven mitts or a towel to protect your hands. See Note 14 in Appendix E for further discussion.

98. Do liquids evaporate more quickly in open pans or pans with a small opening? (I)

99. Do all liquids evaporate at the same rate? *(P)*

100. Does evaporation cause cooling? What is the effect of the rate of evaporation on the temperature of the object? Place alcohol and water on your wrist. *(P)*

101. Why doesn't a liquid evaporate in a closed container? *(U)*

102. How does a vacuum affect the normal rate of evaporation of a liquid? *(U)*

103. Can you determine the rate of sublimation (solid changing directly to a gas) of various substances such as ice, moth crystals, dry ice, and sulfur?[41] *(I)*

104. What happens when water boils? What are the first bubbles that rise? Do the bubbles get larger or smaller as they rise in the liquid? Why do bubbles form? *(I)*

105. Can you determine the heat of vaporization (the quantity of heat per gram required to vaporize the liquid without changing its temperature) of various liquids? *(U)*

106. What is the heat of condensation and how can you measure it? How does this affect weather conditions, particularly hurricanes? *(U)*

107. What effect does the pressure on a liquid have on the temperature at which the water will boil? Boil water in a flask. Remove from the heat and insert a stopper. Turn the flask over and run cold water on the bottom.[42] *(I)*

Lenses

1. Are all lenses the same shape? (Collect lenses from eyeglasses, old cameras, viewfinders, flashlights, etc.) *(P)*

2. How do objects look through different kinds of lenses? *(P)*

3. What are lenses made out of? *(I)*

4. Can a drop of water serve as a lens? Place a drop of water on some Saran Wrap that is covering a page from a newspaper. *(I)*

5. What has the size of the drop to do with the amount of magnification? *(I)*

6. Can you determine the focal length of the various lenses? How far do you hold a lens above a piece of paper in order to focus the light to one point? *(P)*

7. What is the temperature of various lights focused to a point? Try sunlight, artificial light, candle, and match. *(I)*

8. How far from the lens does the image form? How far away do you hold the lens from your eye in order to focus on a distant object? *(I)*

9. Is this distance the same as the focal point or greater? *(I)*

10. How does the image distance (distance from the lens to where the image is in focus) change as the object distance (distance from the object to the lens) changes? Does this formula help solve the problem: $1/D_o + 1/D_i = 1/F$? *(U)*

---

41 See Note 20 in Appendix E about mothballs and Note 14 in Appendix E about dry ice.

42 Glass can shatter during rapid temperature changes; take appropriate precautions.

11. Can you build a smoke box to study the rays of light entering a lens? Cut out part of a cardboard box and cover the hole with Saran Wrap. Burn incense for your smoke. Mount your lens in a hole at one end of the box. Use a small pencil flashlight[43] for a source of light. *(I)*

12. Can you use other liquids besides water as a lens? *(U)*

13. Does a clear marble work as a lens? *(P)*

14. Does a lens have more than one focal point? Try both sides of the lens. *(U)*

15. How does the diameter of the lens affect the amount of light admitted? *(U)*

16. Try using lenses in combinations. How does this affect the focal point, image, and object distance? *(I)*

17. Why are some objects upside down when seen through a lens? How can you make the objects right side up? *(U)*

18. Can you measure the power of different lenses? Make a series of parallel lines about 1/16 inch apart. Focus the lens on these lines. If you see three lines outside the lens for everyone line seen through the lens, the lens would be rated three power. *(I)*

19. Can you use a mirror to reflect light through a lens? Do you get more or less light this way? *(U)*

20. Can you use a concave shaving or vanity mirror to focus light to a point? Hold a pencil in front of the mirror. Move the end of the pencil slowly away from the shaving mirror. The point where the image turns upside down is near the focal point. Use the mirror to focus light or rays on a sheet of paper. Measure the distance from the lens to the point. *(P)*

21. Do concave mirrors act the same as concave lenses? Try experiments mentioned above. *(I)*

22. Can bottles be used for lenses? Are bottles thrown from cars a forest fire hazard? *(I)*

23. Can you measure the amount light bends in different liquids and solids? (This bending is the refractive index.) *(U)*

24. Can you focus moonlight? How does the temperature from moonlight compare with temperature from sunlight when focused through a lens? *(I)*

25. Can you make a sample reflecting type of telescope by using a shaving or vanity mirror, plain mirror, and a small concave lens? Find out how a reflecting telescope works. *(U)*

26. What is the difference between a refracting and a reflecting telescope. *(I)*

27. How are lenses made? Can you make a lens? *(U)*

Black Light

1. What is black light? How was it first "discovered?" *(I)*

2. What rocks and minerals fluoresce when exposed to black light in a darkened room? *(P)*

3. What rocks and minerals retain this fluorescence after the light is removed? *(P)*

4. Can you determine the color certain minerals in rocks will fluoresce under black light? *(I)*

---

43 Or a laser pointer.

5. What effect does the length of time a material is exposed to black light have on the length of time it will retain this ability to fluoresce after the light has been removed? *(I)*

6. What effect does black light have on plant growth? *(I)*

7. Why will certain articles of clothing fluoresce under black light? Try a T-shirt. *(I)*

8. What soap powders fluoresce? What is the "magic blue whitener?" *(U)*

9. Can you paint a picture with fluorescent or phosphorescent paint? Try novelty stores or hardware stores for the paint. This is an excellent project to use around Halloween. *(P)*

10. Can you detect the difference between fresh and old eggs with black light? How old do the eggs have to be before their age can be detected? *(I)*

11. What kind of uses can you devise for black light in crime detection? Can you test your theories? *(I)*

12. How do laundries use black light for identifying clothing? *(I)*

13. Can you use black light to trace the growth of plants and certain animals? *(U)*

14. Can you devise a way of measuring the wave length of different black lights? *(U)*

15. Can a person get a "suntan" from black light? *(P)*

16. Can you experiment with black light on the movement of different animals? *(P)*

17. What effect do color filters have on black light? *(I)*

18. Do plants exhibit a negative or positive tropism (attraction) to black light? *(P)*

19. Are plants more attracted to black light or natural light? *(I)*

20. What effect does black light have on a radiometer? *(I)*

21. What spectrum is cast by black light? Use your spectroscope. *(I)*

22. Can you focus black light rays with a lens? What is the temperature of such focused rays? *(U)*

23. Can you use black light to operate a photoelectric bulb? *(U)*

24. What is the effect of black light on the growth of bacteria, mold, and protozoa? *(U)*

25. Can black light be used to kill germs? *(I)*

26. Does black light travel in a straight line? *(P)*

27. What causes certain minerals to fluoresce? Can you change this property in these materials? *(U)*

28. Can you experiment with different foods such as nuts, butter, meats, and vegetables to determine the effect black light has on them and also to determine if this effect changes with the age of the food? *(I)*

Atomic Energy

3

Radiometric Dating

Purpose: This model clock demonstrates how scientists can tell the age of the Earth by examining radioactive ore. The clock can also be used as a timing device.[1]

Materials: Two straight-sided bottles (such as baby bottles), two rubber stoppers (two-hole), and two very short pieces of glass tubing.

What to Do: Join the two stoppers together with glass tubing through each hole (for baby bottles the stoppers should be #7). Fill one of the bottles with water. Insert the stoppers and the other bottle as shown. Turn the "atomic time clock" over so the water slowly runs into the second bottle.[2]

Operation of Equipment: Time how long it takes the water to flow to a depth of one inch, then two inches, etc. If you use a baby bottle, marks are already calibrated in CC's. You can change your bottle into quite an accurate clock. Place a piece of masking tape alongside the

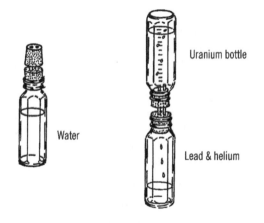

Water

Uranium bottle

Lead & helium

marks on the bottle. Place a pencil mark for the time it takes the water to flow to that height. You will probably have a mark for one minute, two minutes, etc. You might even have short marks for the half minutes between the long marks.

The scientist uses a similar clock to tell the age of rocks. Uranium gives off radioactivity at a

---

1 In this project, originally titled "Atomic Time Clock," you will build a *model* that simulates, by way of analogy, how radiometric dating can reveal the age of a sample. Radiometric dating (e.g., carbon dating or lead-lead dating) works by comparing the relative abundance of a radioactive isotope to its decay products. This is almost the inverse of a modern "atomic clock," which is an extremely accurate timekeeping apparatus that relies upon the consistent properties of stable (non-radioactive) atoms.

2 Note that the actual apparatus is that of the "Water Hourglass" on page 158.

rate as regular as water flows. The uranium gradually changes into helium and lead. The scientist notes the percentage or amount of uranium as compared to lead and helium. By knowing the rate of flow or change, he[3] can tell quite accurately how old the piece of uranium is.

Mark one bottle uranium. Mark the second bottle lead and helium. Now place another tape on the bottle labeled lead and helium. This time, instead of minutes, label the marks in thousands of years.

Can You Work Like a Scientist?

1. Have a student leave the room. This student represents the scientist who wasn't even alive at the time these rocks were formed. At some time after the student has left the room, start your water atomic clock. Note the time you started the water (or uranium) flowing. Call the student back into the room and have him tell you exactly how long ago you started the water flowing. This would be the same as the scientist telling the age of the rock.

2. Why do you use two-hole rubber stoppers? Can you make a clock using white sand?

3. Heat some glass tubing and pull it into a narrow opening. If you used this glass tubing with a narrow opening, would your time clock run slower or faster? Do certain radioactive materials have a slower rate of flow than others?

Spinthariscope

Purpose: This is a small instrument for watching atoms of radioactive materials bombarding a phosphorescent screen.

Materials: Small cardboard box (about 3″ × 3″ × 3″), cardboard tubing, small glass plate to fit into the bottom of the box, small double convex lens (magnifying lens), zinc sulfide, and a source of radioactive material.[4]

What to Do: Remove the radioactive paint from the numbers of an old alarm clock[5] by scraping the paint with a sharp knife. Place the small bits of radioactive material on a clean piece of glass. The glass should be small enough to fit in the bottom of the cardboard box. You can cut glass by following the directions in the chemistry section.

Add about a gram of zinc sulfide powder to the glass plate. Mix this powder and the radioactive material together. Grind the two powders so the mixture contains very small bits of both materials. Remove the mixture from the glass and coat the glass with a layer of shellac.[6] While the shellac is still tacky, the mixture is sprinkled over the glass surface. Care should be taken to spread the mixture evenly. *Don't handle the radioactive material with your bare hands. Wash thoroughly after working with radioactive materials.*

The glass is then glued with airplane cement[7] to the bottom of the box, as shown. Be sure that the coated side of the glass plate is up. Allow the coating on the glass to dry completely.

Cut a hole in the part of the box opposite the glass plate and mount a lens much the same way as you did with the telescope. Over this

---

3 Or she!

4 See Note 28 in Appendix E about radioactive sources, including radioactive paint found on some old alarm clocks.

5 Not all alarm clocks have radioactive paint; see the previous note.

6 Does it need to be genuine shellac? What other materials would work?

7 Model airplane (plastic) cement, Duco cement, or five-minute epoxy.

hole and lens glue a piece of cardboard tubing about an inch in length.

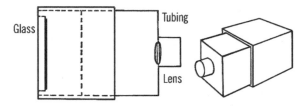

Operation of Equipment: The spinthariscope works the best under total darkness. Therefore, use the instrument in a darkened closet or a room without outside windows. Look through the cardboard tubing toward the glass plate. Focus the lens by moving the top and bottom of the box apart. It probably will take several minutes for your eye to become adjusted to the darkness. Then, you should see bright flashes as the radioactive atoms throw off bits of themselves and slowly change from one element to another.

Modern Safety Practice

Low-level radioactive sources are potentially hazardous, particularly if ingested. Read Note 28 in Appendix E about radioactive sources.

Can You Work Like a Scientist?

1. When your spinthariscope stops twinkling in the dark, the radioactive materials have completely decomposed into non-radioactive materials. Do you notice any weakening of your spinthariscope over a period of a week? A month? How long will your spinthariscope continue to work?

2. Can you devise a way of measuring the strength by counting the number of flashes in a certain time period? Could you observe the flashes under a microscope?

Dosimeter

Purpose: A dosimeter is used to detect the amount of radiation to which a person or area is being exposed. A dosimeter such as the one described can be used to determine the safe limits when working with radioactive material or it can be used to detect radiation levels after a nuclear bomb has been exploded in the area.

Materials: Aluminum foil, nylon thread, large drinking glass, ruler, Scotch tape, and the lid from a tuna or similar-size can.

What to Do: Wrap a piece of aluminum foil around the outside and bottom of the glass. Remove the glass and then carefully work the aluminum form into the inside of the glass. The aluminum should be carefully worked with your fingers and the end of a pencil so that it takes the form of the inside and the bottom of the glass.

Remove the aluminum from the glass. Measure down two inches and then cut a one inch square window out of the aluminum. Cut a similar window on the other side of the aluminum form so the two windows are in line. Put the aluminum back into the glass and check to see that the two windows line up **A**.

Cut a lid off a tuna can. Measure one inch from the center of the tin can lid and punch two small holes side by side as close together as possible. Measure one inch from the center of the lid just opposite these holes and make two similar holes **B**.

Lay two pieces of nylon thread on a table. Do not touch the thread near the middle. The moisture from your hands will affect the results. Measure three inches from the middle of the two threads and fasten the threads down to the table with Scotch tape. Measure 6 inches from this spot toward the middle of the thread and place another piece of Scotch tape over the two threads **C**. Cut out two aluminum circles about the size of a quarter. Slide one of the aluminum circles under the thread between

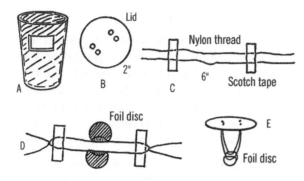

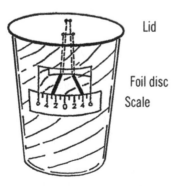

the two pieces of Scotch tape. Carefully bend the edge of the disc over the thread. Do the same with the other thread and aluminum disc.

Tie a knot with the nylon threads at each end of the pieces of Scotch tape. The knot should be retied several times so that it won't come undone later **D**. Next, slip the two loose ends of the threads through one set of two holes in the lid. Immediately tie the two ends so that the knot fastens firmly down near the lid. Remove the Scotch tape. Repeat the steps with the other two loose ends and the other set of holes in the lid.

When you finish, your two discs of aluminum should be suspended from the lid as shown **E**. The discs can be slightly moved so that they will match up. Care should be taken not to touch the thread upon which the aluminum circles hang.

Operation of Equipment: The two aluminum circles suspended from the lid are really a simple type of electroscope. If you bring a material which is charged with static electricity near the piece of foil, the foil becomes charged. Since the charge is the same on both discs, the two circles repel or push each other away. If the lid is carefully lowered over the glass containing the aluminum liner, the two discs will retain their charge. This can be seen through the two windows cut in the aluminum. The lid should be turned so that the discs run perpendicular to the windows. In that way it is easy to tell if the discs are still separated. A scale should be made as shown. This scale is then glued on the outside of the glass just under the window. It is important that the distances on the scale be accurate. Each mark is exactly 2½ mm. apart. Thus the entire scale from 6 to 0 and then to 6 is 30 mm. The scale should be adjusted so that the bottom of each charged disc is the same distance from the 0 mark.

The discs are charged by lifting the lid up and bringing a charged material, such as plastic or a black rubber comb, nearby and touching the discs. Be sure the discs remain charged before you lower them into the glass. The discs should be at least one-half inch apart.

If little or no radiation is present, the discs will remain apart for fifteen or more minutes. If radiation is present, the discs will discharge their static electricity and come together much sooner.

When the discs are placed in the glass, the viewer should mark down the number indicated by the bottom of the disc. The time that it takes the disc to move one mark closer together is then recorded in seconds. The safety limit is then figured by multiplying this time in seconds by ten. The resulting answer is the number of hours a person could stand the degree of radiation present before a fatal dose had accumulated.

Thus, if it takes twenty seconds for the disc to drop from four to three, the time limit would be ten times twenty or two hundred hours.

Can You Work Like a Scientist?

1. Dry air is a good insulator and does not conduct static electricity very well. Would your dosimeter be accurate if you had a lot of moisture present?

2. Moisture in the air makes air a better conductor of static electricity. Thus a static electricity charge does not build up as much. Why should you be sure not to touch the nylon thread between the lid and the aluminum disc?

3. Does it take as long for the discs to discharge outside of the glass as inside the glass? What purpose does the aluminum foil have inside the glass?

4. Radiation turns the air into a good conductor of electricity. Why does increased radiation bring the discs closer together?

5. Moisture will affect the accuracy of this instrument. How can you protect your dosimeter from moisture?

Diffusion Cloud Chamber

Purpose: The cloud chamber is used for detecting and identifying radioactive particles and rays such as Alpha particles, Beta particles, and Gamma rays.

The vapor trails of these radioactive materials can be seen when the chamber is operating properly.

Materials: Gallon jug, aluminum cake tin, 7" × 7" plate of window glass, a ¾" × 48" strip of 24-gauge copper, black felt or black enamel paint, sponge, 240-volt photoflash battery, alcohol (or ditto fluid), 5-pound block of dry ice, two pieces of insulated wire, about 4' of weather stripping, and a radioactive source.[8]

What to Do: Cut the top and the bottom off the jug with the bottle cutter (see chemistry section). Glue a piece of weather stripping around the top and the bottom on the outside of the jug. The stripping should stick just beyond the edges of the glass so that it can form a seal.

A copper strip is fitted around the outside of the jug. The ends of the strip are soldered together. The strip is positioned about one inch from the top of the jug. A second copper strip is fitted and soldered around the inside of the bottom of the jug.[9] This strip should be positioned about an inch and a half from the bottom edge.

The bottom of the aluminum cake or pie tin is either painted black or covered with black felt. When the jug is placed on the aluminum pie tin, the weather stripping should form a seal around the bottom.

The sponge should be cut in strips and glued around the inside of the jug opposite the upper copper band. This sponge is used to hold a supply of alcohol during the operation of the chamber.

A wire is connected to the inside of the lower copper band. The wire is run through the strip-

8 Read Note 29 in Appendix E for an extended discussion about materials for this project.

9 A soldering iron with at least 75 W power is recommended.

ping around the bottom of the jug and is connected to the negative pole of the battery. A second wire is run from the positive pole and is connected to the aluminum tin.

The radioactive source is fastened or glued to the point of a thumbtack. Radium can be scraped off a radium dial wristwatch or clock. The tack containing the radioactive sample is then placed in the center of the aluminum tin.

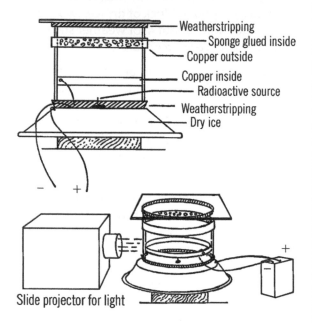

Slide projector for light

Operation of Equipment: Place a block of dry ice on a plywood board. Set the aluminum tin over the dry ice. The bottom of the tin should be in contact with the dry ice. After the radioactive source has been placed in the center of the bottom of the aluminum cake tin, lower the jug down on the tin. The weather stripping around the bottom of the jug will form a tighter seal if you coat the stripping with Vaseline. Pour a small amount of alcohol around the sponge.

Some of the alcohol will drip down to the bottom of the container. The glass plate should be placed over the top of the jug. Again, Vaseline

can be used to make a seal between the stripping and the glass plate. Shine a light source directly across the jug and about two or three inches above the surface of the aluminum tin. A flashlight or slide projector provides a good source of light.[10] Finally, the wires should be connected to the battery.

Modern Safety Practice

1. Dry ice can cause severe frostbite injuries. Handle only briefly, using oven mitts or a towel to protect your hands. See Note 14 in Appendix E about safety with dry ice.

2. Avoid directly touching the radioactive source and wash your hands thoroughly after. Read Note 28 in Appendix E about radioactive sources and safety.

3. Alcohol vapor is flammable. Keep sources of spark and flame away, and have a type ABC fire extinguisher handy.

4. If you are applying a voltage between the copper strips, be sure not to touch the strips or wiring. You could receive a severe electric shock. Additionally, there is some chance of a spark when you connect or disconnect wires under power. Connect the wires to your strips first, and only apply or disconnect power from the other ends of the wire, away from the alcohol vapors.

Theory of Operation: Alcohol evaporates from the sponge and the bottom of the chamber. This forms an alcohol vapor atmosphere. The dry ice cools the air near the bottom of the chamber. The cold temperature of the tin draws the heat energy down the glass. The copper band around the outside of the chamber absorbs heat energy from the room and passes this energy through the glass jug and directly to the sponge. This extra heat energy

10 Slide projectors can often be found used at garage sales and thrift stores. Modern computer projectors will also work well.

causes the alcohol in the sponge to evaporate more rapidly. As the vapor continues to evaporate, the atmosphere inside the jug is soon saturated with alcohol vapor. As the vapor diffuses (spreads) throughout the chamber, the vapor molecules near the bottom of the chamber are cooled. Since there are more molecules in the form of vapor than can normally exist at a very low temperature, we say the atmosphere near the surface of the aluminum tin is supersaturated. The vapor molecules will condense on charged particles passing through the supersaturated vapor.

Unfortunately, many air molecules have either lost or gained electrons and therefore have an electrical charge. Thus the vapor molecules are very likely to form on the charged air molecules (ions) as a visible alcohol fog. Dust particles have an electrical charge and also serve as nuclei or centers for condensing alcohol vapor.

Since we want to see the vapor trail of just the radioactive material, it is usually necessary to remove the ions (electrically charged molecules) and dust particles from the air. When the wires forming the aluminum tin and the copper band are connected up to the high-voltage battery, negatively charged particles are attracted to the aluminum base because of its positive charge. Since the copper band has a negative charge, particles with a positive charge are drawn to it. This removes electrically charged material from the lower part of the chamber. We then can see the vapor trails formed by the passage of radioactive materials.

An Alpha particle is really the nucleus of a helium atom. This nucleus is thought to consist of two neutrons and two protons. Since the protons have a positive charge, the Alpha particle is charged positively. As this radioactive particle is given off by the radioactive source, it travels through the chamber at about one-twentieth of the speed of light. This particle attracts electrons from the neutral air molecules, thus giving the molecules an electrical charge. The alcohol vapor condenses on these air ions.

Thus, the ions formed by the passage of a radioactive particle mark the path of the particle. In the cloud chamber, the trail of Alpha particles is straight and wide.

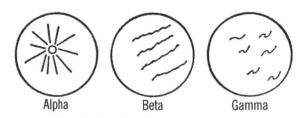

Alpha   Beta   Gamma

Beta particles are electrons given off as neutrons change to protons. These particles are negatively charged and travel at almost the speed of light. The Beta particles leave curly, wispy trails that are much harder to detect than the trail of Alpha particles.

Gamma rays have a great penetrating power and travel at the speed of light. They will appear occasionally as a spot with trails going in all directions.

Can You Work Like a Scientist?

1. Is copper a better conductor of heat than glass? How can you prove this?

2. Why do you need a band around the top of the gallon jug?

3. What is the difference in temperature between the top and the bottom of the glass chamber? Can you measure this?

4. Does heat increase the rate of evaporation for both alcohol and water?

5. Will the cloud chamber work without using the electronic clearing field?

6. Will the cloud chamber work if you use just plain ice instead of dry ice?

7. Will the chamber work as well if you reverse the leads on the battery?

8. Do you have to clear the field by connecting up the battery every time you want to view the path of particles? What happens if you leave the battery connected up? Are the ions that are formed immediately attracted to the metal plates?

9. If you used a water vapor instead of alcohol vapor, would the chamber work as well?

10. Can you keep water vapor from condensing by removing the dust and ion particles?

11. Which form of radiation is the most common?

12. Can you explain why the tracks of the various particles and rays differ?

13. Will any of the radioactive materials go through paper? You might place a paper barrier in the chamber.

14. Does radiation travel in a straight line? Can you prove this?

15. Can you take pictures of these vapor trails?

16. Remove the radioactive source from the chamber. Observe carefully. Do you still see an occasional vapor trail? This could be a cosmic ray.

17. Can you tell from your chamber what form of radiation is a cosmic ray?

18. Can you shield your chamber so that cosmic rays cannot enter?

19. If some forms of radiation are electrically charged particles, can you change the path of the particles by using a strong magnetic field?

20. Can you devise a simpler chamber than the one described?

21. Are we bothered by cosmic rays more in the daytime or at night?

22. What effect do sunspots have on the amount of cosmic rays?

23. Does radiation help discharge an electroscope? Can you explain the reason?

24. Does rain help bring radiation down from the atmosphere?

25. Another type of cloud chamber is the expansion type. The theory behind it is this:

   If air is pumped into a chamber containing alcohol, the temperature is increased as the air is compressed. As the temperature increases, the atmosphere is able to hold more and more vapor. Finally, if the air pressure is released suddenly, the alcohol vapor and air atmosphere expand rapidly. This rapid expansion cools the atmosphere in the chamber. The atmosphere can no longer hold the alcohol vapor at this decreased temperature, so the alcohol condenses on any available nuclei.

   Can you design an expansion type of cloud chamber? You might get an idea by examining the chest cavity mentioned in the biology section.[11]

---

11 Or possibly from the "Cloud Jar" on page 190.

Electricity and Magnetism

4

Making a Magnet (or Recharging a Magnet)

Purpose: If you learn how to make magnets you will not only be able to make magnets without cost, but you will learn something about the nature of magnetism.

Materials: Strong bar magnet[1] and some things made of iron, steel, brass, pot metal, copper, and other metals.

What to Do: Stroke the object to be magnetized with a bar magnet. Try to imagine the object to be magnetized as being made up of many small compasses. You want to line up all the compasses so the needles point in the same direction. You will need to stroke the object many times.

Can You Work Like a Scientist?

1. If you rub the object *back* and *forth* with a magnet, does it magnetize very well? Check with a magnetometer (see "Magnetometer" on page 108).

2. Try rubbing the object in only one direction. Is the result better?

3. Check your magnet with a compass. Does it have one or two poles? Can you explain this on the basis that the object is made up of small materials that act like compasses?

4. Rub two objects the same way with a magnet. Check the poles with a compass. Are the poles alike or different? Do the same poles of the objects push away (repel) or draw together (attract) each other?

5. Which materials magnetize the easiest? Which materials keep their magnetism the longest?

6. Can you make a magnet by placing it in a north and south position and leav-

---

1 A neodymium (NdFeB) magnet is ideal.

ing it? Try this for several days. Why might or might not this work?

7. Can you place an object in a north and south position and magnetize it by hitting it with a hammer?

8. Check cans in a store to see if any are magnetized.

9. Are cans just made of tin?

10. Will a magnet pick up a nickel? Try a Canadian nickel too.

11. Can you make a magnet out of a nail by holding a strong magnet nearby?

12. Will magnetism go through things? Try glass, paper, books, table tops, other metals, etc.

13. Set a thin box on top of your magnet. Shake in some iron filings. If you don't have any, use a file on a nail. Tap the box. Do you see the lines of force? What part of the magnet is the strongest?

14. How can you destroy your magnet? Can you think of five ways?

Magnetometer

Purpose: This instrument is used for measuring the strength of magnets, either permanent or electric magnets.

Materials: Ruler or yardstick, rubber band that will stretch easily, thumbtack, and a paper clip.

What to Do: Stick a thumbtack in the ruler or yardstick near the end. Break the rubber band and tie one end around the tack. Tie the other end to a paper clip.

Operation of Equipment: Hold the ruler or yardstick with one hand. Bring the magnet to be tested near the end of the paper clip. Pull the paper clip with the magnet slowly. The rubber band will stretch to a point where the strength of the magnet is equal to the pull of

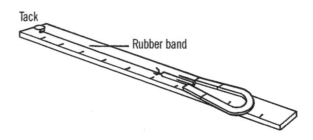

the rubber band. At that point, the paper clip will break loose from the magnet. Note the number on the ruler that the paper clip reaches. This number will tell you the strength of the magnet compared with other magnets.

Can You Work Like a Scientist?

1. How can you make your magnetometer more sensitive?

2. Keep a chart of how many times you rub a screwdriver with a magnet and the strength the magnet has each time. What do you conclude from this?

3. Do both poles of a magnet have the same strength?

4. Try magnetizing different materials in the same way. Which material is magnetized the most? Which the least?

5. Try putting two bar magnets together end to end. Is the magnet stronger?

6. What part of the magnet is the strongest-the end or the middle?

7. What materials hold their magnetism the longest? Keep a graph on the time and the strength of each material.

8. Make an electromagnet out of a coil of wire. Connect it to a battery. How strong is the magnetism around the wire?

Now place a nail in the coil. How strong is the electromagnet? Turn off the current. How strong is it now?

9. Magnetize a screwdriver by rubbing it. Measure the strength. Now hit the screwdriver against a table several times. Measure the strength. Can you make the screwdriver non-magnetic?

10. What effect does heat have on a magnet?

11. If you wrap more wire on your coil, will the magnetism increase? Keep a graph of the turns and the strength.

12. Connect more batteries together. Does it make the electromagnet stronger? Can you measure the strength of batteries with this?

Needle Compass

Purpose: To make a sensitive compass out of simple materials in order to demonstrate the law about magnetic poles.

Materials: Two needles, piece of 3″ × 5″ file card, and a pin.

What to Do: Magnetize the two needles either with your solenoid (see "Solenoid" on page 118) or magnet. Check the poles with a compass. Be sure the north poles of both needles point in the same direction. Bend a piece of file card and stick the needles in as shown. Next put the pin through the rest of the cardboard. Balance the compass needles on the point of the pin.

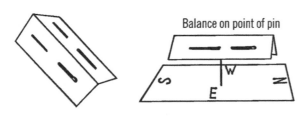

Balance on point of pin

Operation of the Compass: The compass should line up in a north and south direction. Check with your compass. Mark which pole is north and which pole is south.

Can You Work Like a Scientist?

1. Take a magnet. Find its north pole by checking with a compass. Now try your north pole on the north pole of your needle compass. Do they attract or push away (repel)?

2. Try the south pole of the magnet on the south pole of your compass. Do they attract or repel?

3. Now try a north pole on a south pole. Then a south pole on a north pole.

4. Do the same poles attract or repel? Do opposite poles attract or repel?

5. Place some metal nearby to make your needle compass give the wrong reading. Can you correct this by using more metal? Where would you place it? Would this work on a ship?

6. If you were at the North Pole, what direction would be north? What direction would be south?

7. If the Earth acts as a magnet, what must the inside of the Earth be made of?

8. Would a compass work on the moon? How about the other planets?

9. Could you use a compass to locate a meteorite?

10. How far away from your needle compass can you move your magnet and yet affect the compass? Is this distance the same for all magnets?

11. Could you develop a scale for measuring the strength of magnets by using the principle mentioned in question 10?

12. Use your compass to check the polarity (poles) of your refrigerator. Move the compass up and down the outside of the refrigerator.

Watch Spring Compass

Purpose: To detect the North and South Magnetic Poles of the Earth.

Material: Hard steel, preferably the spring out of an old watch.[2]

What to Do: Cut the spring so that it is about an inch and a half long. Using a nail or nail punch and a hammer, try to dent the spring near the middle. Now put a pin through a piece of cardboard and balance the watch spring on the point of the pin. Check with a piece of iron to see if the spring acts as a compass.

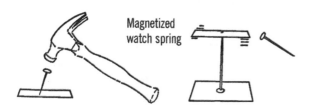

Magnetized watch spring

Columbus Compass

Purpose: To show the type of compass Columbus used.

Materials: Needle, cork, and glass dish.

What to Do: Magnetize the needle. Slice a thin section of cork with a razor blade. Lay the needle on the cork or stick the needle through the center. Place the cork in a dish of water.

Can You Work Like a Scientist?

1. Columbus had trouble with his men. As he sailed west, the compass needle didn't point in the same direction as it did before. Can you show why that would be true by using a globe?

2. Could you use a compass to help you tell longitude?

3. Why didn't you use a metal pie dish? What might act like a metal pie dish on a boat?

4. How do boats and planes protect themselves from metal?

5. What kinds of compasses do ships use now in place of Columbus's compass?

Iron Filings

Purpose: Iron filings are used to detect the magnetic lines of force around a magnet or around a wire conducting electricity.

Materials: Soft nail, file, salt or white sand, small box, and plastic sheet or Saran Wrap.

What to Do: You can make your own iron filings by filing a nail. Magnetite is a very cheap substitute for iron filings. In either case, mix the filings with about one fourth salt or white sand. The white sand is the same kind that is used in hotel containers.[3] Most hotels would give you a small amount. Place the mixture in a shallow flat box and cover the box with a plastic sheet or a glass plate.

Operation of Equipment: Hold a magnet under the box. Tap the box gently. You should be able to see the magnetic lines of force clearly. The sand or salt outlines the iron filings and

---

2 While the spring from an old mechanical wristwatch can be used, there are many other sources. Try straightening a paper clip or the spring from a ballpoint pen. What materials work best?

3 "Hotel containers" in this context refers to a kind of fancy ashtray (once common in hotel lobbies) that would be filled with sand. Kosher salt is a good substitute.

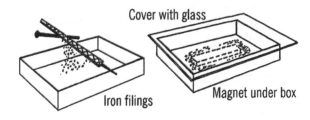

Cover with glass

Iron filings

Magnet under box

makes the lines show up more clearly. If you get filings on your magnet, you can remove the filings with Scotch tape.

Can You Work Like a Scientist?

1. Can you remove the iron filings from the salt or sand mixture? Can you use an electromagnet to do the same thing? Be sure to have a sheet of paper between the filings and the magnet.

2. Does an electromagnet have the same kind of lines of force as a permanent magnet?

3. Run a wire through a card as shown below. Sprinkle the mixture on the card and connect the wire up to a strong source of direct current. Is there a pattern of magnetic lines of force?

4. Wrap a coil of wire through a card as shown. Connect this up to a source of current.[4] Can you see magnetic lines of force?

5. Try this with alternating current.

6. Can you make permanent pictures of magnetic lines of force by forming a pattern of a magnetic field on wax paper and then heating the paper over a strong light bulb?

7. Do filings cling to a bare wire carrying electricity?

8. Can you make Scotch tape magnetic by using iron filings?

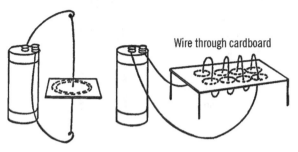

Wire through cardboard

Electroscope

Purpose: The purpose of an electroscope is to detect charges of static electricity and to tell the type of charge present.

Materials: Bottle or clear light bulb, a cork or one-hole stopper, heavy copper wire (size 6, 8, or 10), and two short pieces of tape recorder tape.[5] In place of the tape you could use gold foil or very thin tinfoil. However, tape recorder tape is light and does an excellent job.

What to Do: If you use a light bulb, use a #1 stopper. If you use a bottle, you can use either a stopper or a cork. Remove the rubber insulation from the wire except for the short section that fits into the stopper. Stick the wire through the stopper or cork. Make a loop in the upper end of the wire. Bend the lower end as shown and glue on the tape recorder tape. Insert the wire and stopper in the bottle. Be sure there are no sharp edges on the ends of the wire.

Operation of Equipment: Rub a comb on a piece of wool or flannel. Touch the comb to the loop of the wire. The tapes should become charged. Because the charge on both strips of tape is the same kind, the strips push apart. If you place the charge on the loop correctly, the

---

4 The source of current shown is a No. 6 dry cell, 1.5 V. See Note 30 in Appendix E.

5 E.g., part of the ribbon from an old audio cassette tape.

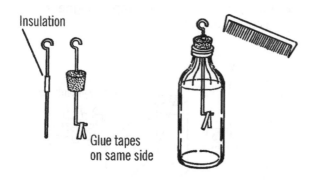

Insulation

Glue tapes on same side

tapes will remain charged after you remove the pencil.

Can You Work Like a Scientist?

1. Touch the loop. What happens to the charge on the tapes?

2. Rub some plastic, like a plastic bag, with your hand many times. Touch the loop with the plastic and see what happens.

3. Rub some paper. When you touch the loop, do the tapes go farther apart or together? Is the charge the same as the plastic? If so, the tapes should go farther apart.

4. Touch the loop. Now rub a glass test tube with flannel, then silk, and finally wool. Test your electroscope each time. Are the charges alike or different?

5. Does your electroscope work better in dry or moist air? What effect does temperature have on the operation of your electroscope?

6. Does the charge seem to leak away? The cork may be your trouble. Try waxing or shellacking the cork. Is this better? Try a rubber stopper.

7. Some of the charge may leak away from the loop. Get an old Christmas tree ornament. Slip this over the rod. The ornament is round. Does the elec-troscope keep its charge longer? Try coating the ornament with aluminum paint.

8. When the air is damp, it is a good conductor of electricity. How does this help to explain why static electricity charges seem to leak away?

Pith Ball Electroscope

Purpose: The pith ball electroscope is used to detect the charge of static electricity.

Materials: Small bottle (ink, olive, maraschino cherry, etc.), cork, silk or nylon thread, stiff wire, and a pith ball. The pith ball can be made from the pith (inside) of an elderberry plant; however, workable substitutes are small cork or styrofoam balls about half an inch in diameter.

What to Do: Wrap aluminum foil around the styrofoam or cork ball. Tie a knot in one end of the thread. Thread the needle and stick it through the ball. Pull the thread through and remove the needle. Tie the end of the thread to the wire loop as shown.

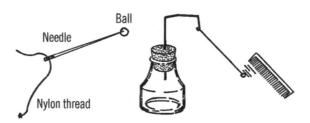

Needle   Ball

Nylon thread

Operation of Equipment: You give a charge of static electricity to an object by rubbing it with another object. Bring a charged object near the pith ball and see what happens.

Can You Work Like a Scientist?

1. Rub a piece of plastic with your hand. Can you transfer the charge on the plastic to the pith ball?

2. Can you charge the pith ball by direct contact? By bringing a charged object near the pith ball?

3. Charge the ball. Give the same charge to another object. Bring the object close to the charged pith ball. What happens?

4. Charge two pith balls with an opposite charge. What happens when the pith balls are brought near each other?

5. Rub a black hard rubber comb with fur or wool. The charge on the comb is negative (-). Charge the pith ball with this negative charge. Objects that will attract this negatively charged ball have a positive charge.

6. Now rub various objects with fur, plastic, nylon, flannel, etc. Record the charge of each. Remember to charge your pith ball electroscope with the comb each time.

7. Rub a balloon. What kind of a charge does a balloon carry?

8. What kind of charge does your electrophorus ("Electrophorus" on page 115) carry?

9. Charge both pith balls the same. Bring a watch with a fluorescent dial near the balls. What happens? Why? (See the dosimeter in the atomic energy section.)

Electronic Electroscope

Purpose: An electronic electroscope is a highly sensitive instrument for detecting faint static electricity charges. An instrument similar to this is used by the weather bureau for detecting and predicting thunderstorms.

Materials: Neon bulb (NE 2), resistor (47 kΩ), high cutoff pentode radio tube (6AU6), transformer (6 volt-½ amp. filament), heavy copper wire for antenna (size 6), and a disc made from aluminum (about 3" in diameter). Two pieces of plastic or wood (about 3" × 6") for the base, and bolts and nuts to hold the base together.[6]

What to Do: The plywood or plastic pieces are separated by nuts used as spacers on the four corner bolts. All the wiring is done on the bottom of the top piece. After the electroscope is completed, the bottom piece is placed in position, and the bolts and nuts are fastened in place. This arrangement prevents your touching any of the wiring and getting a shock. Holes are drilled in the top piece, and all wiring connections are made on the bottom side.

First, fasten the transformer to the base. Next, mount the neon bulb and then the tube on the top side of the base. Tap off the line coming in from the plug (house current) to the transformer.[7] This usually is a black wire. Connect a lead from this wire to one side of the neon bulb. This should be done on the bottom side of the base. The other side of the neon bulb is connected to the resistor located on the bottom side of the base. The other side of the resistor is connected to both the plate and screen connection of the tube. (See wiring diagram.)

The other black lead-in wire to the transformer is connected to the suppressor of the tube and the cathode of the tube. The low-voltage wires from the transformer (green) are connected to the ends of the filament heater in the tube. A heavy copper rod or wire is connected from the grid of the tube and mounted upright through the base. This antenna should be about 12" long. A hole should be drilled in the center of the aluminum disc so that the disc will just fit over the copper wire antenna. The bottom

---

6 See Note 31 in Appendix E for notes on this project, including a low-voltage alternative with LEDs.

7 Exercise *extreme caution* because mains voltage is involved and exposed.

base is then held in place by corner bolts and spacer nuts between the two bases.

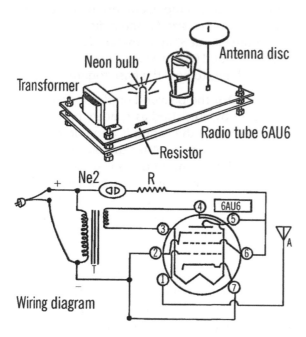

SYMBOLS: Ne2: *Neon bulb*, R: *Resistor* (47 kΩ), 6AU6: *Tube*, A: *Antenna*, T: *Transformer*

TUBE CONNECTIONS: 1: Grid, 2: Suppressor, 3: Filament heater, 4: Filament heater, 5: Plate, 6: Screen, 7: Cathode

Operation of Equipment: In order to understand the operation of the electroscope, we must understand the operation of the tube.

Electricity coming from a house plug is 110-volt alternating current. This is the current that comes into the cathode from socket **7**. The cathode gives off bits of electricity (electrons), and because it has many extra electrons it has a negative charge. The wires from the low voltage side of the transformer bring in low voltage current to the filament heater through sockets **3** and **4**. The filament heater adds extra energy to the electrons coming off the cathode. In fact, a cloud of electrons form around the cathode. The grid connected to socket number **1** normally does not have a charge. This grid is connected to the antenna. A screen (socket no. **6**)

is connected to the other side of the transformer through the resistor and the neon bulb. Since its charge is positive, or opposite to that of the cathode, it pulls the electrons from the cathode. The electrons are then pulled from the screen to the plate (socket no. **5**). The plate is connected on the same side of the transformer as the screen and thus has the same positive charge. Since the plate is larger than the screen suppressor (socket no. **2**), the greater charge pulls the electrons right by the suppressor screen. The suppressor acts to slow down the electrons since it has the same negative charge as the electrons. Like charges repel. When the electrons reach the plate, the electricity flows through the resistor and lights up the neon bulb. Thus, when the neon bulb is glowing, electrons are flowing across the tube.

When the current changes directions, the plate becomes positive and the suppressor negative. Since the electrons are negative, they are repelled by the suppressor and stay on the plate. Thus the tube acts as a gate to let the electricity flow only in one direction.

Modern Safety Practice

Pay careful attention not to touch the exposed wiring: a shock from mains power (AKA line voltage or household wiring) is potentially lethal. Read Note 10 in Appendix E for safety practice around it.

The low-voltage LED electroscope described in Note 31 of Appendix E is a safe and easy alternative to the project described here.

Can You Work Like a Scientist?

1. Normally the grid does not have a charge, so the current flows through the tube and the neon bulb glows. Now, if you rub some object and produce a static electricity charge, let's see what happens. Suppose you bring an object with a positive charge near the

antenna, thus giving the grid a positive charge. Since the cathode has a negative charge, will the electrons flow by the grid to the plate? Remember, unlike charges attract.

2. Will the bulb continue to glow if the charge on the grid is positive?

3. If you bring an object charged negatively near the antenna, the grid becomes negatively charged. The cathode is also negatively charged. Will the grid allow the electrons to flow by? Will the neon bulb continue to glow?

4. Does the electroscope detect positive or negative charges? What does it indicate when the bulb goes out?

5. What kind of charge forms on a black hard rubber comb? Remember, when you rub an object, a positive charge forms on one of the materials, and a negative charge forms on the other.

6. How far away from the antenna will the electroscope detect static electricity charges?

7. Does the distance that a charge can be detected vary with the amount of humidity?

8. Will your electroscope operate as well in a bathroom after a shower as in other rooms of the house?

9. What is the effect of putting hair oil on your hair as far as the generating of static electricity is concerned? Try combing your hair and see.

10. Do you create a static electricity charge when you pet a cat? What kind of charge is formed?

11. Break a lump of sugar in a darkened room. Is static electricity given off?

12. Can you detect the build-up of thunderstorms with your electroscope?

13. Do all cloth materials give off the same type of static electricity charge?

14. What seems to affect the rate at which a static electricity charge leaks off? What kind of material and what shape will retain a charge the longest?

15. Does flowing water have an electrical charge?

16. Is there a static electricity charge built up when a mimeograph or duplicating machine is operating? What causes the charge? What kind of charge is formed? Does it vary with humidity?

17. If you use a 6J7G tube instead of the 6AU6, your electroscope will be more sensitive. The antenna disc is placed on the cap of the tube.

18. Is the distance through which an electronic electroscope will detect static electricity an indirect way of measuring the strength of the charge? Can you devise a scale to measure the strength of static electricity charges?

19. What is the effect of temperature on static electricity charges?

Electrophorus

Purpose: This is a standard piece of equipment used to build up a strong static electricity charge.

Materials: Aluminum cake tin, wood doweling for handle, plastic sheet or rubber sheet (from old inner tube if none other available), piece of fur or flannel.

What to Do: Drill a hole in the bottom of the cake tin. Attach a handle to the tin by nailing through the hole. Lay the rubber sheet on a flat surface.

Operation of Equipment: Rub the rubber or plastic sheet with the flannel or fur for about a minute. Place the cake tin on the rubber sheet.

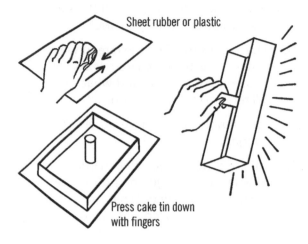

Sheet rubber or plastic

Press cake tin down with fingers

Press down on the inside of the cake tin with your fingers. Then lift the cake tin off the rubber by lifting the tin by its wooden handle. The tin is highly charged.

Can You Work Like a Scientist?

1. Bring your finger near the edge of the cake tin. What do you observe?

2. Does the electrophorus work as well if you rub the rubber with silk?

3. Why did you press down with your fingers? Would it work just as well if you didn't press down with your fingers?

4. Try your electrophorus out in a dry room. Then turn on the shower in the bathroom. Does the electrophorus work as well if there is moisture in the air.

5. Discharge the electrophorus by letting a spark jump to your finger. Press the tin down on the rubber again but don't rub the rubber sheeting. Was the cake tin charged again? Why?

6. If you can't get a good charge, try a handle made out of a candle or some sealing wax.

7. Try your electrophorus on your electroscope. Is the charge negative or positive? Can you get a different charge on the electrophorus?

8. Can you get the electricity to travel from the tin through a wire to the electroscope? Hold the wire with a clothespin.

9. Try your electrophorus on your pith ball electroscope.

10. Drop bits of paper on the cake tin. Does anything happen? Why?

11. A good source for heavy plastic sheets is the plastic typing plate found in each package of mimeograph stencils. School offices and business offices throw these sheets away.[8]

12. Try making an electrophorus by using an old phonograph record in place of the cake tin. A candle or sealing wax can be used for the insulated handle.

Leyden Jar

Purpose: The Leyden jar stores static electricity over a short period of time. The jar can store up quite a powerful charge.

Materials: Mason jar, tin foil or aluminum foil, ¼" plywood for the jar cover, round brass curtain rod with a brass ball on the end, a short chain, and shellac. If you can't get a brass rod, use heavy wire.

What to Do: Clean and dry the jar thoroughly. Coat the inside of the jar with shellac. The tin foil or aluminum foil coating should cover only

---

8 Mimeograph stencils are now rare, but plastic surfaces are *everywhere!* What kinds of plastic generate stronger charges? Can you identify them by manufacturing marks or a recycling number?

the lower two-thirds of the jar. Before the shellac has completely dried, insert the foil and press it smoothly around the inside of the jar. Coat the lower two-thirds of the outside of the jar in the same manner, shellac, and then cover with foil. Cut foil circles for the bottom of the jar, both inside and out.

Cut a piece of plywood into a circle that will just fit the inside of the jar opening. Cut a second circle about half an inch larger than the first. Glue or nail these together. These pieces form the lid of your Leyden jar. Shellac this lid and drill a hole through it for the rod and chain or the wire.[9]

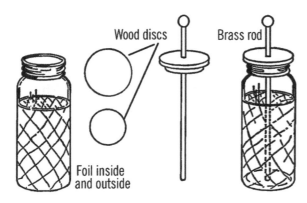

Wood discs   Brass rod

Foil inside and outside

Operation of Equipment: The Leyden jar may be charged by touching an object or wire containing static electricity to the ball or loop on the outside of the jar.

You can discharge a Leyden jar by bringing a piece of wire, which is connected to the outside coating, toward the brass ball on the top of the rod.

Modern Safety Practice

A Leyden jar stores a significant amount of electric charge and can give you a severe (potentially lethal) shock. So it is very important to discharge the jar before touching it. For added safety, mount your discharging wire on a long wooden handle so that you do not touch (or come close to touching) the wire.

Can You Work Like a Scientist?

1. Will other types of metal rods and balls work besides brass?

2. What does the chain do in building up static electricity?

3. Can you charge a Leyden jar positively?

4. Can you charge a jar negatively?

5. Can you think of a way to measure the amount of charge stored by the jar?

6. Can you charge your Leyden jar with an electrophorus?

7. Can you store regular electricity in a Leyden jar?[10] How about direct current? *Be very careful with house current.*

8. Does the static electricity in your Leyden jar affect a compass?

9. You can make a simple Leyden jar with just aluminum foil and a glass. (See the dosimeter in the atomic energy section.)

10. You can store more electricity if you stack glass plates with a layer of aluminum between each. The ends of the

---

9 The chain is intended to make an electrical connection between the rod and the foil on the inside bottom of the jar.

10 That is, "regular" electricity as opposed to static electricity. Exploring AC current is interesting, but can you think of reasons that you shouldn't connect the Leyden jar directly to household (line) voltage?

foil at each of the plates are connected together. This is a capacitor.[11]

Solenoid

Purpose: This instrument is used to make magnets with electricity from batteries.[12]

Materials: Hollow tubing, such as a test tube, cardboard tubing, or glass tubing. Also insulated wire (wire that does not have too much resistance).

What to Do: Wrap the wire around the tubing or test tube many times. Connect each end to a battery. A dry cell or a flashlight battery will work; a storage battery or battery charger is ideal.[13] The wire can be held in place by tying it or by using adhesive tape.

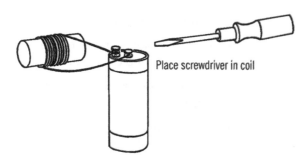

Place screwdriver in coil

Operation of Equipment: Place the object to be magnetized in the coil or tubing. Connect up the electricity. Remove the object.

Safety Tips

Don't leave the electricity on too long. The wire might get hot and cause a burn.

Can You Work Like a Scientist?

1. Does the number of turns have any effect on the strength of the magnet? Test with a magnetometer.

2. What things magnetize the best?

3. What effect does increasing the current have on the magnet?

4. Can you magnetize a watch spring? A file? A screwdriver?

5. Place the nail just inside the coil. Turn on the electricity. What happens to the nail?

6. Try to shake the nail out of the coil while the electricity is on. What holds the nail in?

7. Test the solenoid with a compass. Does it have a north and south pole? Which is the stronger?

8. Does the direction in which you wrap the wire around the tubing have anything to do with the poles?

9. What would happen if you connected the solenoid to alternating current instead of direct current? Use a bell transformer or a train transformer. Your salt water rheostat might work. Check the current with a compass. Place a magnetized object in the coil. What effect does alternating current have on a magnet?

Lemon Battery

Purpose: The lemon battery shows the principle of the battery.

---

11 Original text: "This is a condenser," where "condenser" is an old name for capacitors that has largely fallen out of favor.

12 How does this method compare to magnetizing an object with a bar magnet? How about a rare-earth (neodymium) magnet?

13 Simple battery chargers may work well. However, smarter devices (like smartphone chargers) could be damaged if you connect them this way.

Materials: Strip of zinc, copper strip, paper clips, and a lemon.

What to Do: Connect a wire to each strip of metal with the clips. Insert the copper and zinc strip into the lemon. Connect the lemon battery up to a compass galvanoscope.[14]

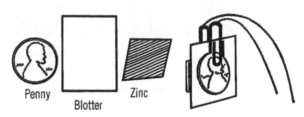

Penny   Blotter   Zinc

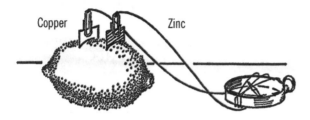

Copper   Zinc

Blotting Paper Battery

Purpose: This simple battery uses simple materials to show the essential parts of the battery.

Materials: Blotting paper, vinegar, paper clips, zinc piece from old flashlight battery,[15] and a penny.[16]

What to Do: Fasten a wire to each paper clip. Soak a small piece of blotting paper in vinegar. Clip the paper clips on the penny and the piece of zinc. Connect the wires up to a compass galvanoscope. Place the blotting paper between the two pieces of metal. A current should flow and affect the compass. Which strip is the positive pole? Can you tell from the compass? Is this the same as the lemon battery?

Electric Cell

Purpose: This liquid cell will produce enough power to light a flashlight bulb.

Modern Safety Practice

Although these are simple electrochemistry experiments, they are still chemistry experiments. Read Note 17 in Appendix E for safety practice around chemicals.

Materials: Mason jar, copper and zinc strip, paper clips, and dilute sulfuric acid. The acid may be purchased from the drugstore, or you may get some from an old car battery. Ask an attendant at a local service station.[17]

What to Do: Clip the wires to the strips with paper clips. Hang the strips in the acid. Connect the wires up to a flashlight bulb. Do you detect an electric current?[18]

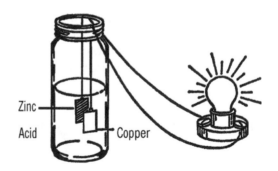

Zinc
Acid   Copper

---

14 One of several galvanoscopes described starting in "Galvanoscope" on page 120.

15 See Appendix A for additional sources.

16 Use a pre-1982 penny, if possible. Newer pennies only have a thin coating of copper.

17 It is better to start with clean acid. See Note 16 in Appendix E.

18 This should be enough to light a small incandescent bulb. Can it light an LED or a neon bulb? Why or why not?

Galvanoscope

Purpose: A galvanoscope is used for detecting very small amounts of electric current.

Materials: For both galvanoscopes shown, you will need very light insulated #22 wire. In addition, you need two needles for the first one, and a razor blade for the second.

What to Do: Wrap about 25" of the wire around a mason jar in a neat coil, leaving a few feet at each end. Tie the wire together with strings as shown. Make a wood frame as shown. In the razor blade galvanoscope the coil is hung from the support. The razor blade is magnetized either with a magnet or electrically. The blade is then hung so that it is parallel to the coil of wire.

In the needle galvanoscope, the coil is mounted on the base. The needles are magnetized and stuck through a small piece of cardboard and are parallel to each other. The poles are opposite. The needles and cardboard are hung by a thread and turn freely through a split in the top of the coil.

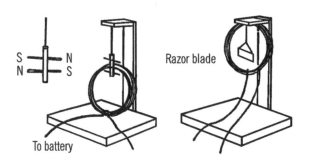

Operation of Equipment: Connect up the power to the free ends of the coil. The coil of wire becomes magnetized when current flows through. The razor blade turns to line up its magnetic field with that of the coil. In the case of the needle galvanoscope, one needle turns to line up with the magnetic force inside the coil. The other needle lines up with the magnetic force outside the coil.

Can You Work Like a Scientist?

1. If you make a simple compass galvanoscope as shown below, can it detect faint currents?

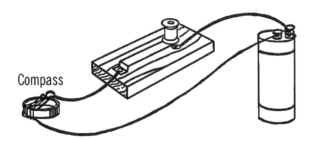

1. What happens when you switch the wires on the dry cell or the transformer?

2. Does increased current affect the galvanoscope more or less?

3. Can you use this idea to make an ammeter?

4. Work the compass galvanoscope with AC.[19]

---

19 The output of a low-voltage ac transformer ("wall-wart") is appropriate; do not use wall power directly.

Storage Battery

Purpose: This is a simple type of storage battery. It can be charged and then the charge used to light a flashlight bulb or to ring a bell. The battery can be charged many times.

Materials: Two strips of lead (lead fishing sinkers will do), acid (sulfuric, hydrochloric, or other strong acid), jar, two wires to connect to charger or bell.

What to Do: Use a gallon jug aquarium (see "Bottle Cutter" on page 18) or a wide-mouth jar. Fill the jar half full of dilute acid (see instructions about batteries, "Electric Cell" on page 119). Hang the fishing sinkers or lead strips from the wires as shown. Connect the wires up to a source of direct current (dry cells, 6-volt car battery,[20] battery charger, DC transformer).

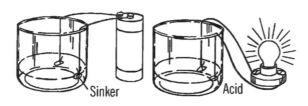

Operation of Equipment: One lead strip or sinker will start to turn brown. This is the positive pole. The brown color is due to the oxygen escaping from the water and joining with the lead to form lead dioxide. One element (sinker) is now lead and the other lead dioxide. Since the poles are different materials, current will flow.

Modern Safety Practice

The acid is potentially hazardous. Take appropriate cautions for working with chemicals (Note 17 in Appendix E), including wearing safety glasses and washing your hands thoroughly afterwards.

See Note 16 in Appendix E about sources of acids.

Can You Work Like a Scientist?

1. What happens after you use your battery to ring a bell?[21]

2. What happens to the brown sinker when you recharge the battery?

3. Can you charge the battery with alternating current?

4. What happens if, after charging the battery with direct current, you switch the wires and continue to charge the battery?

5. Can you measure the amount of charge in any way?

6. Can you use your hydrometer ("Hydrometer" on page 47) to tell how strong your acid is?

7. Can you use one battery to charge another one?

Mercury Switch

Purpose: A mercury switch is used in many automatic electrical devices. When the position of the switch is changed, the liquid mercury flows over to the wires and completes the contact, turning on the electricity.

Materials: Mercury from a thermometer or purchased from a drugstore.[22] A substitute for the mercury could be salt water or some other liquid that conducts electricity easily. A small vial or test tube, and a cork.

What to Do: Stick a needle through the cork and make two holes. Push the bare ends of two wires through the cork. Put enough liquid (salt

20 Or lantern battery.

21 Electric bells were once common, as doorbells, buzzers, and alarms. See "Doorbell" on page 135 about how to build one.

22 Your drugstore won't sell you mercury any more. Note 32 in Appendix E presents alternatives for this project.

water or mercury) into the vial to partially fill it. Insert the stopper in the vial.

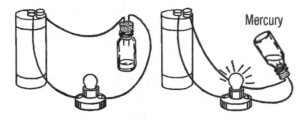

Operation of Equipment: Connect the bare ends of the wire to your circuit of a battery and a small bulb or motor.

Modern Safety Practice

Mercury presents several different health hazards, and is not only harder to obtain, but not recommended for use. See Note 32 in Appendix E for further discussion and methods of building "non-mercury" tilt switches.

Can You Work Like a Scientist?

1. What happens when the vial is in a position so that the liquid contacts both wires?

2. What happens when the vial is turned so that the liquid cannot touch both wires?

3. Mercury is expensive. Can you try other liquids in its place?

4. Would your switch work on current other than that your dry cell produces? Be very careful if you use house current. *Don't touch bare wires.*

5. Does the salt water heat? How about the mercury? If you use salt water, be

careful the gas given off doesn't pop the cork out.

Safety Pin Switch

Purpose: A switch opens and closes a path for electricity. When the path is open, electricity does not flow. When the path is closed, electricity flows and operates lights, motors, and other electrical devices.

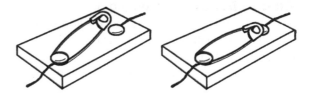

Materials: Safety pin (metal), wire, two tacks, and a board.

What to Do: Remove the insulation from the end of two wires. Wrap the end of one wire around a tack and place the tack through the end of the safety pin. Wrap the bare end of the second wire around the second tack and push the tack into the board so the head of the safety pin makes contact. Wire up your electrical device as shown. Close the switch, and the electricity will flow.[23]

Modern Safety Practice

As this switch is not *insulated* in any sense, it can only be used with finger-safe low-voltage circuitry, e.g., circuits running from one or two battery cells. How could you make a safely-insulated safety-pin switch? See "Push Button" on page 123 for an example of how one might add an enclosure around a switch.

---

23 Switches like these are reliable and easy to make. Large paper clips may also be used instead of safety pins.

Rheostat

Purpose: A rheostat varies the amount of electricity available for a light or other object using current. This nichrome wire rheostat (see chemistry section) can be used on any amount of current up to 60 volts. However, do not connect the rheostat directly to house voltage without having reduced the voltage (see AC Lamp Bank, "Lamp Bank Rectifier and Battery Charger" on page 130).

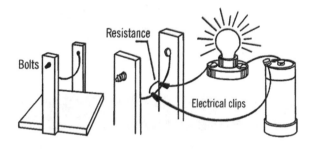

Materials: Nichrome wire (18" of #20), wood for stand, two bolts, four nuts, and four washers. Your bottle cutter can be used for your rheostat.

What to Do: Make your nichrome wire rheostat as shown in the bottle cutter section ("Bottle Cutter" on page 18).

Modern Safety Practice

While this rheostat shouldn't be used at full household line voltage, it does have exposed wiring. Read Note 10 in Appendix E about safety practice around live wires.

Operation of Equipment: Clip a wire from the electrical device to the nichrome wire. The other wire can be connected on any place along the nichrome wire. The farther apart the con-nections, the greater the resistance and the lower the amount of current. The closer the connections to each other, the greater the flow of electricity. If the nichrome wire gives too much resistance, use wire with a lower resistance.

Push Button

Purpose: The push button turns the flow of electricity on and off with a slight movement of the hand. Such a switch can be used with electromagnets, telegraphs, and other devices.

Materials: Wood block for base, two wood screws, a piece of spring brass (other metal will do), and a spool or piece of wood for a handle or knob.

What to Do: Screw the metal strip down as shown. Fasten the knob onto the metal with glue. This knob makes the switch insulated so you won't receive a shock. A slight push closes the circuit. Release the knob and the circuit is broken. See if you can make the second kind of push button as shown. It is insulated.

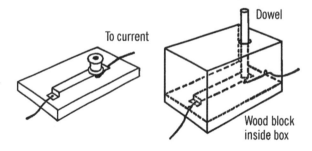

Pencil Rheostat

Purpose: The pencil rheostat reduces the amount of current from a dry cell or small transformer.

Materials: #3 or #4 hardness pencil,[24] bell wire, wood base.

What to Do: Cut halfway through the pencil at both ends. Squeeze the pencil with a pair of pliers so the pencil wood breaks at the cut and the lead is exposed. Fasten the pencil down to a wood base. Connect one wire to the point of the pencil or to a place as close to the point as possible. Strip the insulation off the point of the other wire. Connect up the small motor or light as shown.

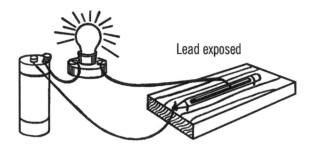

Lead exposed

Operation of Equipment: Slide the bare wire along the lead. The bulb increases in brightness as the two wires get closer together. The amount of current is reduced as the wires are moved farther apart. Pencil lead is made of clay and graphite. Graphite is a form of carbon and is a good conductor of electricity. Clay is a nonconductor.

Can You Work Like a Scientist?

1. Is a #1 or #2 pencil a better conductor than a #3 or #4 hardness pencil?

2. Is there more clay in a #2 pencil or in a #3 hardness pencil?

3. Could pencil lead be used for electrodes? Try them.

4. Make a straight line with a soft lead pencil. Can you use this line as a rheostat?

5. Try welding two pieces of tin foil together, using your pencil rheostat as shown.

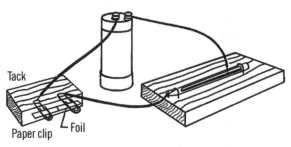

Tack

Paper clip — Foil

Modern Safety Practice

If there's enough heat to weld, there's potentially enough heat to start a fire as well. Take basic fire safety steps, as discussed in Note 3 of Appendix E.

Current Reverser

Purpose: The current reverser changes the poles or direction of the current. When connected up to motors, etc., the motor will reverse directions. This principle can be used on car motors, cranes, etc.

Materials: Five small tin or brass contacts, two brass or copper strips (other metal will work), and a wooden connecting bar with a spool for a handle, plus wood block for the base.

What to Do: Mount the five connectors as shown. A tack can be used for the contact. It can go through the brass or tin. Attach with a screw the two strips as shown. The strips should be tied together with a wooden con-

---

24 Use a pencil made from real wood so that it will split as indicated. Pencil grades B, HB, H, and 2H are equivalent to grades #1, 2, 3, and 4, respectively.

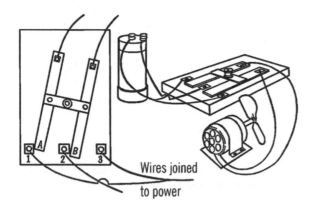

Wires joined to power

necting piece. A handle can be screwed onto the two strips.

Operation of Equipment: When the handle is moved to the left, contacts 1 and 2 should be touched by strips A and B. When the handle is moved to the right, contacts 2 and 3 should be touched by strips A and B. This reverses the flow of current. Wires from 1 and 3 should be connected together. A single wire then runs to the motor. The second wire to the motor comes from connector 2. Strips A and B are connected by wires to the power supply, a dry cell or transformer power supply.

Can You Work Like a Scientist?

1. Try the current reverser out on a small toy motor.

2. Could you use your current reverser on your electrolysis tank? ("Electrolysis Tank" on page 132) What would happen?

3. What would happen if you used your current reverser when electroplating?

4. Connect up your current reverser to your galvanoscope. What happens when you reverse the direction of current?

5. Can you connect the current reverser with the pencil or nichrome wire rheostat and control both the speed and direction of simple motors?

6. Will the current reverser work with alternating current? Be sure to cut the voltage down first with a lamp bank, or with a salt water rheostat.

Conductivity Tester

Purpose: This instrument enables you to test both solids and liquids so that you can find out which will conduct or allow electricity to pass through them. With an electromagnet attachment (see "Magnetometer" on page 108) you can measure the strength or conductivity of the solids and liquids.

Materials: Light bulb socket and bulb,[25] two pieces of tin,[26] source of DC electricity (dry cell, power supply, or battery charger), wire, a bottle with two nails through the cork, and a mounting board.

What to Do: Connect up the materials as shown in the drawing. The pieces of tin should be bent so that they do not touch each other. You do not need to include a push button switch. You can screw the bulb in and out to turn the electricity on and off.

Operation of Equipment: In order to test the conductivity of liquids, pour the liquids into the bottle. If the electricity goes through the liquid, it will make a complete path, and the bulb will light. In order to test bulbs and fuses, lay them across the piece of tin so electricity will flow through the bulb or fuse. If the bulb or fuse is good, the light will glow. Be sure to remove the stopper from the bottle or else remove the liquid when you are testing. You can test the con-

---

25 What would you have to change, in order to build a newer equivalent that uses an LED instead of a light bulb?

26 That is to say, *pieces cut from a tin can*. Other metals will work just as well.

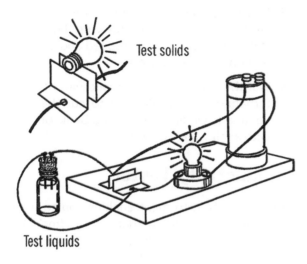

Test solids

Test liquids

ductivity of solids by laying them across the tin strips.

If you connect an electromagnet in the circuit and use your magnetometer, you can tell which solids or liquids are the best conductors by noting the number of pins the electromagnet will pick up.

Can You Work Like a Scientist?

1. Is salt water a better conductor than fresh water?

2. What other materials added to water besides salt makes water a conductor?

3. Can you make a list of which solids conduct electricity and which do not?

4. Which metals are the best conductors?

5. Can you use your tester to test batteries?

6. By adding just a little salt to your liquid, could you use your bottle as a variable load to cut down the voltage?[27]

7. Does the temperature of the liquid have anything to do with the conductivity of the liquid? Of the solid?

AC or DC Power Supply

Purpose: A combination of a small transformer and a rectifier makes up a permanent power supply to replace the customary undependable dry cell. A power supply of this type provides both alternating and direct current.

Materials: A 6-volt, center tap, 1 or 2 amp. filament transformer, 1 amp. full wave rectifier, metal for bracket, wood for a base, four Fahnestock clips, and 3 short wires to make connections. The cost of a suitable transformer and rectifier is about $10.00 - $15.00.[28]

What to Do: Screw the transformer down to the wood base. Connect the two long black leads to rubber-covered electrical cord by soldering the bare ends of the wire and taping the joints with friction tape (zip cord-type on short extension cords). This cord is used to plug into a regular house electrical outlet. Bend a piece of metal (tin, aluminum, etc.) into an L-shape. Drill or punch two holes in this L-bracket, as shown. Screw the bracket to the base. Mount the rectifier to the bracket by sliding the rectifier bolt through the bracket hole and fastening with a nut. Mount your four Fahnestock clips at the end of the wood base with wood screws.

Three wires come out of one side of the transformer. Two wires are green, and the other is yellow. The middle yellow wire is attached di-

---

27 Original wording: "…as a transformer…." See Note 33 in Appendix E about the distinction between the two.

28 Please see Note 34 and Note 35 in Appendix E for specific transformer and rectifier recommendations.

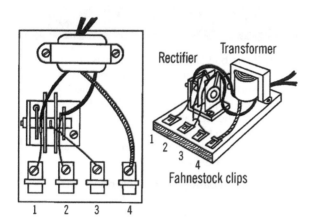

Fahnestock clips

rectly to the outside Fahnestock clip (#4). Solder the green wires to the AC inputs of the rectifier.[29] A wire is soldered from the output terminal of the rectifier[30] to Fahnestock clip #3. A short wire is then soldered one of those two inputs to Fahnestock clip #1. A final wire is soldered to the other of the AC inputs and run to Fahnestock clip #2.

Operation of Equipment: Plug the transformer into house current. Now you may attach small lamps (flashlight bulbs), motors, doorbells, buzzers, etc., to two of the Fahnestock clips. The combinations possible are listed below, and the voltage and type of current is mentioned.

Voltage Combinations:

Fahnestock clips	Voltage	Type
1 and 2	7 V	AC
1 and 3	3½ V	½ wave DC
1 and 4	3½ V	AC
2 and 3	3½ V	½ wave DC
2 and 4	3½ V	AC
3 and 4	3½ V	full wave DC

The transformer reduces house voltage to about 7 volts. This voltage is ideal for most laboratory or classroom use. The power from the transformer is changing direction 120 times a second. In a loudspeaker or earphones, it sounds like a low hum. You can make a direct connection to this alternating current by connecting up to the combinations shown above.

When the rectifier is connected to the transformer, it changes the voltage from alternating (or reversing) to a smoother kind of current that flows in only one direction. This voltage sounds like a high-pitched hum when heard in a speaker or earphones because it is getting twice as many impulses per second as the voltage from the transformer alone.

The rectifier contains semiconductor diodes that permit the current to flow in only one direction, so the voltage flows first from one terminal into the circuit, and then from the other terminal into the circuit. If you are connected only to one terminal, you will receive only half as much current as when connected to both terminals.

Can You Work Like a Scientist?

1. Wind many turns of wire around a piece of hard steel (file, screwdriver). Which combination of connections makes a permanent magnet?

---

29 The rectifier bridge shown in the illustration is an old-school *selenium* rectifier. A suitable modern replacement is the GBU8D rectifier bridge. These instructions have been adapted for that type. See Note 35 in Appendix E for more information.

30 On a modern rectifier bridge, use the output terminal marked with a "+".

Safety Tips

1. Always use some form of direct current on motors having permanent magnets. Alternating voltages will destroy permanent magnets.

2. Be careful not to cause a short circuit to occur for very long as it can cause the transformer to burn out. When either the transformer or rectifier becomes uncomfortably warm to the touch, check for a short circuit or too large a load on the circuit (too much connected to the transformer).

3. There is no danger of shock from this power supply if the cord to the transformer from the outlet plug is properly connected.[31]

Modern Safety Practice

1. While building a power supply is an excellent and educational project, it is worth considering that modern efficient plug-in power supplies will likely cost less and be safer for later use. Related question to investigate: What is the efficiency of this power supply? How can you measure it?

2. This is a basic power supply design that does not have any kind of fuse (or other overcurrent protection) built into it. For what reasons is this problematic, from the standpoint of safety? Could a fuse protect the transformer in the case of the short circuit described previously?

3. When adding a single fuse, it should go on the primary side of the transformer. Why? Would it make sense to also have a second fuse on the secondary side of the transformer? How can you calculate what fuses you need, in terms of current rating?

4. Make sure that you do not leave any exposed wiring, particularly on the "primary" (line voltage) side. Read Note 10 in Appendix E for safety practice around exposed wiring.

2. Which combination of connections will destroy a permanent magnet placed in this coil?

3. Poles 3 and 4 can be used for a source of power for electroplating and electrolysis. Connect the item to be plated to the negative pole. The other electrode (anode) is connected to the positive pole.

4. Can you use a compass to detect which is the positive and which the negative poles of the other combinations?

5. In electrolysis, we can tell where hydrogen is formed because we get twice as much hydrogen from water as we get oxygen. Why? Which pole releases the hydrogen? See "Electrolysis Tank" on page 132.

6. Is a compass affected by alternating current?

7. Can you use your nichrome wire rheostat to vary the amount of electricity? With this can you change the speed of motors, etc.?

---

31 There is never "no danger" when there is exposed wiring at mains voltage level, so make sure none is exposed.

Variable Load

Purpose: To reduce the amount of voltage that comes from a wall plug so that it is safe and can be used to run small motors and other equipment that normally require a dry cell.[32] The dry cell is expensive, is not a dependable source of electricity, and can easily be shorted out. Remember a dry cell is direct current (DC) and the electricity from a transformer is AC.

Materials: Two porcelain light fixtures (one with a pull chain), a 6' extension cord, and light bulbs with different wattages.

What to Do: Buy two porcelain light fixtures from a hardware store, one plain and one with a pull chain. Cut the socket end off a short extension cord. Divide the wires. Use short pieces of the wire to connect the porcelain fixtures in series as shown in the drawing. The two ends of the wire at A and B are left unconnected.[33] These ends can be connected to a motor or piece of electrical equipment. They can be clipped together, and the socket of one of the fixtures can be used for the amount of electricity needed.

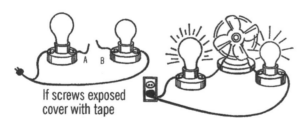

If screws exposed
cover with tape

Operation of Equipment: Turn off the electricity by pulling the pull chain. Connect wires A and B to the piece of equipment you wish to operate. Put a large bulb in one fixture. Put the smallest bulb you have in the other fixture (7- or 15-watt size). Plug in the extension cord to a wall plug. Now turn on the electricity with the pull chain. If the current isn't strong enough, put a larger bulb in the fixture instead of the small one. Increase the size of this bulb until you get the desired amount of electricity. If you wish to plug directly into the line, connect wires A and B. Screw a two-hole plug into one of the fixtures. Now connect directly to this plug.

Safety Tips

1. Be sure you do not work anywhere near a sink or a radiator that might be grounded.

2. Do not touch any bare wires when electricity is turned on.

3. Screw the porcelain fixtures on a board after they are connected up.

4. If the screws on the fixtures are exposed, cover them with electrical tape.

5. Be sure you have one light bulb connected on each side of wires A and B. Then even if you accidentally touch both wires, you will receive only a small amount of electricity.[34]

6. Never work with electricity if you are standing on damp ground or on a cement floor. Both are good grounds. Electricity could pass through you to the ground.

Can You Work Like a Scientist?

1. You have connected the two lights plus wires A and B in series. The electricity, even though alternating current (going back and forth many times a second), has to go through each one of the bulbs in order to make a complete

---

32 This project was originally titled "Power Supply Transformer." See Note 33 in Appendix E for further discussion.

33 Cover any exposed screws with electrical tape.

34 Keep in mind that even a "small" shock can still be dangerous.

Modern Safety Practice

In addition to the safety tips listed above, never operate this equipment without easy access to a electrical cut-off switch (e.g., wall switch or circuit breaker panel) that can disconnect power if needed. Read Note 10 in Appendix E for safety practice around exposed wiring.

Better, follow the advice from "Modern Safety Practice" on page 20 and use an isolated variac to power the equipment.

path. Do the bulbs light up as brightly as they normally do in a lamp?

2. Normal house current has a voltage of about 115 volts. In series, each bulb uses part of the voltage, or electrical push. If you have one bulb in the line, it would get all 115 volts. If you have two bulbs in the line, what is the voltage?

3. You can check your answer. Put in two bulbs of the same size. Connect wires A and B together. Do both bulbs light up to full brightness? Are both bulbs getting their share of the voltage? How do you know?

4. The amount of electrical current flowing is called the amperage, or amps. The electricity has to flow through each lamp. The bulbs act like valves or water faucets. The small bulb lets only a small amount of current through. A large bulb lets more. Test this. Connect wires A and B to a fan. How fast does the fan go when you use a little bulb? How fast does it go when you use a big bulb?

5. Does it make any difference if you use one large and one small bulb or two small bulbs? Does the large or small bulb control the amount of electricity?

6. If you connect a fan up to wires A and B, does the fan increase in speed with the larger-size bulbs? Does the amount of wind given off by the fan also increase?

7. Can you use this wind to turn an anemometer? (See "Anemometer" on page 175.)

8. Can you use this principle to measure the amount of current flow (amps)? An instrument for this purpose is called an ammeter.

9. If you connect another porcelain fixture in series with the two you already have, what voltage will each receive? Will this have any effect on the fan? Can you devise a voltmeter from this?

10. As you add bulbs in series, do the lights become brighter or weaker? Is there a change in amperes (current flow) or voltage (push behind current)?

11. How much voltage does each light get in a string of Christmas tree lights if they are wired in series? There are eight bulbs in a string. How much voltage would each get if there were ten in the string? Would the string last longer if there were ten instead of eight bulbs in the string? In which case would the bulbs glow the brightest? Try this and see. You may be able to save much money on Christmas tree bulbs.

12. Connect up your fan measuring device to the Christmas tree string. Can you determine the flow of electricity and the voltage?

Lamp Bank Rectifier and Battery Charger

Purpose: The lamp bank rectifier changes normal house current (AC) into direct current (DC) which then can be cut down to any desired voltage by the nichrome wire rheostat. The rectifier can be used to recharge storage batteries.

Materials: Two porcelain fixtures, two light bulbs, large jar (bottom of gallon jug—see "Bottle Cutter" on page 18), wood for base, ¼" plywood for jar lid, and a strip of aluminum and lead.

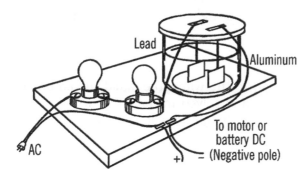

What to Do: Wire up the fixtures as shown. The lid to the jar can be made by cutting one piece of plywood to just fit the jar and a second piece a little larger. These are glued or nailed together. Cut slots for the aluminum and lead strips. This lid should be coated with paraffin to protect it from chemical action. The electrolyte is a strong solution of sodium bicarbonate or baking soda. A package costs about $3.00 at a grocery store.

Operation of Equipment: This rectifier allows current to flow one way but not the other. When the current is such that the lead electrode is positive and the aluminum negative, the current flows through. When the current reverses, the aluminum pole becomes positive. Oxygen gas is formed, coating the electrode with aluminum oxide. Aluminum oxide is a very poor conductor of electricity, so this stops the passage of current back again. This action happens in an instant. This works like a valve, changing the current to direct current.

When charging a storage battery, always connect the negative pole of the battery to the aluminum electrode. The rectifier works well un-

less too much current is passed through. The electrolyte then becomes too hot.

Modern Safety Practice

This apparatus also involves live AC power from the wall (Note 10 in Appendix E) and requires a cut-off switch. Again, it would be ideal to power it through an isolated variac. (See "**Modern Safety Practice**" on page 20.)

Can You Work Like a Scientist?

1. This lamp bank rectifier uses only one-half of the alternating current. Can you design four such rectifier cells in order to use completely the alternating current?

2. Can you use this rectifier to electroplate?

3. Can you use this rectifier to run your electrolysis tank?

4. You get a small amount of current at high voltages with the lamp bank. Try a small bell transformer or train transformer in its place. This will lower the voltage and still give you enough current to run simple motors, etc.

Carbon Rods—Carbon Electrodes

Purpose: Carbon rods are used in electrolysis tanks as electrodes. They are excellent conductors of electricity and do not break down in acids as metal does. Carbon rods are also used for arc furnaces, sensitive microphones, and as resistors.

What to Do: Saw off the ends of a flashlight battery.[35] Cut the cover open with tin snips.

---

35 A specific type of battery is needed. Please see Note 12 in Appendix E.

This metal is probably zinc. Save this zinc, as it comes in handy for experiments. The black rod in the center is a carbon rod. You will get a very large carbon rod out of a dry cell battery.

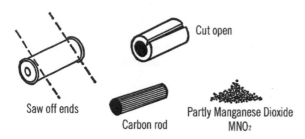

Saw off ends

Cut open

Carbon rod

Partly Manganese Dioxide MNO₂

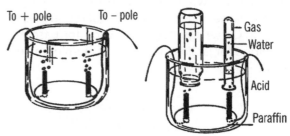

To + pole   To – pole

Gas

Water

Acid

Paraffin

Electrolysis Tank

Purpose: Electricity is passed through a liquid, and gases are given off at the poles or electrodes. This is a good method of collecting hydrogen and oxygen.

Materials: Wide-mouth jar, two carbon rods from flashlight batteries,[36] zip cord (wire covered with rubber),[37] and paraffin (you can use melted candle wax).

What to Do: Cut off the top of the jug (see "Bottle Cutter" on page 18). Divide the zip cord into two pieces. Wrap the bare end of each wire around the end of a carbon rod. Place the rods in position as shown and melt paraffin into the bottom of the jug. (You might use a piece of cardboard to help support the rods while the paraffin cools.) The loose ends of the zip cord should come up on opposite sides of the jug.

Operation of Equipment: Connect a source of power (dry cell, storage battery, DC transformer) to the wires. The liquid should be an acid or a dissolved material that will conduct electricity. Suggestions are: sulfuric acid, hydrochloric acid, washing soda, baking soda (sodium bicarbonate).

Gases will come off the ends of the rods. Hydrogen will come off one rod and oxygen off the other. In order to collect the gases, place a baby bottle or a test tube filled with water over the rod. As the gases rise, they force out the water, and the water is replaced with the gas being collected. The bottle is then filled with the gas collected.

Try lighting a match to the gas in each bottle. Which bottle contains hydrogen? Why do you have water in the bottle and funnel before you collect the gas? Which bottle will fill faster? Why? What will happen if you switch the poles when the bottles are half filled? If you test with a match, be sure to *wrap a towel around the bottle containing the gas.* This is just in case the bottle should crack.[38]

---

36 Carbon rods can also be purchased; see Appendix A.

37 Specifically, two-conductor insulated lamp cord that can be pulled apart.

38 Why should it crack? Remember that hydrogen gas *explodes*. Use a long match, safety glasses, common sense, and caution. Read Note 3 in Appendix E about safety around fire.

Test Tube Electric Motor

Purpose: This type of motor was one of the earliest to be built. From this working model the junior scientist can discover the principles of an electric motor.

Materials: Three 6" spike nails, 20 feet of cotton-covered wire (or bell wire),[39] a cork, a test tube, a thin copper sheet (about 1" × 2"), four short nails, wood for the base, and a ¼" bolt and nut about 6" long.

What to Do: Drill a hole in the center of the board and then drive one of the spikes through the hole. Work a hole in the end of the cork so that the bottom of a test tube will fit tightly in the hole. A second hole should be worked in the side of the cork so that the bolt will fit tightly in the hole as shown. Place the test tube, cork, and bolt over the spike. Drive the other two spikes for the field magnets into the wood base. The spikes should be spaced so that the ends of the bolt will just miss the heads of the spikes.

Wrap about forty turns of wire around each end of the bolt. Notice the direction the wire is wound. This is very important. The copper strip is cut in half to make two pieces ½" × 2". The strips are placed on the opposite sides of the test tube near the opening of the test tube. The two strips can be held in place by wrapping a strip of tape around the bottom edge. Insert the bare ends of the wire from the armature under the copper strips as shown. Wrap a piece of tape around the top of the copper strip to hold the wires firmly in place.

Now wrap the two field magnets. Again, be sure you wrap the wire in the direction shown in the drawing. The ends of the wire from the small nails to the copper sheet are bare. They

should be bent so that each comes in contact with one of the copper strips. When the armature is turned crossways to the two spikes, the bare wires should not touch the copper strips.

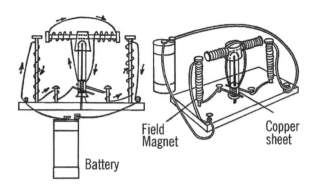

Operation of Equipment: Connect the ends of the wires to a dry cell battery or a three-volt direct current power supply. You may have to adjust the bare wires touching the copper strips or the distance from the heads of the spikes to the bolts in order for the armature to turn. You may have to bend the ends of the two field magnet spikes toward or away from the bolt.

Can You Work Like a Scientist?

1. Can you use a compass to check to see whether the heads of the spikes are north or south electromagnetic poles? Do they ever change? Try turning the armature and checking again.

2. What are the poles of the armature? Again check the ends of the bolt with a compass. Do these poles change when the armature turns?

---

39 *Insulated* wire, ideally solid-core. Cotton insulation is not necessary; most modern plastic insulation is perfectly fine for this application. Commercially made motors always use *magnet wire*, a specific type of insulated wire with a very thin coat of transparent and shiny insulating varnish (usually red, yellow, or brown in color). If you have access to magnet wire, you'll probably need to mechanically scrape the insulation off of the ends with a blade or file in order to make an electrical connection to it.

Bolt and Nut Motor

Purpose: This type of motor can be used to lift objects and to drive small machines.

Materials: Three ¼" bolts about 2¾ inches long, five nuts, eight washers about ¾" in diameter, twenty feet of cotton-covered or bell wire, heavy metal for brackets, finishing nail about four inches long, eight-inch piece of size 8 copper wire, short piece of ½" wood dowel, copper strip about 1" × 1", screws, and wood for the base.

What to Do: Have a hole drilled in the center of one of the bolts just large enough for the finishing nail to fit in. A machine shop will do this for you.[40] Make the metal brackets to hold the armature. Drill a hole through the end of the wood dowel so that a short piece of dowel will slide over the finishing nail. Fasten two strips of thin copper sheeting on each side of the dowel. These strips can be glued or fastened on the dowel by wrapping thread around the ends. Wrap about forty turns of wire around each end of the armature bolt. Be sure the armature is wrapped exactly as shown in the drawing. The direction of the turns is important. The free ends of the armature wire are fastened to the two copper strips on the dowel. It is best if they are soldered.

The heavy copper wire serves as brushes. The two pieces are held in place by wood screws in the base. The heavy wire brushes should be bent so that they make contact with the copper strips on the dowel.

The field magnets are wrapped as shown. The brackets can be moved back and forth until the heads of the bolts from the field magnet and the armature just miss each other. One wire from the field magnet goes to the battery. The other wire is soldered to the heavy wire brush.

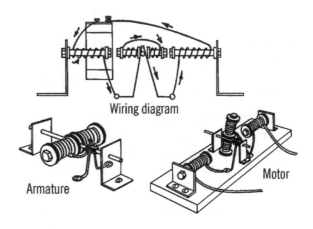

Wiring diagram

Armature

Motor

There should be about 100 turns around each field magnet (bolt). It is important that this coil be wrapped in the right direction.

Operation of Equipment: A piece of rubber tubing can be slipped over the end of the center nail to serve as a pulley. A rubber band can be used to connect the motor to toys. A cord can be wrapped directly around the shaft in order to use the motor to lift things.

Can You Work Like a Scientist?

1. What happens to your motor if you connect the current reverser switch to the motor?

2. Can you vary the speed of the motor by using some kind of resistance or rheostat?

3. How much weight can your motor lift?

4. Will your motor start without your help, or do you have to turn the armature?

---

40 You may also be able to do it yourself, if you have access to a drill press and suitable vise.

Doorbell

Purpose: An electric bell not only demonstrates one of the uses of an electromagnet, it also can be used to indicate by sound when an electric current is flowing. The bell can be used in biology for conditioning experiments.[41]

Materials: Twelve feet of cotton-covered or bell wire, ¼" bolt about three inches long, two large nuts and washers, hacksaw blade for the clapper; nuts, bolts, and wood for the rest of the doorbell, a small metal funnel for the bell, and a long bolt.

What to Do: Drill a hole in a block of wood large enough for the bolt. Insert the bolt in the block of wood and fasten in place with

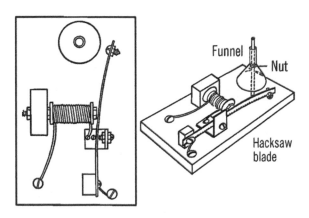

the nuts as shown. Be sure you have the washers in place. Wrap about 140 turns of wire around the bolt. Nail the block of wood to the wooden base.

Drill a hole in the end of a hacksaw blade and fasten the blade to a block of wood with nails and a screw. A small bolt and nut should be fastened in place near the striker end of the blade. Another small bolt and nut is used for a contact point. This bolt and nut is held in place by a tin bracket.

Figure the best location for the bell. Drill a hole in the wood base and insert a long bolt. Thread a nut on the bolt and then hang a small funnel over the bolt.

Operation of Equipment: Connect the bell up to a push-button switch and a dry cell. Will the bell work on alternating current? Try connecting the bell up to your power supply.

Electric Buzzer

Purpose: The buzzer can be used as a sounder for a telegraph set. It has other uses when connected with a push-button switch.

Materials: Three-inch bolt, two large washers, two nuts, twelve feet of bell wire, one-half of a hacksaw blade, heavy copper strip, screw and nuts for the contact point, and wood for the base and supports.

What to Do: Wrap the coil around the bolt as you did in making the doorbell. Fasten the bolt and coil through a hole in the base. Fasten the part of the hacksaw blade to a support with a screw. Drill a hole in the end of the copper strip and fasten the contact screw as shown. Fasten the other end of the copper strip to the tall support. Connect up the wires and push button as shown.[42]

Operation of Equipment: When you press on the push button, electricity flows through the switch and then through the coil. The electricity goes through the copper strip, the contact screw, the hacksaw blade, and then back to the battery. The coil is magnetized and pulls on the hacksaw blade. This breaks the complete path. The coil loses its magnetism, and the blade flies

---

41 Perhaps a more humane use is as an actual doorbell, or alarm bell.

42 Hacksaw blades are normally painted. You may need to scrape away the paint in the area where you wish to make an electrical connection.

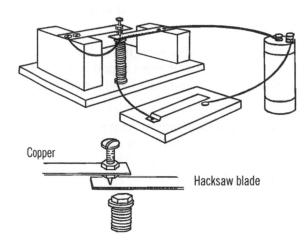

Copper

Hacksaw blade

used for a contact. A wire is also connected to this screw. When you press down on the end of the metal strip, a complete electrical path is made and the bulb will light. A dot is made by a short blink of the light. A dash is a long blink of the flashlight bulb.

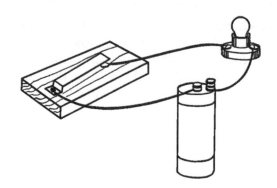

up to the contact screw. The circuit is then complete and the process is repeated.

Can You Work Like a Scientist?

1. Hold one end of a ruler over the edge of the table. Pluck the ruler. Does the ruler make a sound? See if you can make the ruler give off a very high-pitched sound. What causes the sound?

2. How does the movement or vibration of the ruler compare with the movement of the hacksaw blade? Why does the blade produce a buzz?

Telegraph Key

Purpose: A telegraph key is used to make or break a complete electrical circuit or path. A light or sounder plus a source of electrical power are usually connected in the electrical circuit.

Materials: Tin or copper strip, two round-head wood screws, wood for base, cotton-covered wire, flashlight bulb and socket.

What to Do: Bend the tin or copper strip as shown. Punch a hole in the short end of the strip and then fasten the strip and a wire lead with a wood screw. A second wood screw is

Telegraph Sounder

Purpose: A sounder is connected in a path with a telegraph key. As the complete path is made and broken with the key, sounds are produced. From this sound code a telegraph operator can send or receive a message.

Materials: Cigar box, three-inch bolt, two washers, and two nuts, for the magnet coil. Long narrow bolt, two nuts, and a screw eye for the clapper. The sounder is made from a block of wood and a glass microscope slide, 12' of bell wire.

What to Do: Make your set as shown. The clapper is held against the glass plate by a rubber band. The other end of the clapper swings back and forth in the screw eye.

Telegraph Solenoid Sounder

Purpose: The solenoid sounder is sometimes used with a telegraph key instead of a regular sounder and flashlight bulb. Its advantage is that the sound can be heard for a greater distance.

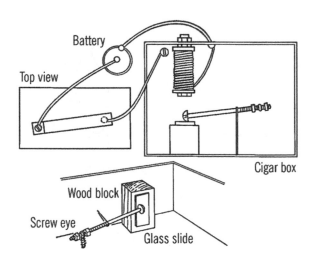

Top view

Battery

Cigar box

Wood block

Screw eye

Glass slide

Materials: Narrow steel or cardboard tubing, cotton-covered wire, microscope slide or other glass, wood for the base and support, and a ¼" bolt.

What to Do: Wrap the wire fifty or more times around the tubing. Attach the tubing to the wood support. Glue a microscope slide on the base directly underneath the opening of the tubing. Place the bolt in the tubing with the head of the bolt down.

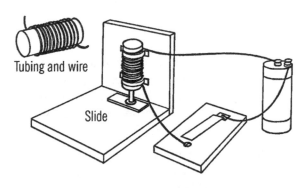

Tubing and wire

Slide

Operation of Equipment: As the telegraph key is pressed down, the electrical current going through the bolt creates an electrical field that pulls the bolt upward. When the key is released, the bolt drops and produces a large sound by striking the microscope slide which has been attached to the wood.

Can You Work Like a Scientist?

1. How far can you send messages with your telegraph set?

2. How many words a minute can you receive when someone else is sending International code?

3. Can you devise your own code?

International Code

1. A dash is as long as three dots.

2. The time between parts of a letter is equal to one dot.

3. The time between two letters is equal to the time required to make three dots.

4. The time between two words is equal to the time required to make five dots.

Current Flow Indicator

Purpose: This current flow indicator can analyze electrical current to determine whether the current is direct or alternating, the direction of flow, the frequency or times the electricity changes direction, and the strength of the current. Many things such as light, sound, and temperature changes can be changed into electrical energy and thus can be monitored.[43]

Materials: Two red LEDs, a resistor (470 Ω), and wire. Two LEDs and one resistor may be purchased for about $1.00 at stores that sell electronic components.

---

43 This project has been updated to use modern components, and the original project has been moved to the appendix. Please see Note 36 in Appendix E for more information.

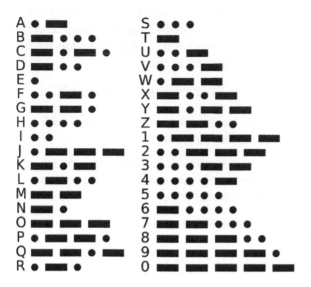

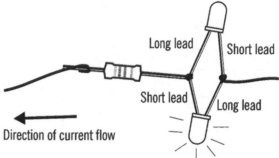

Long lead   Short lead

Short lead   Long lead

← Direction of current flow

What to Do: Connect the components as shown, ideally by soldering. You may be able to twist the wires together sufficiently well to make a solid connection or (better) use a solderless breadboard. The two LEDs are wired facing opposite directions, long lead to short lead, and short lead to long lead. Connect the two end terminals (the loose wire ends, shown as open circles in the diagram) go to the object that you are going to test.

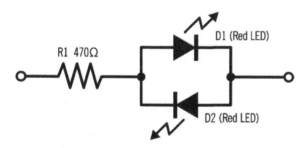

R1 470Ω

D1 (Red LED)

D2 (Red LED)

Operation of Equipment: Each LED conducts electricity in only one direction, and lights only when conducting electricity. When current is flowing in one direction, one LED lights up. When the current flows in the other direction, the other LED lights up. If the current alternates or changes directions, the two LEDs alternate, producing a blinking effect. With red LEDs and

the given choice of resistor (in the range of 470 Ω), this circuit can be used to test circuits from 1.5 V–9 V AC or from 2 V–12 V DC. Using a higher voltage may damage the LEDs.

Can You Work Like a Scientist?

1. Electric current from a battery is normally described to flow from the positive pole through to the circuit and back through the negative pole into the battery. Which LED lights up when you connect the negative pole of the battery to the side of the current flow indicator with the resistor, and the positive pole to the side with the LEDs? Will a single battery cell provide enough voltage?

2. Which LED lights up when you connect the positive pole of the battery to the side of the current flow indicator with the resistor, and the negative pole to the side with the LEDs?

3. Does it affect the direction of flow of the current if you change the wires on the battery? Try a car battery.

4. Connect up the current reverser switch with the current flow indicator and the battery. What happens to the bulb when you reverse the switch? Why?

5. Connect up your power supply. What are the positive and negative poles of

the direct current part of the power supply? See "AC or DC Power Supply" on page 126.

6. Connect up the current flow indicator to the AC clips on the power supply. What difference do you notice? Is there a difference if you use 3 or 6 volts?

7. Connect up the current flow indicator to a hand crank generator. What determines the number of flashes a second? Remember each blink is caused by the current changing direction.

8. Connect up your current flow indicator to the field magnet. Does the current change in the field magnet ("Test Tube Electric Motor" on page 133)? How about on the armature?

9. Why is there a minimum voltage necessary to light up the LEDs? What is that minimum voltage, and what sets that limit? Is the minimum different for other colors of LEDs?

10. Why does the current flow indicator work for a different range for AC or DC voltage?

11. What changes would be necessary in order to use the current flow indicator with household current, to see how it flickers? Could it be done with a simple change in resistor? In addition to electrical safety, you will need to find as the voltage and current that LEDs can tolerate, as well as the power dissipation in the resistor. How can these things be calculated?

Analog Computer

Purpose: An analog computer is a type of electronic brain that changes or converts numbers into something else (such as an electrical current) that can be worked easier than the numbers.[44] A slide rule is an example of an analog computer since numbers have been changed to distances on the slide rule.[45]

Materials: Two 50-ohm potentiometers, one 1000-ohm potentiometer, galvanometer,[46] 3 volt direct-current source (two 1½ volt batteries or power supply), three dial knobs, bell wire, a piece of masonite (1' × 2'), a 2' piece of 1" × 2" wood.

What to Do: Cut the piece of 1×2 in half and nail each piece along the end of the piece of masonite. Drill three ¼" holes in the masonite. These holes should be evenly spaced as shown. Remove the top nut off each potentiometer and slide the potentiometer through the correct hole from the bottom side. Fasten the nuts in place on the top side of the board. The middle potentiometer is the 1000-ohm size.

All of the wiring is done beneath the board. The schematic diagram is given below as well as an actual diagram of the appearance of the bottom of the board. The number 1 on the potentiometer is the side of the potentiometer from which no current will flow. No. 2 is the variable voltage controlled by turning the knob. No. 3 is the position in which the potentiometer is wide open and all possible current flows.

Operation of Equipment: Potentiometers R1 and R2 are wired together in such a way that an electrical current that goes through R1 also goes through R2 and then to one side of the

---

44 This particular analog computer is modeled after commercially available analog computers of the 1960s, such as the Calculo Analog Computer.

45 What modern examples of analog computers can you think of? And, if you've never used a slide rule, it's worth trying out, to understand the principle of operation. You can find online simulators (and even mobile apps) to let you try one out, as well as download-and-print patterns to make a paper slide rule.

46 Either a digital ammeter or multimeter in DC current mode will work fine.

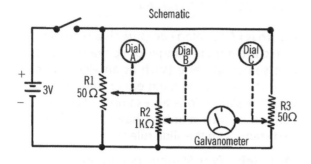

Schematic

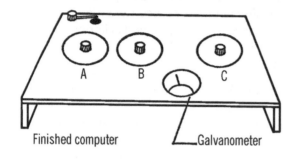

Finished computer ___ Galvanometer

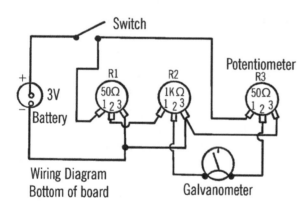

Wiring Diagram
Bottom of board

Galvanometer

galvanometer. If the dial on R1 is set half way, the voltage from the battery is cut in half. This voltage is sent to R2. If the setting on R2 is one-third of the dial, this voltage is cut by one-third. Thus 3 volts is cut to 1½ volts at R1. The voltage at R2 is cut to ½ volt.

R3 is connected directly to the battery and the other side of the galvanometer. When R3 is

turned to such a position that the current flow through it is exactly equal to the current flow through R1 and R2, the two pushes exactly balance each other, and the galvanometer does not register any current flow. The position of the pointer connected to R3 would indicate the correct answer.[47]

In order to program your computer, slip cardboard discs over the stems of the potentiometers on the top side of the board. Fasten the cardboard discs with the nuts or with glue. You should cut out and glue cardboard pointers to the bottoms of the knobs. The knobs and pointers are then fastened on the stems of the potentiometers.

You can make a multiplication scale on the cardboard discs. Turn R1 half way. Label the spot the pointer indicates as 5. Turn R2 half way and label this 5. Now turn R3 so that the galvanometer indicates no current flow. Since 5 times 5 is 25 you should label the position of the R3 pointer as 25. Now turn R1 so that no current flows. Label this 0. Do the same for R2. Turn R3 so that the galvanometer indicates there is no current flow. Mark this 0. Turn both R1 and R2 wide open. Mark 10 on each dial. Turn R3 so that the galvanometer indicates no current flow. Since 10 times ten equals a hundred, label the position of the pointer of R3 one hundred. Then make your subdivisions from one to ten on both R1 and R2. Each time turn R3 and mark the answer on the dial as you did before.

Can You Work Like a Scientist?

1. If you made the dials as just described, you can multiply up through the ten tables. Can you make subdivisions between each number so that you can multiply such numbers as 3.5 times 6.8?

---

47 How does this work? Can you use Ohm's law together with Kirchhoff's rules to understand *exactly* what is going on?

2. How can you multiply numbers larger than ten? Think of it this way. Three could also be thirty and 3.4 could be 34.

3. Can you divide with your computer? If you set R2 and R3, the answer should be on R1.

4. Computers can be used for many things. Can you make new dials for your computer which will help you predict weather? R1 could be labeled with air pressure readings. R2 could be labeled with the seasons of the year. R3 then would indicate the type of weather you generally would expect. When you make out your dials, you are doing what is called programing your computer. The more accurately you program your computer, the more accurate the results.

5. Can you program a computer to indicate the total number of calories you will eat in a day? One dial can list various lunches. Another dial can list various dinners. The third dial can indicate the total number of calories.

6. Can you program a computer to give automatically the correct distances between various combinations of cities?

7. Can you program a computer that will automatically give you the area for a circle of known diameter?

8. Can you program a computer so that if you know the time and distance it took to get to a certain place, the computer would give you the average speed?

9. Can you program a computer to give you the square and the square root of various numbers?

10. Can you program a computer to give you the cube and the cube root of various numbers?

11. Can you program your computer so that you can figure the distances from the Earth to different planets at various times throughout the year?

12. How can you multiply three numbers by three numbers?

13. Can you program a computer so that if the velocity of a missile and the distance to a target is known, you automatically can figure the proper angle to fire the missile?

14. Could you use a compass galvanometer instead of a commercial galvanometer for your computer? What are the limitations?

15. What effect does the amount of electricity have on the operation of the computer?

16. Could you use a voltmeter instead of a galvanometer?

17. Below is a wiring diagram using an audio signal generator and a head set instead of a galvanometer. When you do not hear a sound, the current flow is balanced. Could you make such a computer?

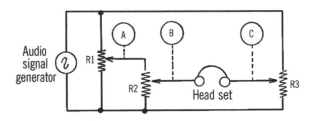

Digital Computer

Purpose: A digital computer is a type of electronic brain. It consists of involved wiring patterns. In an analog computer numbers are converted to electrical current. In the digital computer exact answers are possible because an electrical circuit has been wired that will furnish only that answer. Answers are sometimes given by certain bulbs that light up.

Materials: Masonite for dials and base, small ½" bolts and nuts, bell wire, dry cell battery or power supply, five flashlight bulbs, sockets, and washers.

What to Do: Cut out six masonite discs about six inches in diameter. Drill holes in the center of each disc. Mount the discs on the board as shown. Washers should be used as spacers between the masonite disc and the board. Drill a hole in the outer edge of the disc. Place a pencil in the hole and turn the disc. Remove the disc and drill holes every half inch around the circle drawn on the board. Insert bolts into each hole and fasten with nuts on the reverse side of the board.

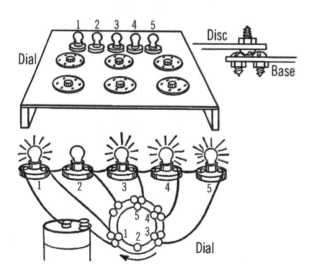

Drill other holes in the disc. Each hole should be the same distance from the center hole and a half inch apart. The necessary bolts are placed with the head of the bolt down. When the disc is rotated, the head of the bolt on the disc fits

and makes firm contact with the heads of the bolts coming through the board.

A series of five flashlight bulbs should be mounted on the base in sockets. These can be connected with any number of bolts to form electrical circuits for different problems.[48]

Operation of Equipment: A simple circuit should give you an idea how this computer can be programed. You can then use your imagination for making other circuits.

PROBLEM 1—Can you wire a dial so that you can choose to light up any of five different bulbs? (For the sake of clarity, the heads of the bolts that would be directly under the bolt on the dial are shown alongside the dial.) Connect up the wires as shown. As you rotate the dial, bulb one will light up. Turn the dial again and bulbs one and two will light up because of the complete circuit. As you continue to turn the dial more and more bulbs will light. Can you trace the complete circuit?[49]

Problems to Investigate in the Study of Electricity and Magnetism

(P)-Primary

(I)-Intermediate

(U)-Upper

1. Are all magnets the same shape and size? Collect different types of magnets. *(P)*

2. Do magnets attract each other? *(P)*

3. What things do magnets attract? Will all magnets attract the same things? *(P)*

---

48 Could you build a "modern" version of this, using LEDs instead of light bulbs? How small could you make it?

49 Are you wondering how to get from here all the way to something that you might recognize as a computer? Read Note 37 in Appendix E.

4. What things will magnets not attract? *(P)*

5. Will magnets attract things through paper? Through tin? What materials will magnetism pass through? *(P)*

6. Do all magnets attract things with the same amount of force? Can you measure the attracting force of different magnets? *(P)*

7. Does magnetism travel through the air? *(P)*

8. Will both ends of a magnet attract things? *(P)*

9. Do both ends of a magnet attract the end of another magnet? *(P)*

10. Does a magnet affect a compass? *(P)*

11. Do both ends of a magnet affect a compass in the same way? Are both ends of a compass alike? *(P)*

12. What end of a magnet attracts the point or arrow of the compass? Mark this end with a piece of tape and write N on the tape. *(P)*

13. What end of a magnet attracts the back end of the pointer or arrow? Label this S. Does the magnet act this way with all compasses? *(P)*

14. Mark another magnet in the same way as you did in the two problems mentioned before. Does this magnet make different compasses act in the same way as the first compass? *(P)*

15. If you bring the two ends of the magnets marked N together, will the magnets attract? *(P)*

16. If you bring the two ends of the magnets marked S together, will the magnets attract? *(P)*

17. Is there a force between the two N ends that seems to push the magnets away from each other? Is there a force between the two S ends that seems to push away (repel) from each other? *(P)*

18. Can you measure the force that pushes away or repels two like poles? *(I)*

19. Is this force of repulsion the same for all magnets? *(I)*

20. Do both north poles and both south poles push away or repel with equal force? *(I)*

21. If you bring the end marked N of one magnet and the S end of the other magnet near each other, do the magnets push away or attract? Try the other two ends of the magnets (marked S and N). Do they act the same way? *(P)*

22. Try these experiments on other magnets. Do all magnets act in a similar way? *(I)*

23. Will magnets pick up all kinds of metal? What kinds of metals will magnets attract? *(P)*

24. Will magnetism pass through metal and attract something else? Try touching a strong magnet to a nail and then picking up a pin with the nail. *(P)*

25. Will magnetism pass through the air and then through a piece of metal and make the metal magnetic? *(I)*

26. Will magnetism go through a vacuum? *(I)*

27. Are both ends of a magnet just as strong? Be sure to try many different magnets. *(I)*

28. Does the strength of the magnetic force increase or decrease as you move the magnet away from a piece of iron or steel? Can you measure and compare the strength and the distance? *(I)*

29. Does a magnet attract a nail? Is the nail a magnet? *(P)*

30. Does a nail affect a compass? Do both ends of a nail affect the compass the same way as a magnet? *(P)*

31. Do other pieces of iron and steel affect the compass in the same way? How about other metals? *(P)*

32. Is the needle or pointer of a compass magnetized? *(P)*

33. Will the compass needle always point in the same direction if magnets and other metal objects are not nearby? *(P)*

34. Will a compass point in the same direction both inside and outside a building? *(P)*

35. Does the compass always point in the same direction? What direction is this? *(P)*

36. What causes a compass to face always toward the north? *(P)*

37. In what directions will compasses point in different parts of the world? *(I)*

38. Where are the Earth's magnetic poles? Have these poles always been in the same place throughout the history of the Earth? *(U)*

39. Can you magnetize a nail by rubbing it with a magnet? What ways do you rub the nail to make the strongest magnet? *(P)*

40. What materials are easy to magnetize? What materials are difficult to magnetize? Can you keep track of how long these materials retain their magnetism? *(P)*

41. Can you magnetize a piece of iron or steel by laying it near a magnet and leaving it a few days? *(P)*

42. Can you magnetize a piece of iron or steel by using the Earth's magnetic poles? *(I)*

43. What materials around you (pipes in your home, cans on a shelf, steel beams in a building) have been magnetized? Check with a compass. *(I)*

44. What effect does heat have on a magnet? Magnetize a nail and then heat it in an alcohol burner.[50] Test the strength on a magnetometer. *(P)*

45. What effect does hitting a magnet sharply have on its strength?[51] *(P)*

46. Do magnets get weaker while they are not being used? Can you recharge a magnet? *(I)*

47. What can you do to keep a magnet from growing weak while you aren't using it? *(I)*

48. What parts of a magnet are the strongest? What part is the weakest? Measure with a magnetometer or place a small steel ball in the center of the magnet. *(P)*

49. Could you make a magnet stronger by changing its shape so the strong parts of the magnet pull together? *(P)*

50. What shapes of magnets are used by industries? How does the shape and power of the magnet fit the particular need? *(U)*

51. If a compass needle is a magnet, could you make a magnet by magnetizing a needle? *(P)*

---

50 Only try this with a magnet that is made of steel or iron, not a plastic or rubber-coated magnet, nor rare earth or ceramic magnets.

51 Rare earth and ceramic magnets will shatter; if you decide to try this, do it with a different kind of magnet.

52. How did Columbus make a floating compass to guide his ships? *(P)*

53. If you make a floating magnet, should you use a glass or steel dish? *(P)*

54. What happens to a compass if there is a piece of iron or steel nearby? Could you offset this pull by using more iron or steel? *(I)*

55. How do ships offset the pull on the compass of ship parts made of iron and steel? *(U)*

56. What is the advantage of having magnets float in oil or other heavy liquids? *(U)*

57. How can you see the pattern of force given off by a magnet? Use iron filings and sprinkle over a piece of cardboard. What hypothesis have you proven or verified by observing the magnetic lines of force? *(P)*

58. What is the pattern of magnetic field for two magnets with their N poles placed so that they face each other? *(P)*

59. What is the pattern of magnetic field for two bar magnets with their S poles facing each other? *(P)*

60. What is the pattern of magnetic field for two bar magnets with the N pole of one magnet facing the S pole of the other magnet? *(P)*

61. What is the pattern of magnetic field for two bar magnets placed so they form a T? *(I)*

62. What is the pattern of magnetic field for two horseshoe magnets placed so that the N poles face each other and the S poles face each other? *(P)*

63. What is the pattern of magnetic field for two horseshoe magnets when the N pole of one magnet faces the S pole of the other magnet? *(P)*

64. Are there any particles in dirt that are attracted by a magnet? Can you try dirt from many areas? *(I)*

65. Are there any rocks that are attracted by a magnet? Do these rocks have magnetic poles? *(I)*

66. What effect does electric current flowing through a wire have on a compass if the wire is laid over the compass? *(P)*

67. Does the Earth have a magnetic field around it? Where are the poles? *(I)*

68. Are there lines of magnetic force surrounding the Earth? How can we map these? *(U)*

69. What causes the magnetic force inside the Earth? Is this force the same as gravity? *(U)*

70. What effect does the Earth's magnetic field have on radio communication? *(U)*

71. Do other members of the Solar System have magnetic fields surrounding them? *(U)*

72. What effect does the Earth's magnetic field have on energy from the sun? On energy from other sources in the universe? *(U)*

73. Do alternating (electricity from a house plug) and direct current (electricity from a battery) have the same effect if they flow through a wire that is placed on a compass? *(I)*

74. What effect has direct current on a compass needle if the wires connecting to the battery are reversed? *(I)*

75. If you coil wire around a compass, is the magnetic field strengthened? Can you measure the strength of this magnetic field? *(I)*

76. If you place a nail in this coil of wire, will it be magnetized? *(I)*

77. What other objects can be magnetized by being placed in a coil of wire? *(I)*

78. Does a magnet increase in strength if placed in a coil of wire connected to direct current? *(P)*

79. Does a magnet increase in strength if placed in a coil of wire connected to alternating current? *(I)*

80. How does increasing the strength of the current (adding more batteries) affect the strength of the electromagnet? *(I)*

81. How does the number of turns of wire in a coil affect the strength of an electromagnet? *(P)*

82. Does an electromagnet have a north and south pole? *(P)*

83. Does an electromagnet have the same effect on a compass as a regular magnet? *(P)*

84. What materials in a coil retain their magnetism after the current has been turned off? How long do they retain their magnetism?

85. Is it better to use alternating or direct current in an electromagnet? *(I)*

86. What uses are made of electromagnets in industry? Can you experiment with some of these uses? *(U)*

87. What effect has different kinds of wire in the coil on the strength of an electromagnet? *(I)*

88. Can a piece of iron or steel always be made a stronger electromagnet by increasing the number of coils or the strength of current? What limits the strength of an electromagnet? *(U)*

89. How strong can you make a magnet by rubbing a nail or piece of steel with another magnet? Can you increase the strength by rubbing the object longer? *(I)*

90. Does an object increase in magnetic strength if it is left in the Earth's magnetic field for a long period of time? What is the limit of the strength of an object magnetized in this manner? *(U)*

91. Can you make a magnet by magnetizing bits of iron filings in a tube? Does each piece of iron filing act like a magnet? *(I)*

92. If a magnet is broken, is each piece as strong as the original magnet? *(I)*

93. If a magnet is broken, do both pieces still have north and south poles? *(P)*

94. Why does steel retain its magnetism longer than objects made of iron? *(U)*

95. Are magnets made of iron, nickel, and cobalt stronger than magnets made of steel? Why? *(U)*

96. Can you experiment with various types of iron, steel, and alloys to determine the ease of magnetism and the ability of each to retain its magnetism? *(U)*

97. How is a magnet used in a telephone? In a telegraph? In a doorbell? *(I)*

98. How are magnets used in engines? Radios? Television? *(U)*

99. Does the ground contain bits of iron? Try using a strong magnet on different kinds of soil. *(I)*

100. Can you use a magnet to separate various mixtures? Try mixing filings with different materials. *(P)*

101. Can you detect the magnetic field around a wire by using iron filings? Run a wire through a piece of cardboard. Connect the wire to a dry cell battery. Sprinkle iron filings on the cardboard near the wire and then tap the cardboard. *(I)*

102. If an electric current can be used to make a magnet, can a magnet be used to make an electric current? Attach a coil of wire to a galvanometer. Move a magnet in and out of the coil of wire. *(I)*

103. Is it the magnet that generates the current, or is it the magnet breaking the lines of force around the coil of wire? *(I)*

104. Will materials that are not magnetized help generate a current in a coil? *(U)*

105. What kind of current is generated by moving a magnet back and forth through a coil? *(U)*

106. What effect is the rate of moving the magnet back and forth through the magnetic field having on the amount of current produced? *(I)*

107. Can a current be produced in a coil if the coil is revolved around a permanent magnet? *(I)*

108. What effect does the size of the coil have on the amount of current produced? *(U)*

109. What effect has the strength of the magnet in the coil on the amount of current produced? *(U)*

110. Can you make alternating and direct current generators? What is the difference between the two? *(I)*

111. How does the voltage vary in an alternating generator? What is the effect of frequency (number of complete turns of the coil over the permanent magnet) on the amount of current produced? *(U)*

112. What frequency is necessary in order to avoid flickering when alternating current is used for lighting? *(U)*

113. What is the difference between a generator and an electric motor? Can you make a generator and measure the output? Try various experiments to improve the performance. *(U)*

114. Through what kind of wire will electricity flow? *(P)*

115. What kind of wire is the best conductor of electricity? *(I)*

116. Does electricity need a complete path? Why doesn't electricity flow if there is a break in the wire? *(P)*

117. Can you make a switch in order to control the flow of electricity? *(P)*

118. What materials will conduct electricity? What materials will not? *(P)*

119. Can you measure the conductivity of different materials? *(I)*

120. What materials resist the flow of electricity? Can you measure the resistance of various materials? *(I)*

121. What effect has the size of the wire on the resistance to the flow of electricity? What effect has the length of the wire on the flow? *(U)*

122. How is resistance to electrical flow used to produce heat and light? Can you measure the amount of heat and light given off by various resistors? *(I)*

123. How do different kinds of switches make a complete path? Examine knife switches, flashlights, house light switches, outlet plugs, starter buttons, doorbells, mercury switches. *(P)*

124. Is there electricity that is not made by man? What is electricity? *(P)*

125. Do different materials contain electricity? Rub a balloon on some wool. Hold the balloon over the hair on your head. Is this electricity? *(P)*

126. Does paper contain electricity? Rub a sheet of newspaper with a ruler while holding the sheet of newspaper

against a smooth surface. How do you know electricity affects the paper? Can you see a spark? *(P)*

127. Does wool contain static electricity? Place a piece of wool (such as a wool blanket or shirt) in a clothes dryer. Turn the dryer on and leave the wool in until the material is very warm and dry. Is the wool charged with static (electricity at rest) electricity? *(P)*

128. What causes static electricity? Try rubbing wool, silk, nylon, rubber. Do these materials become charged? Try other experiments with these materials, such as heating, cooling, hitting, etc. What causes the static electricity charge? *(P)*

129. How do you create a static electricity charge when you walk across a rug and then touch a doorknob? *(P)*

130. What causes a static electricity charge when you reach for the door handle of the car from inside the car? *(I)*

131. What causes dust to collect on a phonograph record? *(U)*

132. Does dust have a positive or negative charge? How about the phonograph record? *(U)*

133. What materials can be charged with static electricity? What materials do not become charged? *(P)*

134. Will material that can be charged also be affected by a magnet? *(P)*

135. Are all static electricity charges alike? Charge different materials. Bring different combinations together. Are materials charged with static electricity always attracted to each other? *(P)*

136. What charged materials seem to attract each other? What charged materials seem to push away (repel) each other? *(P)*

137. Are materials that are charged in the same way attracted or repelled by each other? Charge two pieces of Saran Wrap by rubbing with your hand. Are the pieces of Saran Wrap attracted or repelled? *(P)*

138. Are all static electricity charges alike? Make an electroscope and record whether the leaves are attracted or repelled. *(I)*

139. Do like charges (two materials charged in the same way and affecting an electroscope in the same way) attract or repel? How about unlike charges? *(I)*

140. Does static electricity affect a compass? *(P)*

141. Is static electricity affected by a magnet? *(P)*

142. Is static electricity affected by the humidity of the air? *(P)*

143. Is static electricity affected by temperature? *(P)*

144. Can static electricity be made to jump through the air? Use an electrophorus, Van de Graaff generator, or charge wool and then observe it in a darkened room. *(P)*

145. Can you measure the strength of static electricity you generate? *(U)*

146. What affects the length of time a material will retain its static electricity charge? *(I)*

147. Is lightning the same as static electricity? What conditions cause lightning? *(P)*

148. Can you predict a lightning storm? *(U)*

149. Will static electricity flow through a wire? Use a Van de Graaff generator or some other generator of static electricity. *(U)*

150. Can static electricity be stored? Use a Leyden jar. *(I)*

151. What materials will conduct a static electricity charge? Try discharging a Leyden jar or Van de Graaff through different materials.[52] *(I)*

152. What materials are non-conductors of static electricity? *(I)*

153. How fast does static electricity travel through different materials? Through gases? *(U)*

154. In what ways is static electricity useful? *(P)*

155. Is static electricity harmful? What has man done to eliminate or control hazards from static electricity? *(P)*

156. What objects will store static electricity? *(I)*

157. What is a capacitor or condenser? Can you experiment with making different kinds of condensers and measuring the charges stored by them? *(U)*

158. Can you experiment with different materials for lightning rods? Be sure you use a safe source of static electricity and not lightning to test your theories. *(U)*

159. What is a battery? Can you make a simple battery out of a lemon, copper wire, and a steel knife? Make fruit batteries using different fruits and metals. Can you measure the current flow in each? The current will operate a headphone. *(I)*

160. How does a flashlight battery work? Can you make your own? *(I)*

161. How does an automobile storage battery work? Can you make one? Be sure to experiment with different materials and electrolytes. *(U)*

162. Is the electricity from a battery the same as the electricity from a generator? Be sure to do many of the experiments mentioned with both. *(P)*

163. Will water conduct electricity? *(P)*

164. What materials added to water will make water a better conductor? Try salt, baking soda, sugar, etc. Be sure to measure the current flow of each. *(I)*

165. What gases are given off when current from a battery is sent through water? *(I)*

166. What happens when you reverse the wires on the battery? Which pole gives off hydrogen? Remember, water is made of two atoms of hydrogen to every atom of oxygen. *(I)*

167. How is electroplating done? Can you do various types of electroplating? *(I)*

168. What effect does the strength of the electric current have on the amount of metal deposited by electrolysis? *(I)*

169. What effect does the length of time the current flows have on the amount of metal deposited during electrolysis? *(I)*

170. What determines the voltage of a battery? What determines the amperage or current flow of a battery? *(U)*

171. Can you detect changes in heat energy produced by various amounts of voltage and current? Measure the temperature change by using a coil of wire in a container of water. *(I)*

---

52 See "**Modern Safety Practice**" on page 117 about safety practice around Leyden jars. Read up on safety about Van de Graaff generators separately before using one.

172. What effect does temperature have on the conductivity of various metals? *(U)*

173. How is a fuse used to protect against too much current flow? Can you make different types of fuses and determine the electrical load they will carry? *(P)*

174. How does a circuit breaker work? What determines the danger level for different circuits? *(U)*

175. What is the difference between an alternating and direct current motor? Can you build an example of each? *(I)*

176. How does a voltmeter work? Can you build a voltmeter and use it to measure the voltage in various circuits? *(U)*

177. How does an ammeter work? Can you build an ammeter and measure the amount of current in various circuits? *(U)*

178. How does an electric clock keep time? How is the voltage regulated by power companies? *(U)*

179. Can the magnetic field around one wire induce or start a current in another wire? Try both alternating and direct current. *(I)*

180. What is a transformer? Can you make a transformer by placing an electromagnet inside a coil of wire? *(I)*

181. How does the number of turns of wire around the iron core in the primary (the wire with the current flowing through) affect the voltage of the outside coil of wire in which the current is induced (secondary)? *(U)*

182. How does the size of wire used in the primary and secondary affect the voltage? *(U)*

183. Can you devise transformers for different voltage uses? *(U)*

184. What is a step-up transformer? What is a set-down transformer? *(U)*

185. How does an induction coil enable direct current to be stepped up? Can you experiment by making and breaking the direct current in the primary? Is there a current induced in the secondary? *(U)*

186. How does a light bulb work? Can you experiment with different kinds of materials as the filament? *(I)*

187. Can you make a light bulb that won't burn out? Compare your light bulb with standard commercial light bulbs. Record the life and brightness of each. *(U)*

188. What gases under low pressure conduct electricity? How is this the basis for neon lights? Can you experiment with the pressure as compared to the brightness? *(U)*

189. What colors do various gases give off when an electrical current is passed through the gas at low pressure? *(U)*

190. How does a fluorescent light work? How can invisible ultraviolet rays give off visible light? *(U)*

191. Can you make a carbon arc light? How much light is given off? Can you measure the heat at the point of contact of the two carbon rods? *(U)*

192. Can you devise a carbon arc projector? A carbon arc searchlight? Remember to use a good mirror to collect and focus your light. *(U)*

193. Can you collect and focus the light from a carbon arc projector through a lens onto a piece of wood? Does focusing the rays increase the temperature? *(U)*

194. How does a carbon arc furnace work? Can you make one and carry on experiments with different metals? Can you determine the melting points of various metals and alloys? *(U)*

195. What is the difference between lights wired in series and those wired in parallel? Can you measure the voltage and amperage at various points in the circuit? *(I)*

196. What is the effect on the voltage if additional lights are added in series? *(I)*

197. How is electrical current changed from alternating to direct current? How does a rectifier work? Can you build a rectifier? *(U)*

198. How does a burglar alarm work? Can you devise a burglar alarm? *(I)*

199. How do two-and three-way switches work? Can you devise a wiring circuit that can do many things? *(I)*

200. What is a digital computer? Can you make a digital computer that can do certain tasks? Be sure to use the computer after you make it. *(U)*

201. Can you make an analog computer? Can you calibrate your computer for a particular purpose? For instance, can you design the computer to show the area of a circle when the radius is known? *(U)*

202. Can you use the heating effect of an electric current to measure electricity? Send an electric current through a cold iron wire. Note the sag in the wire as it gets hotter. Is the amount of sag related to the strength of the current? *(U)*

203. What effect do the material, length, cross-sectional area, and temperature have on the resistance of a wire? Set up an experiment using different kinds of wire. Keep the voltage constant. You can measure the amperes or amount of current. From the voltage and amperage you can determine the resistance. *(U)*

204. How do fuses protect against an overload of current? *(P)*

205. Can you make a wheatstone bridge in order to determine the resistance of various materials? *(U)*

206. How does an induction coil work? Can you measure the amount of voltage at the spark gap? *(U)*

207. What are eddy currents? Can you demonstrate the uses of eddy currents by making a repulsion coil? Try experiments to see if you can measure these currents. *(U)*

208. Can you magnetize needles by building up a static electricity charge in a Leyden jar and then discharging the current through a coil containing a needle? *(U)*

209. Can you use alternating current in the electrolysis of water? Can you use static electricity in the electrolysis of water? *(U)*

210. Will a magnet have any effect on an electric current flowing through a wire? Stretch a size No. 24 wire about 6 feet long between two insulated supports. Connect the wire to an alternating current supply.[53] Use a rheostat to vary the current. Bring one end of a strong bar magnet near the wire while the current is flowing. *(U)*

---

53 A *low voltage* ac supply.

211. Can you design and build a Tesla coil? Can you develop a practical use for such an instrument? *(U)*

212. Can you build different kinds of radio sets? Be sure to trace the principles behind radio and show how they apply to your work. *(I)*

213. Can you devise different kinds of television antennas? Try these out with your TV set. Be sure to test your own designs instead of copying those already on the market. *(I)*

214. Can you experiment with TV reception problems in your area? *(U)*

215. Can you devise a new use for a photoelectric cell? *(U)*

216. Can you experiment with different kinds of condensers? You should measure the charge that is stored. *(U)*

217. What effect has static electricity on plant growth? *(I)*

218. What effect has DC and AC current on plant growth? *(I)*

219. What are the problems of tuning in radios? Can you devise your own tuner? *(I)*

220. How does a speaker on a radio or television set operate? Can you design and make your own? *(U)*

221. Can you design and build a computer that can forecast weather from data you feed it? This might be either a digital or analog computer. *(U)*

222. Can you design and build a computer (analog) for learning the multiplication tables? Be sure to calibrate your instrument experimentally. *(I)*

223. What is the difference between AM and FM radio broadcasting? Can you devise a means of modulating radio carrier current so loudspeakers can respond to it? *(U)*

224. How fast do radio waves travel? Can you measure the velocity, frequency, and wave lengths of radio waves you generate? *(U)*

225. What are rectifiers? Can you experiment with using different kinds of rectifiers in place of diode and triode tubes?[54] *(U)*

226. How does the sound track on a movie film work? Can you devise a means to put a sound track on film? *(U)*

227. Can you devise a means to put a sound track on film strips? *(U)*

228. Can you design and build your own oscilloscope? *(U)*

229. What are the problems of color television? Can you devise your own method of solving some of these problems? *(U)*

---

54 Experiment with different kinds of diodes. What are (*were*) diode and triode tubes?

Force, Measurement, and Motion | 5

Spring Scale

Purpose: A spring scale is used to determine the force or pull an object causes or exerts.

Materials: Wood ruler (or half a yardstick), tack, paper clip, and a rubber band.[1]

What to Do: Cut the rubber band in half. Stick a tack in one end of the ruler. Tie one end of the rubber band around the tack. Bend a paper clip as shown. Tie the other end of the rubber band to the paper clip. Part of the paper clip serves as a marker. The other part of the paper clip serves as a hook from which objects to be weighed can be suspended.

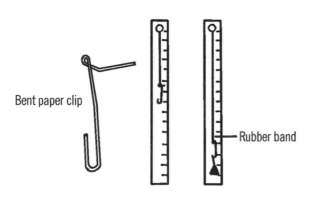

Bent paper clip

Rubber band

Operation of Equipment: Hang a known weight from the end of the paper clip. Place a piece of adhesive or masking tape at the spot the paper clip indicates. Write the weight of the object on the tape. Do this for several known weights until you have a complete scale.

Can You Work Like a Scientist?

1. Do all rocks that look the same size weigh the same?

2. Do things weigh the same in water as in air?

3. What would you have to change on your rubber band spring scale if you had to weigh very heavy weights?

4. How would you weigh very light-weight objects?

Rubber Band Weighing Scales

Purpose: This type of scale is a simple measuring device to determine weight or the pull of gravity on objects. The scale has certain limitations as to accuracy.

Materials: Rubber band, lid, wood.

---

1 Could you use a metal spring instead of the rubber band?

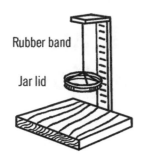

Rubber band

Jar lid

the scale in place, cut the lath so that it is just 1000 millimeters long.

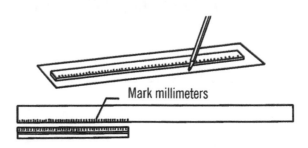

Mark millimeters

Meter Stick

Purpose: A meter stick is similar to a yardstick except that the divisions are much finer and all the measurements are expressed in the metric system which is used by scientists all over the world.

Materials: Yardstick or wood lath, plastic ruler (type with both millimeters and inches), white butcher paper, and glue.

What to Do: A meter stick is almost forty inches long.[2] You can make part of a meter stick by just putting metric measurements on the back of a yardstick. A better and more complete meter stick can be made from a piece of smooth lath. Lay such a piece of wood down on a piece of butcher paper. Trace around the wood with a pencil. The outline of the wood on the paper forms your scale. Lay your ruler down on the paper and copy the metric system as well as the English system of measurement on the butcher paper scale. The very small divisions on your ruler are in millimeters, the next largest division is in centimeters. It takes 1000 millimeters or 100 centimeters to make one meter. Therefore, you will have to move your ruler four times along the butcher paper in order to make the complete scale.

When you have finished making the scale, glue the scale on the wood lath. Use rubber cement or Elmer's glue. After you have finished gluing

Operation of Equipment: The metric system's main advantages are finer divisions for more accurate measurement and the ease of changing one measurement into another. For instance, if you want to change centimeters into millimeters you just multiply the centimeters by ten. In actual practice you move the decimal point one place to the right. For example, 36 cm (centimeters) would be written 36. Therefore 36 multiplied by 10 equals 360 mm. (millimeters).

Can You Work Like a Scientist?

1. How would you change millimeters into centimeters? Remember, your answer should be less.

2. How would you change meters into millimeters? Into centimeters?

3. What are the equivalents in the metric system for the inch? Foot? Yard? How many centimeters equal one inch? One yard?

4. What measurement is used for the "mile" in the metric system?

---

2 To slightly higher precision, 39.37 inches. What is the *exact* conversion?

Optical Micrometer

Purpose: A micrometer is used to measure the thickness of very thin objects. The optical micrometer is accurate to a thousandth of an inch.

Materials: Two glass microscope slides, two finishing nails, piece of black art paper, small rubber band, block of wood (4" × 1½" × 1"), wood for the base (8" × 16") and cardboard for a scale.

What to Do: Cut a 1" × 1¼" notch off the block of wood as shown. Use masking or Scotch tape to fasten one microscope slide to the side of the block of wood. Glue a piece of black art paper over all the slide except for ½ inch near the notched end of the block. The finishing nail near the slide should be driven into the wood base very near the slide at the edge of the notch in the block. A second finishing nail is nailed on the other end of the base about an inch in from the side.

Tie a piece of white thread around the finishing nail near the slide. Fasten another slide to the first slide with a rubber band. The slide should be so positioned that if you press on "B" the slides will open up at "A."[3] Finally glue or tape a piece of white cardboard to the base for a scale.

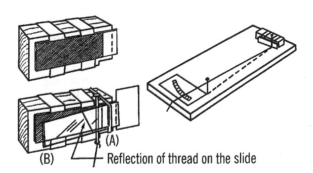

(B) (A) — Reflection of thread on the slide

Operation of Equipment: Line up the two finishing nails with your eye so that they lie in a straight line. Now move the thread from one side of the base to the other. You should be able to see the reflection of the thread on the microscope slide. Move the thread so that the reflected image of the thread disappears behind the nail between the slides. Mark this position of the thread "0" on your scale. Slip a piece of newspaper between the two slides at "A." Move the thread again so the reflection lines up with the two nails. Since a newspaper is .0025 of an inch (2½ thousandths) thick, place a mark on the scale and label this mark 2.5. Now insert two pieces of newspaper between the slides. Mark the correct position 5 for five thousandths of an inch. Repeat this until you have a scale that goes up to fifty thousandths of an inch.[4] You can make smaller divisions by dividing the distance between zero and 5 into five equal distances. Then each small mark indicates one-thousandth of an inch.

Can You Work Like a Scientist?

1. Can you measure the thickness of different metals? Hair? Paper? Glass?

2. Can you change your scale into metric measurement?

3. Does glass get thicker when it is heated? How about other things?

4. What is the effect of freezing temperatures on various things?

Centrifugal Force Indicator

Purpose: This instrument is used to measure the force necessary to balance the outward pull of a mass that is whirled at different speeds and at different distances (radii). With this indicator

3 That is to say, the two slides are hinged by the nail that is between them, like the two sides of a spring-loaded clothespin.

4 You may want to consider using other types of objects—possibly better known than newspaper thickness—to define your scale. A good reference point is that a (shiny, new) US dime is 1.35 mm (0.053 inches) thick.

you can measure the increased gravitational force exerted on an object that is whirling in a circular path.

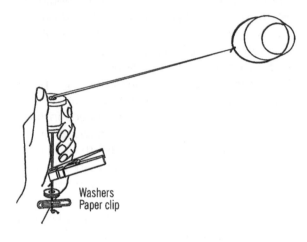

Washers
Paper clip

Materials: Small bucket or can with a handle, strong cord about four feet long, large spool, clothespin, paper clip, and several washers for weights.

What to Do: Tie one end of the cord on the handle of the small bucket or can. Slip the other end of the cord through the spool. Tie a paper clip on the other end of the string in such a manner that washers can be slipped over the paper clip and yet won't fall off. Clip a clothespin onto the cord at a certain distance from the handle of the can. This distance is the radius for your particular experiment. Add washer weights till the pull of the can balances the weights.

Operation of Equipment: Whirl the can by holding the spool in your hand and turning the spool as you whip the can around your head. When the clothespin is pulled up to the bottom of the spool the outward force is too great. Add washer weights till the outward pull just balan-

ces the inward pull.[5] The amount of weight required to balance this outward pull is a measure of the "centrifugal force" or artificial gravity on the bucket.[6]

Can You Work Like a Scientist?

1. Does it require more or fewer weights to balance the whirling bucket if you increase the speed at which you are rotating the bucket?

2. Time the number of rotations you make in fifteen seconds. If you know the radius, can you figure the miles per hour the object is moving?

3. Can you keep a graph of the weight required to balance the outward pull as the bucket increases its speed around you?

4. What effect does increasing the radius have to do with miles per hour and total force needed to balance the outward pull?

5. What effect does adding a weight in the bucket have on the outward pull at various speeds? Does increasing the speed of rotation seem to increase the weight of the object?

6. Try whirling small animals in the bucket. Put a lid on the bucket. What effects does an increase in gravitational force have on small animals?[7]

7. Can you whirl an animal in such a way as to equal the force that would be exerted on the animal on the planet Jupiter?

---

5 Depending on the weight of your bucket, you may want to use a second bucket that you fill for weight, rather than just more or heavier washers.

6 Does centrifugal force really exist? See Note 38 in Appendix E.

7 See Note 1 in Appendix E about working with animals.

8. The pull of gravity is measured in "G's." One G is the regular weight of the object. If it takes twice the normal weight of the object to balance the object when you whirl it, the object is subjected to two "G's." How many "G's" can you cause by whirling an object?

Hand Stroboscope

Purpose: This instrument is used for measuring and examining objects that rotate or reappear up to sixty times per second. The stroboscope seems to stop the motion of a moving object by matching its speed with that of a rotating object.

Materials: Plywood or masonite disc as shown, wood handle, screw, and washer.

What to Do: Cut twelve slots in the wooden disc as shown. Attach the handle by using the washer and the wood screw. Drill a hole next to the handle. The hole should be large enough for your finger. The disc should turn easily when you rotate it with your finger.

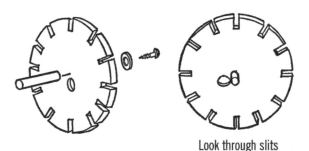

Look through slits

Operation of Equipment: The stroboscope operates on the principle that if you always see a moving object in the same place, it looks as if it is standing still. Move your finger back and forth in front of your face. Close your eyes before you move your finger to one side. Open your eyes when your finger is in its original position in front of your face again. It looks as if your finger is always in the same position even though you know it is moving back and forth.

This is how the stroboscope works. A vibrating or rotating object is viewed through the slits in the disc. If the object is vibrating slowly enough, all the slits but one can be covered with masking tape. Then if you rotate the stroboscope at the exact speed the object is vibrating or rotating, you will always see the object in the same place, and the object will appear as if it were not in motion. By counting the rate at which you rotate the hand stroboscope, you know the rate at which the object is rotating or vibrating. If the object is vibrating too fast, uncover a second slit. Then each half rotation of the disc is equal to one rotation or vibration of the object. If you have two slits uncovered and you rotate the stroboscope thirty times in a minute to stop the motion, you have seen the object thirty times two or sixty times in the minute at the same spot. Therefore, the object has rotated sixty times in a minute.

Can You Work Like a Scientist?

1. Fasten a hacksaw blade in a vise. Pluck the blade so it vibrates. Can you determine the rate of vibration?

2. Can you stop the motion of an engine with your stroboscope and determine the revolutions per minute? If not, what do you conclude?

3. If your stroboscope is moving too slowly, the slot won't quite get around for each rotation of the object. Will the object seem to be moving slowly forward or backward?

4. If your stroboscope is moving too rapidly, the object will not quite make one complete turn each time you see it. Will the object seem to be moving backward or forward?

5. Why do the wheels of wagons in the movies appear to be moving backward?

Motorized Stroboscope

Purpose: The stroboscope is used seemingly to stop the motion of rapidly rotating or vibrating objects in order to examine them at high speeds or determine the speed at which they are rotating. The motorized stroboscope is capable of stopping the movement of high-speed motors. When the stroboscope is hooked up with a rheostat, the speed of the stroboscope can be matched up with the unknown speed of the object to be viewed.

Materials: Small motor such as a phonograph motor, a rheostat, and a plastic phonograph record or round piece of masonite.[8]

What to Do: Cut six slits in the record or masonite as shown. You can drill holes in place of the slits if necessary. Attach the record to the shaft of the small motor. You may need to wrap tape around the shaft in order for the record to fit tightly on the shaft. Connect the rheostat to the motor and the plug as shown.

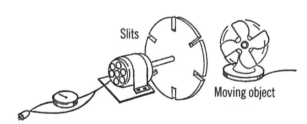

Slits

Moving object

Operation of Equipment: Cover up five of the six holes with masking tape. Look through the remaining hole of the rotating record at a fan that is turning at a known number of revolutions per minute.[9] Use your rheostat to control the speed of your stroboscope. When the speed of the stroboscope matches that of the fan, the blades of the fan will seem to stand still. Make a mark on your rheostat and label it

with the number of revolutions the fan is making. You can determine other speeds for your stroboscope by using similar rotating motors of known speed.

If you want to double the speed of your stroboscope, uncover a second hole on the disc. The hole should be opposite the first hole. Three holes triple the rate of your stroboscope reading. Six holes will increase the rate six times. Thus, if your rheostat gives a rate of 1200 revolutions per minute and you were using six holes, the total number of revolutions would be six times 1200 or 7200 revolutions per minute. An intense light on the object being viewed makes examination much easier.

Can You Work Like a Scientist?

1. Can you use your stroboscope on fluorescent lights? Do fluorescent lights give a steady light?

2. If the stroboscope turned twice as fast as a motor being examined, would the motor still appear to be stopped?

3. If the stroboscope turned four times as fast as the motor, what pattern would you see through the stroboscope disc?

4. How is a stroboscope used to help time a car engine? Ask a mechanic.

Water Hourglass

Purpose: This water glass clock is used to measure a certain length of time. The amount of time is controlled by the amount of water used.

Materials: Two bottles, two rubber stoppers, and two short pieces of glass tubing. Baby bottles are very good since they have measuring

---

8 Please read Note 39 in Appendix E about materials for this project.

9 How would you know the rpm speed of a fan? Are there other consistent speed references that you can find? What about LED lights, or an orange neon "pilot" bulb in a power strip?

marks on them. Baby bottles require size #6½ stoppers. (You can use clay for a stopper and soda straws for tubing.)

What to Do: Cut two short pieces of glass tubing about an inch long (see "Cutting Glass Tubing" on page 6). Use the tubing to connect the stoppers. If you use soap suds, the tubing will slip into the stopper more easily.

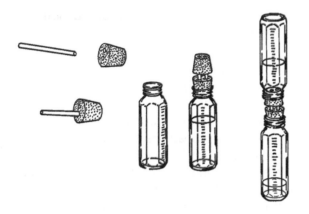

Operation of Equipment: Baby bottles are marked in milliliters. Fill one bottle with 200 ml of water. Place the stoppers in both bottles. Turn the bottles over and time how long it takes for the water to go from the upper to the lower bottle. If you want only half the time for a measurement, put in only half the water.

Can You Work Like a Scientist?

1. How many milliliters of water would be necessary for three minutes? Is this long enough to soft-boil an egg?

2. What clock is the most accurate, the water clock or the phonograph stop clock ("Stop Clock" on page 159)? Can you think of times when one would be better than the other?

3. If a student is to give a two-minute talk, can you set your water clock for just two minutes? Will this help classmates keep track of the time?

4. How many baskets can you shoot in five minutes (use your water clock)?

5. How many arithmetic problems can you do in fifteen minutes?

6. What effect does the amount of light have on plant growth? Time the experiment with your water clock.

7. Try making your hourglass with just one-hole stoppers.

8. Put water in a baby bottle. Use a two-hole stopper. Will the water run out if one hole is covered? Will it run out if neither hole is covered?

Stop Clock

Purpose: There are many uses in science for the accurate measurement of time. Often it is necessary to measure time in parts of a second. With this stop clock you can break time up and measure it down to one-hundredth of a second.[10]

Materials: Phonograph turntable (should have 45 rpm speed), old phonograph record, spring clothespin, and a large empty spool for thread.

What to Do: With a protractor divide a phonograph record into a hundred sections as shown. Mark these divisions by using tape. Mount a clothespin as shown on a block of wood. The block should raise the clothespin to the same height as the record on the phonograph.

How to Operate Equipment: Place the phonograph record on the turntable. Turn on the

---

10 At first glance, this may seem absurd: most of us already carry around *at least one* digital device that has a fast stopwatch function. But *how could you do it* if you couldn't use modern digital electronics? Here, one plausible method is illustrated.

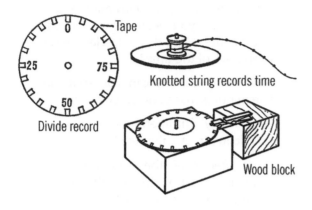

Tape

0

25   o   75

50

Divide record

Knotted string records time

Wood block

2. Time the fall of objects. See how fast they fall as compared with the distance they fall.

3. Time a student running a set distance. Figure their speed in miles per hour.

4. What clock is the most accurate, the water clock or the phonograph stop clock?

5. Time the complete swing of a pendulum (see "Foucault Pendulum" on page 75).

record player at 45 rpm. Keep the record from revolving by pinching it with the clothespin. The record should be turned so that "0" is at the spot marked by the clothespin. To start the time clock, press on the clothespin. To stop it, release the pressure on the clothespin. In order to figure the time, count the total times the record went around.[11] Add the part of the turn as shown by the clothespin. An example would be: 9 full turns and 25/100 of a turn. Since the record turns only 45 times a minute, it makes only ¾ of a turn each second. To change your reading to seconds, multiply by 4/3. On the example, 4/3 times 9.25 gives an answer of about 12 and ⅓, so the time would be 12.33 seconds.

If you wish to record the time without watching the record player, tie a long piece of string onto a spool. Glue the spool to the center of the record. Tie knots on the string to show one turn of the spool as it winds the string. Mark every fifth knot with some color for easy counting. After the record has stopped, count the number of knots on the spool. Add the part of a record and multiply by 4/3.

Can You Work Like a Scientist?

1. Use your timer clock to time reactions of boys and girls.

Harmonograph

Purpose: Harmonic motion is a movement or motion that repeats itself. A pendulum is a good example of harmonic motion. When two forces work together, the result is a pattern of movement that combines the movements of both forces. The harmonograph records such movement in pleasing and unusual designs.

Materials: Wood for stand, piece of plywood, two wooden rods about 5' long, two metal collars, two cans, two clamps for the rods, four spikes, cement for the cans.

What to Do: File four slots in each metal collar or ring. Two slots should be on the top side of the collar and two on the underneath side as shown. Build the stand. Make the two pendulums out of the wood rods or doweling. Drill a hole about one foot from the top of each rod and drive a spike through each hole. Cut off the head of the spike. Attach a plywood board to the top of one rod and an arm with a hole for a ball point pen on the top of the other rod.

The cans are used for weights. These should be large fruit juice cans. Cut a hole in the top and bottom of each can. Slip in a piece of plastic pipe that is just large enough to slip over your wood rods. Pour cement in the cans and let the

---

11 Do you notice a flaw in this method? See Note 40 in Appendix E.

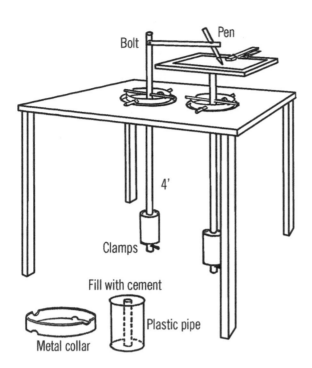

Bolt

Pen

4'

Clamps

Fill with cement

Metal collar

Plastic pipe

The short pins that the collars ride on can be spikes fastened to the table of the stand. Slip the rods into position through the holes. Clamp on the weights.

Operation of Equipment: Fasten a sheet of paper to the plywood board at the top of one rod. Start both pendulums going by releasing the weights. The pen will draw various patterns depending on the direction and motion of the pendulums. Try adjusting the weights in various positions until you develop patterns that suit you.

cement harden. The weights are then slipped over the wood rods and kept in position by an adjustable clamp or a pin through the rod.

Geology

S

Geology is often thought of as a field science. Many people feel that geology is limited to only what one can see and collect. To them geology is collecting interesting rocks and perhaps fossils.

Geology is not a spectator type of science. It is rather a participation type of science filled with nearly as much experimentation as any other branch of the physical sciences.

Geology is often defined as the study of the history of the Earth, what it is made of, and how it was formed. The branch of geology that deals with the study of rocks is called petrology. The branch that deals with minerals is called mineralogy.

Identification is important in either the study of rocks or minerals. The equipment needed to perform various identification tests is very simple and inexpensive. Most of the equipment has been described in other sections of this book, and if this is the case, the reader will be referred to the appropriate page.

General Equipment Needed for Geology

- Beam balance ("Sensitive Gram Scale" on page 30)
- Magnet ("Making a Magnet (or Recharging a Magnet)" on page 107)
- Test tube holder ("Ring Support for Support Stand and Test Tube Holder" on page 16)
- Alcohol lamp ("Alcohol Burner" on page 8)
- Blowpipe ("Blowtorch Type of Alcohol Burner" on page 11)

Other equipment you will need:

- Hand lens (tripod magnifying glass 5 to 10 power)
- File
- Test tube (closed tube)
- Jackknife

Chemicals: [1]

- Borax

- Ammonia (kitchen cupboard)

- Sodium Bicarbonate (baking soda)

- Hydrochloric acid-concentrated and dilute

- Nitric acid-concentrated and dilute (optional)

Additional Equipment to Make or Improvise

- Streak plate (unglazed tile)

- Glass rod (seal both ends of a piece of glass tubing)

- Nichrome wire loop (nichrome wire and glass tubing)

Rock and Mineral Worksheet

It is very helpful to make several copies of the following worksheet. This record sheet will aid you in your investigation of the specimen.

Test for Hardness

The standard test for hardness is the Mohs scale. This scale rates the minerals from 1 to 10, with ten being the hardest. Sample minerals for each degree of hardness on Mohs scale are:

Talc: 1	Orthoclase: 6
Gypsum: 2	Quartz: 7
Calcite: 3	Topaz: 8
Fluorite: 4	Corundum: 9

Apatite: 5	Diamond: 10

Rock and Mineral Worksheet

Location from which specimen was collected _____
Date _____ Comments about surrounding area _____
_____
_____

Tests:
Color _____
Hardness _____
Luster _____
Streak _____
Fusibility _____
Magnetism _____

Acid Test _____
Bead Test _____
Chemical test *(Specify)* _____
Chemical test *(Specify)* _____

Specific Gravity Calculations:
Weight in air _____
Weight in water _____
Sp. G. _____

$$Sp.\ G. = \frac{Wt.\ Air}{Wt.\ Air - Wt.\ Water}$$

Hand Lens Investigation:
Composition _____
Crystal Structure _____
Granularity _____

Special Properties Noted:
_____

Other Notes:
_____
_____

Conclusions:
_____
_____
_____

You may substitute many common materials for some of the minerals which you do not pos-

---

1 Cobalt nitrate was originally in this list, but may be omitted from a standard kit. Certain substances become colored when moistened with a cobalt nitrate solution and heated.

sess. Some common materials[2] and their degrees of hardness are:

Fingernail: 2.5	Quartz/Flint: 7.0
Penny: 3	Emery paper: 8.0 to 9.0
Window glass/Knife blade: 5.5 to 6.0	Corundum paper: 9.0
Steel file: 6.5 to 7.0	Carborundum: 9.5

The hardness test is based on the fact that a harder material will scratch a softer piece of material. For instance, a penny will scratch a piece of talc (hardness 1) but not a piece of window glass (hardness 5.5). The penny is harder than 1 but softer than 5.5. By trying the penny on gypsum and fluorite we would find the penny would scratch the gypsum but not the fluorite. Therefore, the penny is somewhere between 2 and 4 on the hardness scale. By trying the penny on calcite we find it just about makes a faint mark on the calcite. We would conclude the hardness is just about that of calcite.

You must observe the mineral you are testing very closely to be sure that the surface you are testing is being marked. Many times a mineral of lesser hardness will leave a streak of its own on the mineral being tested. This is sometimes mistaken for a scratch.

Research suggestions:

1. Can you determine the hardness of various materials found around the house? Sample materials might include buttons, various metals, wood, ceramics, and others.

2. Can you determine the hardness of various rocks and minerals found around your neighborhood? You might compare the average hardness with that of rocks and minerals found in other locations.

3. Can you develop your own complete scale of hardness based on common materials instead of those suggested for Mohs scale?

Color Test

A careful visual examination of a specimen is usually sufficient to give you the correct natural color of the material. Be sure, however, that you have a fresh sample, as some specimens tarnish or change color from weathering.

Luster Test

Another characteristic or property of a mineral is the *luster*.[3] Luster is the appearance of the mineral in ordinary light. Quartz and glass are vitreous or glassy in appearance. A diamond has adamantine luster.[4] Stibnite has a metallic luster. Other terms are: earthy, dull, greasy, silky, pearly, and resinous.

Streak Test

The *streak* of a mineral is in some cases one of the most important tests. The streak is determined by any of three ways. First, the specimen is rubbed across a streak plate of unglazed porcelain (such as unglazed tile). The resulting color of the mark on the plate is the streak color. A second way streak can be determined is by scratching the mineral with a knife and observing the color of the powder. A third method for determining the streak is by observing the color of a crushed mineral.

---

2 Quartz may be found as common agate rocks. Corundum paper is more often labeled as aluminum oxide sandpaper. Carborundum is the abrasive on silicon carbide sandpaper. Most "wet or dry" sandpaper is silicon carbide.

3 Note that the word is more commonly spelled *lustre*, in the context of mineralogy

4 Adamantine luster means *diamondlike*: as in, transparent and sparkly.

If the mineral is harder than the streak plate, it will scratch the plate and leave a powder residue on the plate. If your streak is white, check the plate with your hand lens. The white mark may not be from the mineral. It may be caused by scratching the plate.

The streak is in many cases a different color from that of the mineral being tested and is more constant than color as a property of a mineral.

Fusibility Test

In order to test the fusibility of a mineral, a thin piece of the specimen is held in a pair of forceps or tweezers and placed in the flame of an alcohol burner or a blowpipe. If the mineral seems to melt, it is fusible at that temperature. A scale of fusibility is as follows:

1. Stibnite: fuses in the flame from an alcohol burner (980 °F)

2. Chalcopyrite: fuses easily in a blowpipe flame (1475 °F)

3. Almandine: fuses less easily in a blowpipe flame (1920 °F)

4. Actinolite: thin edges fuse in a blowpipe flame (2190 °F)

5. Orthoclase: thin edges fuse with difficulty (2374 °F)

6. Enstatite: thinnest edges fuse with a blowpipe (2550 °F)

7. Quartz: will not fuse in a blowpipe flame.

Acid Test

For your acid bottle, use a small nose drop bottle or other bottle containing an eyedropper. This bottle should be labeled, "Dilute hydrochloric acid" (Dil. HCl). Your druggist will make up a dilute solution for you.[5] You need the acid

Tweezers holding sample

Modern Safety Practice

Follow the safety guidance given previously when using an open flame. See Note 3 in Appendix E. Wear your safety glasses!

diluted in a ratio of one part acid and three parts water. If you are going to dilute the acid yourself, you must be sure to pour the acid into the water and not pour the water into the acid.

Be sure to carry this acid bottle with you in your field kit when you go exploring for rocks and minerals. This acid test offers one of the simplest means of determining carbonate rocks and minerals. If you place a small drop of dilute HCl on a specimen of a carbonate mineral, it will effervesce or give off bubbles.

Specific Gravity Test

The specific gravity of a mineral is its weight relative to the weight of an equal volume of water.

In the chemistry section you were given instructions for building a beam balance. You may use the same instrument for determining specific gravity if you make a few changes. Remove one of the paper cups and replace it with a small jar lid attached to the yardstick by string

---

5 No longer; see Note 16 in Appendix E about acids.

or thread as shown. Punch holes in the bottom of the jar lid with a nail. Place the beam balance on a stack of books or a block of wood about 4" high. The pan should be low enough so that it can be completely covered in the glass of water.

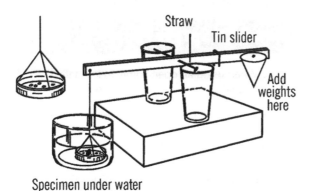

Specimen under water

Procedure: Without the glass of water under the pan, balance the beam scale by using your slider on the right-hand side of the beam. Place a small piece of the specimen to be investigated on the pan and weigh it while it is dry. Record this weight.

Then, place the glass about half full of water under the left side of the beam and weigh the same specimen completely under water. Note this weight. Subtract the weight in water from the weight in air. This will give you the loss of weight, or the displacement in water. Then, divide the dry weight by the loss of weight, which will give the specific gravity.

The formula for specific gravity is:

$$\text{Sp. Gr.} = \frac{W_A}{W_A - W_W}$$

where W_A is the weight in air, and W_W is the weight in water.

Example: The weight of a specimen is 6 grams in air (its dry weight). In water the specimen weighed only 4 grams. The specific gravity would be determined as follows:

$$\text{Sp. Gr.} = \frac{W_A}{W_A - W_W} = \frac{6}{6 - 4} = \frac{6}{2} = 3$$

Bead Test

Purpose: Many minerals give off a characteristic color when placed in the tip of a very hot flame. The color in the flame is a clue to the chemical composition and thus to the type of mineral.

Materials: Nichrome wire about 3" long, glass tubing about 6" long, double wick alcohol burner, and a blowpipe.

What to Do: First make a loop about ⅛" in diameter in one end of the nichrome wire. You can form the loop around the tip of a pencil. Next, heat a 6" piece of glass tubing in your double wick alcohol burner until the glass becomes soft and begins to seal over. Push the straight end of the wire into the sealed end of the glass tube and then remove the tubing from the flame. The glass will then cool and harden around the wire.

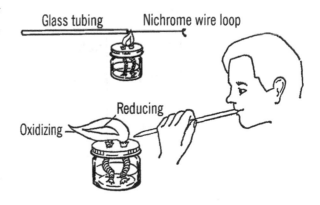

Operation of Equipment: Grind the mineral that you want to investigate into a fine powder. You can do this by breaking a small piece off the sample, covering the piece with a cloth so that the bits of rock and powder won't scatter, and then hammering it into a powder.

Place the tip of the blowpipe in the edge of the flame of the alcohol burner. Blowing gently and continuously, you will produce a flame which is composed of two parts: the outer tip of the flame is called the oxidizing flame. The inner part is called the reducing flame.

With a little practice you will find that you can breathe in through your nose and blowout gently through the blowpipe at the same time. This produces a good continuous flame.[6]

Place the loop of wire in the oxidizing flame until it becomes red hot. Dip the loop quickly into powdered borax and pick up all that will stick to the wire. Place the loop with the borax back into the oxidizing flame until a bead is formed. This bead will be red hot. Pick up a small amount of the powdered mineral to be tested with the bead. Place the loop back into the oxidizing flame and note the color when hot. Then remove the loop from the flame and note the color when the bead is cold.

The following is a list of colors of the bead in the oxidizing flame and when cold.[7]

Substance	Hot	Cold
Antimony (Stibnite)	Yellow	Colorless
Copper (Azurite)	Green	Blue
Iron (Limonite)	Yellow	Green

Be sure the loop is clean before each test. Heat the bead in the blowpipe flame until the bead is red. Take the wire out of the flame and rap the bead sharply on a hard object to remove the bead. Dip the wire into HCl and then heat the wire in an alcohol flame. Repeat these directions until the flame is no longer colored.

Modern Safety Practice

1. When using an open flame, follow safety guidance in Note 3 of Appendix E.

Special Properties

Magnetism: Some of the iron minerals are naturally magnetic. Magnetite is one of these. Touch a magnet to a specimen of magnetite and you will find it attracts the specimen. Some specimens are very weakly magnetic. You can check these by bringing a compass near the specimen. If the rock contains iron, the compass needle is attracted or repelled. Some iron minerals become magnetic when heated. You will find this true of limonite, pyrrhotite, and pyrite.

Odor: Some minerals give off a characteristic odor. Pyrite, when heated, smells like sulfur. Arsenopyrite gives off sparks and a garlic odor when struck. Kaolinite has an earthy odor when breathed upon. Examine your specimens to see if other minerals give off odors.

Solubility: Minerals which dissolve in water are said to be water soluble. The following have this special property: halite, Epsom salt, potash, nitrates, borax, and soda.

Feel: Some minerals have a characteristic feel when rubbed with the fingers. Examples of this special property are:

---

6 If you do this, be sure to practice *without* the flame!

7 How can you find the bead color for other materials?

- Talc feels greasy
- Kaolinite feels greasy
- Graphite feels greasy
- Meerschaum feels smooth
- Molybdenite feels greasy

Chemical Tests: Some minerals are soluble (they dissolve) in hot or cold acids. As the mineral dissolves, it gives a characteristic color to the solution. If you then add some other chemical to the solution, the color may change again. These chemical tests are sometimes very helpful in the identification of a mineral.

In all chemical tests, try to keep the mineral being investigated as pure as possible. The associated minerals in a specimen might affect the weight and the chemical reactions of the specimen in question.

Only a very small amount of the mineral sample is necessary for these tests. The reactions are highly sensitive so a few grains of the solid with a drop or two of liquid is all that is needed.

You can study the mineral identification chart and try some of these tests mentioned under the heading, "Special Properties."

Seismograph

Purpose: seismograph is used to record shock waves from disturbances such as earthquakes. A modern use of seismographs is to detect nuclear and thermonuclear bomb tests. The seismograph operates on the principle of a stationary pendulum. A suspended pendulum remains motionless while the Earth moves beneath it. Since the recording drum is attached to the Earth, the drum moves back and forth and the record of the movement is inscribed by a pointer attached to the motionless pendulum.

Materials: The base is made of either a heavy 2″ × 12″ plank or a slab of concrete. Since the concrete is heavier, it will make a firmer support and will not vibrate with slight movements caused by local sources such as cars. You will also need a pipe flange, a ¾″ pipe about three feet long, a metal rod about three feet in length, a steel wire about five feet in length, and an old alarm clock.[8]

What to Do: Drill a small dimple (the beginning of a hole) in the ¾″ pipe as shown. Grind one end of the rod to a point. This point should fit in the dimple drilled in the pipe. Attach weights to the rod. Heavy bricks or iron can be wired onto the rod. Attach the pipe flange to the base with bolts. Screw the pipe into the flange. Run a wire from the top of the pipe to the end of the rod after the pointed end of the rod is placed in the dimple.

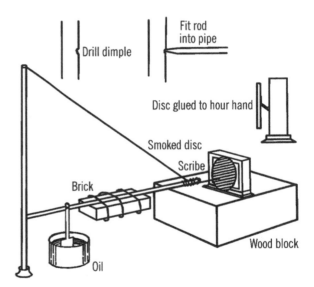

Bolt a stand supporting an alarm clock to the base in such a position that the end of the rod almost touches the outside edge of the clock dial. The minute hand of the clock should be removed. A thin cardboard disc is glued to the hour hand. The disc is smoked by holding it

---

8 Specifically, a clock with moving hands. A clock that runs on wall power (one driven by an internal AC motor) is ideal. *An alarm function is not actually needed.*

MINERAL IDENTIFICATION CHART

Mineral	Hard.	Color	Fusibility	Streak	Luster	Sp. Grav.	Formula	Acid	Bead	Special Properties
Quartz	7	Clear, white, pink, smoky	No	None	Vitreous	2.65	SiO_2	No	No	Crystals form in hexagonal prisms and pyramids. Colors due to impurities as quartz is colorless.
Feldspar	6	White, pink, gray	5	White	Vitreous	2.57	$KAlSi_3O_8$ / $NaAlSi_3O_8$ / $CaAl_2Si_2O_8$	Insoluble	No	
Hornblende	5-6	Dark green to black	4	None	Vitreous or Silky	3.2	Variable Silicate	No	No	Yields water when heated in a test tube.
Calcite	3	Clear, white, brown	No	White	Vitreous to Earthy	2.72	$CaCO_3$	Strong Bubbles in HCl	No	Transparent to opaque. Hexagonal crystal. Crystal has property of double refraction.
Limonite	5-5.5	Brown	5-5.5	Yellow-brown	Dull-earthy	3.6-4.0	$FeO(OH)\cdot nH_2O$	No	Yellow-Hot Green-Cold	Yields water when heated in a test tube. Magnetic after heating. Massive, fibrous, or porous.
Bauxite	1-3	White, gray, yellow, red	No	Brown	Earthy	2.0-2.55	$Al(OH)_3$ (and others)	Insoluble	No	Yields water when heated in a test tube. Moisten with cobalt nitrate and specimen gives off a blue flame upon burning. Chief ore of aluminum. Earthy odor.
Pyrrhotite	4	Brownish-bronze	3	Black	Metallic	4.58-4.65	FeS to Fe_4S_5	Soluble in conc. HCl	Yellow-Hot Green-Cold	Decomposes in conc. HCl. Gives off hydrogen sulfide gas.[a] Only magnetic sulfide.
Cinnabar	2.5	Red	No	Red	Adamantine	8.10	HgS	No	No	Black sublimate in a test tube. When mixed with sodium carbonate in a tube, it gives off globules of mercury.[b] Cinnabar tastes chalky.[c]
Stibnite	2	Gray to black	1	Gray to black	Metallic	4.52 -4.62	Sb_2S_3	No	Yellow-Hot Colorless-Cold	If heated in a test tube and then cooled, it gives one ring of yellow above and one ring of red below. Tarnishes black.
Gypsum	2	White	3	White	Vitreous and Pearly	2.32	$CaSO_4\cdot 2H_2O$	Hot	No	Heated it test tube, it turns white, yields much water.
Azurite	3.5-4	Blue	3	Blue	Vitreous	3.77	$Cu_3(CO_3)_2(OH)_2$	Yes	Green-Hot Blue-Cold	Gives green solution in HCl. Add ammonia and it turns deep blue.
Tourmaline	7-7.5	Black, or brownish to bluish-black	Difficult	Uncolored	Vitreous to Resinous	2.98-3.2	Variable Cyclosilicate	No	No	Crystals prismatic, slender to barrel-shaped. Becomes electric with friction. In granite, gneiss.
Chalcopyrite	3.5-4	Brass-yellow	No	Greenish-black	Metallic	4.1-4.3	$CuFeS_2$	Decomposed by Nitric acid	Magnetic bead	Often mistaken for gold but brittle. Important copper ore. Crystals commonly tetrahedral.
Mica (Muscovite)	2-2.5	Yellowish-white	No	White	Vitreous to Pearly	2.8-3.0	$KAl_2(Si_3AlO_{10})\cdot(OH)_2$	No	No	Monoclinic, thin plates, flexible, clear.
Pyrite	6-6.5	Brass-yellow	2.5-3	Greenish/Brownish-black	Metallic	5.0	FeS_2	Insoluble in HCl	No	Magnetic when heated. Fine powder solution in strong nitric acid. Commonly in cubes. "Fool's gold."
Halite	2.5	White	Deep yellow flame	White	Vitreous	2.1-2.6	$NaCl$	See properties	No	Dissolves in water. Salty taste. Usually in cubes.
Galena	3.0	Lead-gray	Easily	Lead-gray	Metallic	7.4-7.6	PbS	Nitric acid	Magnetic Globule	Soluble in strong nitric acid. Commonly in cubes as crystal. Note specific gravity and color.

[a]Hydrogen sulfide gas is a poisonous gas with the characteristic smell of rotting eggs.
[b]Generating metallic mercury is not recommended. [c]DO NOT taste cinnabar!

over a candle flame. The soot will blacken the face of the disc in a few minutes. A ball point pen or a stiff wire is fastened to the rod. This point makes a scratch mark in the black soot as the hour hand turns the disc.[9]

Operation of Equipment: When an earthquake or severe shock occurs, the base shakes. The pendulum point (end of rod with the ball point pen) does not move, and therefore the point of the pen moves back and forth across the dial. The greater the shock the farther the point will seem to move across the dial. If the outside of the dial is marked with the hours, the time of the shock can be read by close observation of the marks on the dial. Minor shocks can be eliminated by fastening a metal plate from the bottom of the rod so that the plate hangs vertically in a pan of oil. Then the oil will slow the movement of the rod (dampen it) and you will have one or two marks across the dial instead of a large number. You can determine the direction of the earthquake if you make two seismographs and have them face crossways to each other.

An earthquake sends out shock waves that travel at different speeds. The first wave to reach your seismograph is called a pressure wave. If you note the time that lapses between the first and second wave (shear wave), you can tell the approximate distance at which the earthquake occurred. The second wave travels about 5/9ths as fast as the first wave. Charts are available which give the time lapse and the distance for each time lapse.

Can You Work Like a Scientist?

1. A seismograph works much better if the base is fastened to bedrock (the solid layer of rock underneath the topsoil and subsoil). Can you find out how far down bedrock is in your neighborhood? Does the depth to bedrock vary from place to place?

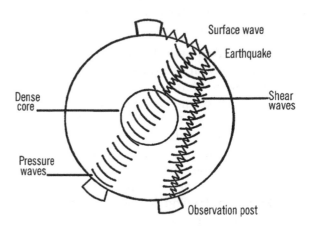

2. Earthquakes occur more often in certain places around the Earth. Can you find out where the earthquake belts occur around the Earth?

3. The Earth is often likened to a huge ball filled with molten liquid rock and metals which surrounds a dense core. Fill a balloon with water. The outside skin of the balloon is similar to the crust of the Earth. What would happen if a hole were punched in the balloon? What occurs when this happens on the surface of the Earth?

4. Tie the neck of the water-filled balloon to a bent nail inserted in a hand drill. Turn the hand drill rapidly so you whirl the balloon. Can you see why, as the Earth was cooling, the equator bulged out and the poles of the Earth seemed to flatten? The Earth contained molten materials. These materials were thrown out away from the equator as the Earth rotated on its axis.

5. Earthquakes are often caused by a pressure or push somewhere else on Earth. Push on one side of the water balloon with your finger. Does this ex-

---

9 There is probably not enough pressure to write on paper with a ballpoint pen, but a rollerball pen might work on plain (unsmoked) paper or cardboard. Would a felt-tip pen work instead? Why or why not?

tra pressure affect the balloon in another place?

6. Earthquakes help the geologist to determine what is inside the Earth and the layers of the interior part of the Earth. There are observation posts all around the Earth. When an earthquake occurs, it sends out three different types of waves. The shear waves go through the molten rock but not through the dense core of the Earth. From the drawing can you see how reports from different observation points would help pinpoint the location of the dense core?

7. Get several knitting needles (or wires) and an avocado. Stick the knitting needles through the avocado from different positions. Can you locate the core without seeing it?

8. The pressure of the Earth often pushes layers of rocks together. Since the rocks can't compress, they fold either up or down. Open a book. Push on the top and bottom of the page. Can you see an example of an upfold? A downfold? Can you find examples of these Earth activities around where you live?

9. Mix several sheets of colored paper together. These represent layers of Earth. Cut the stack and let one pile drop about a quarter of an inch. Can you see how a sudden dropping of a section of the Earth can be detected by observing rock layers? This dropping is called faulting. There are many such weak areas around the Earth where faulting occurs. Can you find where the major fault lines run in your area? Faults send out waves, which are called earthquakes.

Volcano

Purpose: This model volcano erupts and demonstrates the building of mountains.

Materials: Two pieces of cardboard (the larger the pieces the bigger the volcanoes), two small cans (baby food or fruit juice),[10] and plaster of Paris.

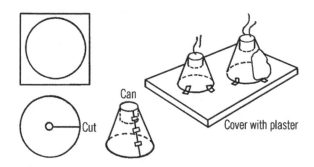

Can

Cut

Cover with plaster

Safety Tips

1. Don't handle the chemicals with your bare hands. Some people are allergic to certain chemicals of this type. You should never directly handle any chemicals. Use spoons, etc.

2. Don't inhale the fumes. You should never inhale gas fumes of any type.

3. Don't try to plug the volcano to get an explosion. Any gas fumes that are trapped build up pressure. If these fumes can't get out, they explode. The baby food can would burst, and it could be dangerous. There is no danger at all if you follow these simple precautions.[11]

---

10 Cat food or tuna cans might be good choices as well.

11 Correct in spirit. However, think it through as well: There is no danger *other* than the fire-spitting pile of flaming poisonous chemicals.

Modern Safety Practice

This project has chemicals, fire, flying sparks, and toxic fumes. If at all possible, do it *outdoors* in a place where sparks and fumes won't be a problem.

1. Follow safe-handling precautions for chemicals (Note 17 in Appendix E), including the use of protective gloves so that you do not handle the chemicals directly.

2. There will be fire. Awesome, fun fire, but fire nonetheless. Follow fire safety practice as described in Note 3 of Appendix E.

3. The fumes are dangerous, so excellent ventilation is required. *Ammonium dichromate is both toxic and a carcinogen.* Again, it is best to perform this experiment *outdoors*.

What to Do: Make as large a circle as possible on the cardboard. Cut the circle out. Cut the cardboard as shown. Make the hole in the center so that the cardboard will fit tightly around the baby food can. Bend the cardboard into a cone and insert the baby food can. Hold the cardboard together with tape or staples. Make a second, similar cone. Fasten them down to a wooden base with masking tape. Mix up plaster of Paris into a thick paste. Cover the cones with plaster of Paris. Shape the outside of the cones like two mountain peaks, making a rough finish. Put some plaster of Paris inside the cans so the opening in the can is only about an inch or an inch and a half deep. Allow the plaster to dry and then paint the volcano with a mixture of colors.

Operation of Equipment: Fill the opening of the volcanoes with about two spoonfuls of a chemical, ammonium dichromate. This chemical is used for photography and is available in photographic supply houses.[12] It can also be purchased from scientific supply houses. Lay a lighted match on the top of the ammonium dichromate. The chemical will ignite, and the volcano will start to erupt. Bits of magnesium powder, added to the ammonium dichromate, gives a more spectacular effect.

Can You Work Like a Scientist?

1. Does the volcanic ash go way up in the air? How high does the ash go on a real volcano? Find out about the volcano that erupted on the island of Krakatoa in the South Pacific.

2. Could such action of a volcano have caused the start of the ice ages?

3. Can you find the cinder cone? Does your mountain grow? Where is the lava?

---

12 That was true back in the days *before* digital photography! See Appendix A for current sources.

Anemometer

Purpose: The anemometer is used to measure wind speed from any direction. The weather bureau uses one with three cups. As these cups turn, a dial connected electrically to the anemometer shows the wind speed to the weatherman inside the weather station.

Materials: Milk carton, four paper cups, medicine dropper or short piece of glass tubing, staples or paper clips, and a board with a large nail for the base.[1]

What to Do: Cut the four corner strips off a milk carton as shown with the dotted lines. Slip the folded ends of each pair of strips together. This should make two long strips. Lay one of the long strips on the table. Place the glass part of an eyedropper or a sealed end of a piece of glass tubing on the middle of the long strip. Lay the other long strip over the top of the eyedropper and fasten the two strips around the eyedropper with staples or paper clips. Bend the four arms out as shown. Cut a slot in each of the four paper cups and insert the end of one arm in each cup. Drive a spike through a

piece of board and then drop the eyedropper over the end of the spike so that the cups will turn freely on the spike. This wooden base can be nailed to the top of a post outside. Color or mark one cup so that it is easy to see.

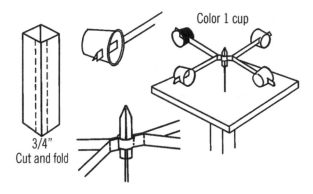

3/4"
Cut and fold

Color 1 cup

Operation of Equipment: In order to measure the wind speed, count the number of turns in 30 seconds and divide by five. This will give you the wind speed in miles per hour. Each anemometer is a little different. A more accurate way to calibrate the wind speed is to take a ride in a car on a calm day. Stick your anemometer out the car window[2] and count how many

---

1 Could you design a version of this instrument that could be 3D printed instead?

2 In the modern era, we are normally trained not to stick appendages out of a car window, because it's an excellent method of creating a rapid unscheduled amputation. *Use appropriate care.*

times it turns in 30 seconds at, say, 5 miles an hour. Then count the number of turns when the speedometer shows ten miles an hour. You can do this for any speed and make an accurate table.

Can You Work Like a Scientist?

1. Can you think of a way you could connect up your anemometer to show the speed electrically on a dial?

2. What area near your home or school is the windiest? The calmest?

3. How does the wind vary for you and your friend who lives several miles away?[3]

4. What has wind speed to do with humidity (remember your hygrometer)?

5. What month of the year is the windiest? (Keep a chart for several months.)

6. How does your knowledge of wind speed help you predict weather?

7. With your wind-chill chart ("Wind Chill Table" on page 179) can you use your anemometer to help you decide what kind of clothes to wear on a given day?

Rain Gauge

Purpose: A rain gauge is used to measure the amount of rainfall over a period of time, for instance, 24 hours. The difficulty in measuring rainfall is that it seldom rains enough to be measured accurately with a ruler. Rainfall is usually measured in hundredths of an inch. This rain gauge magnifies the readings, so that the amount is measurable.

Materials: Baby bottle, milk carton, and a three-inch funnel.

What to Do: Cut a slot about ½ inch wide down the middle of one side of the carton. Fasten a piece of adhesive tape along this slot. This will serve as your measuring scale. Cut out part of the top of the carton so you can slip a baby bottle into it. Then a funnel is placed on top of the carton so that the rain, going into the funnel, will drain into the baby bottle.

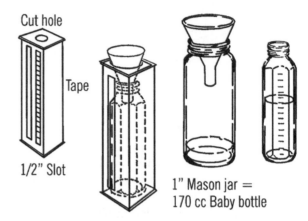

Cut hole

Tape

1/2" Slot

1" Mason jar =
170 cc Baby bottle

Operation of Equipment: The funnel is larger than the baby bottle in diameter, so an inch of rain going into the funnel will be much deeper than an inch in the baby bottle. The large funnel and the small baby bottle magnify the reading. In order to place accurate measurements on the scale, we must find out how high one inch of rain through the funnel will show in the baby bottle. To do this, we need a bottle whose opening is the same size as the funnel. Then one inch of water in this bottle would be equal to one inch of rainfall going through the funnel. We then pour this water into the baby bottle and mark the height as one inch. A wide-mouth mason jar has the same opening as a three-inch funnel. One inch of water in this jar when poured into the baby bottle comes up to 170 cc. Mark 1" on the tape at this spot. Divide the inch into tenths. These tenths can be redivi-

---

3 Related question: How detailed are the wind speed maps available online?

---

ded into ten parts so your rain gauge will measure hundredths of an inch.

Can You Work Like a Scientist?

1. How can you magnify your readings more?

2. Can you keep a chart of rainfall for the month and record the rainfall each day? Or what day does it rain the most?

3. Can you compare months with each other? In what month does it rain the most?

4. Does it rain as much at a friend's house many miles away as it does at your house? Does it rain as much in the country as the city?

5. Do you know someone who lives in another state? Could you compare your readings with theirs?

6. In order to keep a daily record, would you have to record the rainfall at the same time each day?

7. How do your readings agree with those of the weather bureau?[4] Could you both be right?

Wind Vane

Purpose: The wind vane detects the direction from which the wind is blowing close to the surface of the ground.

Materials: Paper milk carton, paper clips or staples, short piece of glass tubing or an eyedropper, spike, and wood for the base.

What to Do: Cut off a corner strip of the carton as shown. Shape the cardboard into an arrow. Be sure the tail is very large. Seal the end of a short piece of glass tubing or use a medicine dropper. Place the tubing between the double

shaft of the arrow and keep it in position with staples or paper clips. The base to hold the wind vane can be a large nail (spike) driven through a board.

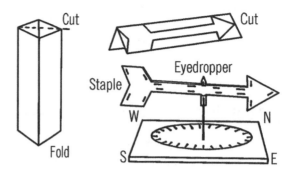

Operation of Equipment: You can use the wooden base with your wind vane almost any place. Just set the glass tubing or medicine dropper over the nail. This type of base can also be mounted on the top of a post by nailing down through the wood. A permanent base can be made from a coat hanger. However, this cannot be moved easily. If you do not wish to leave your weather instruments outside, you need only set up the stand. Then you can quickly set up your wind vane or anemometer to take a reading.

Can You Work Like a Scientist?

1. Can you make a simple wind vane with a pencil, thread, and a tack?

2. On the base you can write down the type of weather to expect if the wind is blowing from a certain direction. From what direction does most of the rain come?

3. Does your wind vane show the same direction as your school flag?

---

4 Check online precipitation maps.

4. Why is the direction of the wind important to ships and planes? Can you think of an experiment that will show this?

5. Record the direction of the wind and the temperature over a period of time. What does the direction of the wind have to do with temperature?

Nephoscope

Purpose: The wind vane detects the direction of the wind near the ground. High above the earth the wind may be blowing in a different direction. The weather service uses weather balloons to find out this information. The clouds can serve as your weather balloons. The nephoscope helps you detect the slightest movement of the clouds.

Materials: Shaving mirror with stand (found in most variety stores for several dollars)[5] and some gum labels[6] or other paper.

What to Do: Write the directions on paper and glue onto the mirror as shown. Glue a small dot in the center of the mirror.

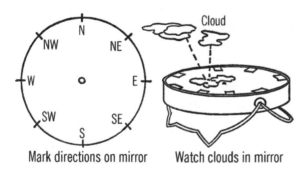

Mark directions on mirror   Watch clouds in mirror

Operation of Equipment: Set your mirror up on its stand (as shown). Turn the mirror so that north on the mirror points to north. Check your directions with a compass. Locate the tip of a cloud overhead in your mirror. Center this tip

with the center dot on your mirror. Follow the movement of the cloud in your mirror. The place where the cloud moves off your mirror glass is the direction *toward* which the cloud is moving. Since the mirror is so small, any slight movement is noticeable on the mirror.

Can You Work Like a Scientist?

1. Since most weather reports give the direction the wind is coming from, your nephoscope readings might confuse you. Could you switch directions on your nephoscope so it will give the direction the wind is coming from?

2. Record the direction of the winds for a period of time (several weeks at least). Also record weather conditions. Can you find any connection between weather conditions and the direction of the wind?

3. Could you use the nephoscope to give you the speed of the wind high above the ground? Could you make a nephoscope that could give you wind speed? What are the problems in making it accurate?

4. Record the winds high above the ground with your nephoscope. Record the winds close to the ground with your wind vane. How often are they the same? How often are they different?

5. Could you use your nephoscope on high-flying airplanes?

6. Could you use your nephoscope to observe the movement of the sun and the moon? Why do these move across the mirror? *Don't look directly at the reflec-*

---

5 Check your local dollar store. The mirror surface itself needs to be *flat* (not curved for magnification).

6 I.e., *self-adhesive* labels.

---

tion of the sun (see "Sunspot Viewer" on page 76).

7. Record both the wind direction and the temperature over a period of time. What effect on temperature does the direction of the wind have?

Beaufort Wind Scale

Wind speed	Description	Beaufort #
< 1 MPH	Smoke rises vertically	0
1-3 MPH	Smoke shows direction. Wind vane does not.	1
4-7 MPH	Vane moves. Wind felt on face.	2
8-12 MPH	Leaves and twigs in constant motion.	3
13-18 MPH	Dust and loose paper rise. Small branches move.	4
19-24 MPH	Small trees wave. Waves on small bodies of water.	5
25-31 MPH	Large branches move. Wires whistle	6
32-38 MPH	Whole tree in motion. Walking difficult.	7
39-46 MPH	Twigs break off.	8
47-54 MPH	Damage to chimneys and roofs	9
55-63 MPH	Trees uprooted. Considerable damage evident.	10
64-75 MPH	Widespread damage. Rarely experienced.	11
>75 MPH	Hurricane.	12

Can You Work Like a Scientist?

1. How accurate is this scale? Check your observations with an anemometer.

2. Can you devise your own wind scale?

3. Make a wind tunnel with an air blower or fan for your source of wind. Test the Beaufort scale with your wind tunnel.

Wind Chill Table

Scientists, at work in the Arctic and Antarctic, have found that the faster the wind blows, the colder the skin gets. When wind speed rises, the exposed skin will become much colder than when there is no wind. Scientists call this wind chill. You can use the table following to find wind chill just as scientists do.

1. Find your present temperature and wind speed.

2. Find the number at the left of the chart (below) that is the same as, or closest to, your wind speed in miles per hour.

3. Read across the top row until you find the number that is closest to what your temperature is.

4. Read the number that is in the row and column matching your wind speed and temperature. This number is the equivalent temperature reading, or what the temperature would be equal to, if there were no wind.

Example: Suppose the temperature is 16 degrees. The wind speed is 28 miles an hour. Find the number closest to what your wind speed is in the wind speed column at the left. It would be 30. Read across the top row to find the number that would be nearest the 16 degrees. This would be 15. Follow down to the intersection between that row and column. The number you find is -5 degrees. Minus 5 degrees means 5 degrees below zero. This shows that with a temperature of 16 degrees and a wind of 28 miles an hour, the effect on all exposed skin would be the same as -5 degrees with no wind. We can say that the higher the wind speed, the more wind chill there will be. Wind chill compares the temperatures at various wind speeds to the temperature when no wind is blowing.

Balloon Barometer

Purpose: The barometer measures the pressure, or push, of the air molecules around us.

Changes of pressure affect weather conditions, so the barometer is a tool to help forecast weather.

Materials: Mason jar, balloon, rubber band, and a broom straw[7] or soda straw.

What to Do: Cut a balloon open as shown. Stretch the rubber over the opening of the mason jar. Fasten the balloon down with a rubber band. Fasten a broom or soda straw to the balloon with a piece of Scotch tape. The tape should be attached to the middle of the balloon so that when the center of the balloon is forced downward, the tip of the straw will go up.

Operation of Equipment: Set the barometer on a table or window sill. Have the straw pointing to the wall or sill. Place a piece of tape on the wall or sill and make a mark each eighth of an inch. Number the mark the straw points to with the number "ten." The bottom of the scale should be zero and the top of the scale about

Temperature (°F)

Calm	40	35	30	25	20	15	10	5	0	-5	-10	-15	-20	-25	-30	-35	-40	-45
5	36	31	25	19	13	7	1	-5	-11	-16	-22	-28	-34	-40	-46	-52	-57	-63
10	34	27	21	15	9	3	-4	-10	-16	-22	-28	-35	-41	-47	-53	-59	-66	-72
15	32	25	19	13	6	0	-7	-13	-19	-26	-32	-39	-45	-51	-58	-64	-71	-77
20	30	24	17	11	4	-2	-9	-15	-22	-29	-35	-42	-48	-55	-61	-68	-74	-81
25	29	23	16	9	3	-4	-11	-17	-24	-31	-37	-44	-51	-58	-64	-71	-78	-84
30	28	22	15	8	1	-5	-12	-19	-26	-33	-39	-46	-53	-60	-67	-73	-80	-87
35	28	21	14	7	0	-7	-14	-21	-27	-34	-41	-48	-55	-62	-69	-76	-82	-89
40	27	20	13	6	-1	-8	-15	-22	-29	-36	-43	-50	-57	-64	-71	-78	-84	-91
45	26	19	12	5	-2	-9	-16	-23	-30	-37	-44	-51	-58	-65	-72	-79	-86	-93
50	26	19	12	4	-3	-10	-17	-24	-31	-38	-45	-52	-60	-67	-74	-81	-88	-95
55	25	18	11	4	-3	-11	-18	-25	-32	-39	-46	-54	-61	-68	-75	-82	-89	-97
60	25	17	10	3	-4	-11	-19	-26	-33	-40	-48	-55	-62	-69	-76	-84	-91	-98

Figure 7-5 **Wind Chill Table**

---

7 I.e., a bristle from a straw broom.

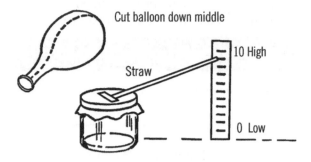

Cut balloon down middle

Straw

10 High

0 Low

twenty. Make a chart and mark down where the straw points and the time of the reading. Check your barometer again after an hour or so. Again record the findings. If the barometer seems to be rising, the weather should be fair. If the readings are lower, we have low pressure, and you may expect cloudy weather and perhaps rain.

Can You Work Like a Scientist?

1. High pressure means the air is heavier. If the air pushes more on the center of the balloon, in which direction will the end of the straw move?

2. Low pressure means the air is lighter. If the air does not push as much on the center of the balloon, in which direction will the end of the straw pointer move?

3. Pressure is caused by the weight and movement of the air molecules. Would heavy air contain more or less molecules than light air (low pressure)?

4. High pressure is caused because the air molecules are all pushed together (packed or very dense). Low pressure is caused because the molecules move farther apart. When the air molecules

are far enough apart, there is room for water molecules (vapor). Squeeze a sponge together. This represents molecules squeezed together (high pressure). Can you pick up much water with the sponge squeezed together? Now let the sponge spread out. Will the sponge hold more?

5. Take the barometer outside. Do you detect a change in pressure?

6. Check the barometer at home when you wash[8] clothes. Do you notice any difference?

7. Take the barometer into a warm room. Note any pressure changes.

8. Place the barometer in the refrigerator and leave for about an hour. Does cold air affect pressure? How? Can you explain why?

Aneroid Barometer

Purpose: The barometer detects changes in air pressure. An aneroid barometer is made from a metal container in which nearly all the air has been removed. The container is kept from collapsing by a spring coil inside the container.

Materials: The container can be made from a condensed milk can. It also can be made from a pint jar, a coiled spring, plywood, and a balloon. In either case, additional materials needed are a two-quart milk carton, a long needle or a piece of coat hanger wire, a penny, a broom straw, and some thread.

What to Do: If you use a milk can for your container, empty the milk can[9] and wash it out as best you can through the two holes. Leave about one inch of water in the bottom. Solder

8 The 1960s version of this book had "your mother washes"(!) here.

9 Use a pointy *piercing* can opener, not one that cuts the lid off.

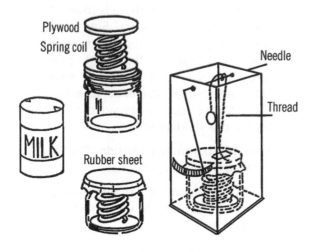

Plywood
Spring coil
MILK
Rubber sheet
Needle
Thread

glue fasten a straw pointer to the wire or the eye of the needle. Make a scale as shown.

Modern Safety Practice

If soldering the condensed milk can, the can may become very hot. Take caution to avoid burns.

Operation of Equipment: When the pressure increases, it should push down on the top of the container. This should cause the needle or wire to turn slightly. This slight movement should make the tip of the straw pointer move a great distance.

Can You Work Like a Scientist?

One reading means little. Take readings every two hours and record your findings. Compare your findings with actual weather conditions the next day or so.

Hair Hygrometer

Purpose: A hygrometer measures the humidity or moisture in the air. [11]

Materials: Two paper clips, milk carton, needle with a large eye, a human hair about ten inches long, glue or cement, broom straw, and a blank file card (any cardboard will do).

What to Do: Wash the hair in alcohol or soapy water to remove the oil. The oil coats the hair and keeps it from absorbing moisture. Rinse and allow the hair to dry completely. Cut two flaps near one end of the carton. Make holes in the two flaps as shown. Work the needle back and forth in the holes so that the needle turns easily. Cut a slot for a paper clip at the other end of the carton. Stick the end of the straw through the eye of the needle. Glue or cement

one of the holes shut.[10] Heat the can on a stove or over your alcohol burner. When the steam from the water in the can has removed most of the air, quickly remove the can from the source of heat and solder the other hole shut.

If you use a pint jar for your container, fasten one end of a coiled spring to a piece of plywood which is almost the same size as the bottom of the jar. Fasten a second piece of plywood to the other end of the spring. Place a little water in the jar. Insert and compress the spring and cover the jar with rubber from a balloon or rubber sheeting. Leave a small corner open for air to escape. Heat the bottle, and when most of the air has escaped, seal the jar by placing rubber bands over the rubber cover.

Cut off the top of the milk carton. Set the container in the bottom of the carton. Make two small holes near the top of the carton and insert the needle or coat hanger wire. This wire should turn easily. Attach a thread with tape to the top of the container inside the carton. Wrap the thread around the wire and hang a weight (penny) from the other end of the thread. With

---

10 See Note 41 in Appendix E about soldering to a can.

11 This particular hygrometer is interesting but more challenging to build and get working than the "Wet- and Dry-bulb Hygrometer" on page 183.

---

the straw in place. Make a dial on your file card as shown. You can hold the file card in place with Scotch tape or tacks. Slip one end of the hair into the paper clip and place the clip in the slot. Do not touch the hair with your bare hand as you will get oil on it. Slip the needle in place and then wrap one turn of the hair around the needle. Slip the other end of the hair into the second paper clip. Hang this clip over the end of the carton.

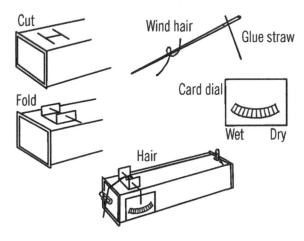

Operation of Equipment: When the hair is dry, it shrinks or gets shorter. When there is a lot of moisture in the air, the hair stretches or gets longer. You can't see this change, but because a small movement of the hair moves the needle and the end of the straw a greater distance, you are magnifying the change in the length of the hair. In order to set your hair hygrometer to make accurate readings, place it in a bucket that contains a wet towel. Cover the bucket with another wet towel, thus making the air as moist as possible. After about 20 minutes, remove the hygrometer from the bucket and quickly set your dial on 10, or 100% humidity.

Can You Work Like a Scientist?

1. What is the humidity when you take a shower?

2. Does the humidity change when there are many people in a room?

3. What effect does humidity have on static electricity?

Wet- and Dry-bulb Hygrometer

Purpose: This instrument is used to measure both the temperature and the amount of moisture in the air.

Materials: Two thermometers, milk carton, two rubber bands, and a piece of clothesline rope about 5" long. The thermometers should read alike, and the bulb should not be covered. Thermometers with cardboard backs are ideal. The clothesline rope should be the kind with soft cotton fibers inside.

What to Do: Cut a hole about three inches from the bottom of the carton. Stick the clothesline rope through the hole so only about one inch sticks out. Slip this end over one of the thermometer bulbs. Fasten the thermometers to the outside of the carton with rubber bands. Pour about two inches of water into the carton.

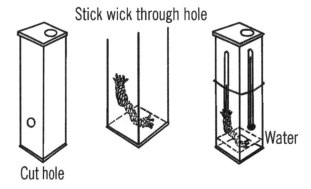

Operation of Equipment: The fine hairlike strands in the cotton rope suck up the water. Soon the water will climb up the rope and dampen the bulb of the "wet" thermometer. In order to understand what happens, wet the back of your hand. Blow across it. The water on

Dry °F	1°	2°	3°	4°	5°	6°	7°	8°	9°	10°	11°	12°	13°	14°	15°
50	93	87	81	74	68	62	56	50	44	39	33	28	22	17	12
52	94	88	81	75	69	63	58	52	46	41	36	30	25	20	15
54	94	88	82	76	70	65	59	54	48	43	38	33	28	23	18
56	94	88	82	77	71	66	61	55	50	45	40	35	31	26	21
58	94	89	83	77	72	67	62	57	52	47	42	38	33	28	24
60	94	89	84	78	73	68	63	58	53	49	44	40	35	31	27
62	94	89	84	79	74	69	64	60	55	50	46	41	37	33	29
64	95	90	85	79	75	70	66	61	56	52	48	43	39	35	31
66	95	90	85	80	76	71	66	62	58	53	49	45	41	37	33
68	95	90	85	81	76	72	67	63	59	55	51	47	43	39	35
70	95	90	86	81	77	72	68	64	60	56	52	48	44	40	37
72	95	91	86	82	78	73	69	65	61	57	53	49	46	42	39
74	95	91	86	82	78	74	70	66	62	58	54	51	47	44	40
76	96	91	87	83	78	74	70	67	63	59	55	52	48	45	42
78	96	91	87	83	79	75	71	67	64	60	57	53	50	46	43
80	96	91	87	83	79	76	72	68	64	61	57	54	51	47	44
84	96	92	88	84	80	77	73	70	66	63	59	56	53	50	47
88	96	92	88	85	81	78	74	71	67	64	61	58	55	52	49
90	96	92	89	85	82	78	75	72	68	64	62	58	56	53	50

Difference between Dry- and Wet-Bulb Thermometers

Figure 7-10 **Table Showing Relative Humidity**

your hand evaporates, and your hand seems cooler. This is what happens to the wet bulb thermometer. The rate or speed of evaporation depends on how much water is in the air and the temperature of the air. To use this instrument, fan the wet bulb with a piece of cardboard. Check the temperature. Note the temperature of the dry thermometer and the difference in temperatures between the two thermometers. Use this information with your humidity chart to find the relative humidity (amount of moisture in the air compared to the amount that the air could hold at that temperature). The relative humidity is given as a percent.

Can You Work Like a Scientist?

1. Place a drop of water and a drop of alcohol on the back of your hand. Blow across each. Which feels cooler? Which one evaporates quicker? What has the rate of evaporation to do with cooling?

2. Note the readings on your thermometers when you use water. Then, instead of water, use alcohol. Could you make a humidity chart based on the evaporation of alcohol?

3. Try your gauge in the bathroom when someone is taking a shower. As the humidity increases, what happens to the mirror?

4. How is humidity related to temperature? What effect does wind have on your wet-and dry-bulb hygrometer?

What Determines Weather

Daily Weather Record[12]

Date	Air Pressure (barometer)	Temperature (thermometer)	Humidity (Hygrometer)			Wind Direction (wind vane)	Weather (use map symbols)
			Wet Bulb	Dry Bulb	Relative Humidity		

Figure 7-11 **A Daily Weather Record Chart**

12 You can use Figure 7-11 as a "template" to make your own daily weather chart. What other things have you learned to measure that might have some bearing on predicting the weather?

Mercury Barometer

Purpose: A barometer is used to measure air pressure (the push exerted on an object by the weight and movement of all the air molecules above it). A mercury barometer is quite accurate because it is not affected by changes in temperature as are many other types of barometers.[13]

Materials: A piece of 6 mm glass tubing at least 32 inches long, six ounces of mercury,[14] one-half-inch piece of rubber tubing, test tube, yardstick, rubber bands, wood to make the supporting stand as shown, and an eyedropper.

What to Do: Build the stand as shown. The stand should be tall enough to support the tube of mercury and the yardstick. Heat and seal the end of the glass tube in the flame of a gas burner. The double wick alcohol burner ("Broad Flame Alcohol Burner" on page 9) will produce enough heat to seal the tubing. Be careful to turn the tubing while you are heating it so that the glass will not crack. A slight leak will make the readings of your barometer inaccurate.

Use an eyedropper to fill the tube with mercury. The smaller the hole in your dropper, the easier it is to fill the long glass tubing with mercury.

Tip down the sealed end of the long glass tubing so that the mercury will run slowly down the tube. Use your eyedropper to add the mercury to the open end of the tube as shown. If the drops of mercury are small enough, the mercury will flow down to the bottom of the tube. If the drops are not small enough, you will have to shake them down to the bottom of the tube. Fill the long tube completely with mercury. This is to remove all the air.

Slice the short piece of rubber tubing lengthwise so that it will fit over the opened end of the tubing. Press the test tube down on the rubber tubing and turn the glass tubing over. Add a few drops of mercury to the bottom of the test tube and then release the pressure on the glass tubing. The glass tubing will rise slightly, and some mercury will drop down into the bottom of the test tube from the tubing. The level of the mercury in the glass tubing should be between 29 and 30 inches above the level of the mercury in the test tube. Fasten the test tube and the glass tubing to a yardstick or meter stick with rubber bands. The yardstick or meter stick is then attached to the wooden upright with small brads or screws.

Safety Tips

Mercury is very poisonous. Never touch it with any part of your skin. Mercury will enter the body through any cuts or opening in the skin.

Modern Safety Practice

Mercury is no longer considered safe to work with, even if you don't touch it with your skin. See Note 42 in Appendix E about building a water barometer instead.

Operation of Equipment: Compare your reading with that of the weatherman or of a standard barometer. If the correct reading is 30.2 and your reading is 29.5, either raise the glass tubing slightly or lower the yardstick until the level of the mercury matches the correct mark on the measuring stick. Always tap your barometer before you take a reading. This will

---

13 Mercury may be affected *less* by changes in temperature than other types, but it is not unaffected. If mercury did not change volume/density with temperature, would mercury thermometers work?

14 For better or worse, that's no longer so easy to get. See *Modern Safety Practice* for this project.

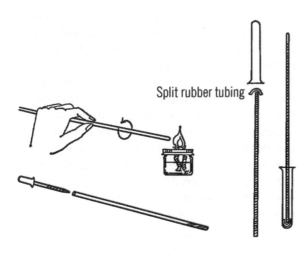

Split rubber tubing

allow the column of mercury to move slightly up or down and settle at the correct level.

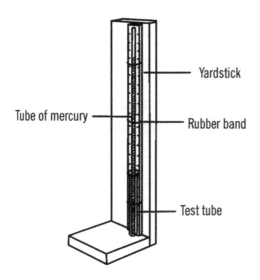

Yardstick

Tube of mercury

Rubber band

Test tube

Can You Work Like a Scientist?

1. The atmosphere will hold up nearly 30 inches of mercury. The density of mercury is about 13 times that of water. How high a column of water will the pressure of the atmosphere support?

2. Can you find out the density of other liquids such as alcohol and then determine how high a column of the liquid you would need in order to make a barometer?

3. Could you make a water barometer similar to the mercury barometer by using plastic garden hose for your column?

4. Is your mercury barometer affected by altitude? Take readings at various floor levels in a tall building. Can you calibrate your barometer for different floor levels? This instrument might be helpful for an elevator operator.

5. Why do you need a vacuum at the top of the glass tubing? Make one barometer and let some air bubbles in through the mercury. Bring a heat source near the top of the barometer. Does the mercury level change? Try the same experiment after you have created a vacuum at the top of the tube by turning the tubing over and not allowing air bubbles to enter.

6. Does mercury expand (get larger) or contract when heated?

7. What material will mercury wet or stick to?

8. Can you make a large drop of mercury by pushing small drops of mercury together with a pencil? Try the same experiment with drops of water on a piece of wax paper.

9. Can you compare the readings of your mercury barometer with actual weather conditions over a period of a month or more?

Chemical Weather Glass

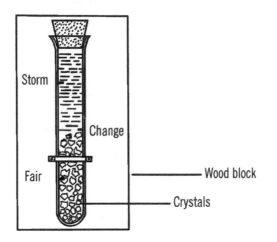

Tube full of crystals	Storm
Dim liquid	Storm
Crystals rising and nearing change mark	Weather change (probably a storm)
Crystals falling below the change mark	Weather change (probably better weather)
Crystals near the bottom	Fair weather (heavy frost in the winter and high humidity in the summer)
Clear liquid	Fair weather

Purpose: The chemical weather glass is another instrument that helps you predict the weather.[15]

Materials: Test tube, about two ounces of water, two ounces of pure alcohol, ½ dram of potassium nitrate, ½ dram of ammonium chloride, and two drams of camphor. A dram is about an eighth of an ounce. (A baby bottle is marked off in ounces.)[16]

What to Do: Mix the chemicals in the test tube. Fill the test tube up to about one inch from the top. Seal with a cork. Mount the test tube on a wood block with either rubber bands or strips of tin for brackets. Make a scale on the wood block as shown. The mark at the bottom should be about 1¼" from the bottom of the test tube. Measure up another inch and a quarter and label the mark CHANGE. Measure another inch and a quarter up and label the mark STORM.

Operation of Equipment: Can you predict the weather from the following chart?

Modern Safety Practice

Note 17 in Appendix E discusses safety when working with chemicals.

Can You Work Like a Scientist?

1. How does the chemical weather glass work?

2. Why should crystals form when a storm is coming?

3. What affects the weather glass, humidity or pressure? Can you test your guess (hypothesis)?

4. Keep a weather chart. How accurate is the chemical weather glass?

---

15 This is a traditional weather forecasting device also known as "FitzRoy's Storm Glass." But *does it actually work?* See Note 43 in Appendix E.

16 See Appendix A for sourcing information.

Air Current Detector

Purpose: To detect slight movements of air, such as warm or cold air currents that are constantly moving.

Materials: Pencil with eraser, soda straw, pin, paper clip, sheet of paper about 2½×4 inches, and a piece of Scotch tape.

Helpful Hints for Building: Scotch tape a piece of paper on one end of the straw as shown. Turn the end of the straw so that the paper is horizontal (like the top of a table). Pinch the other end of the straw and slip a paper clip on it about one inch from the end. The paper clip should be vertical (straight up and down). Now stick the pin through the side of the straw and into the eraser of a pencil. The paper tail and the paper clip should just balance. Move the paper clip until the balance is as perfect as possible.

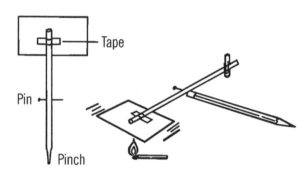

Operation of Equipment: Hold a match under the paper tail of the warm air current detector. Be sure the match is not close enough to burn the tail. The detector should move if it is properly balanced. Faint currents of air push against the tail of the detector and upset the balance. The tail will either go up or down depending on the direction of the air current.

Can You Work Like a Scientist?

1. Does the detector tail always go up, or does it sometimes go down?

2. What kind of air makes the tail go up? What currents make it go down?

3. Hold the detector near a cold windowpane. Can you explain what happens?

4. Hold the detector above a light bulb or a candle. How far away can you get from the bulb and still have the detector move?

5. Open the refrigerator door a few inches. Hold the detector near the bottom of the door. Can you explain what happens? What about the top of the door?

6. Place a sheet of white paper and a sheet of black paper on a window sill that gets lots of direct sunshine. Move the detector slowly over the black paper and then over the white. Can you explain what happens?

7. From question 6, can you explain air currents over a field of dark soil and over a light-colored concrete road?

8. Hold your detector over your bare forearm. If your detector is finely balanced,

the detector will rise. Why? Where did the heat come from?

9. Try your detector over cold water and then over warm water. Can you explain the difference?

10. Can you use your detector to plot the warm and cold air currents in your room?

11. Can you make the detector more sensitive by increasing the size of the tail? Would the detector be more sensitive if you used something longer than a straw?

Air Thermometer

Purpose: A thermometer is used to measure temperature, or the heat energy of air molecules. The one you are about to make is the same kind that Galileo, an Italian scientist, made over 350 years ago.[17]

Materials: Light bulb or baby bottle, one-hole rubber stopper, glass tubing about 1½' long, and a pan holding water.

What to Do: You can make either of two kinds of air thermometers. The easiest to make is the one with the baby bottle or bulb on the bottom. If you use an old light bulb, remove the inside of the bulb as explained in the light bulb chemistry flask section ("Light Bulb Chemistry Flask" on page 5). Fill the bulb or bottle about half full of cold water. Put some food coloring or ink in the water. Insert the glass tubing through the stopper and all the way into the bulb or baby bottle. Insert the stopper as shown. Place a cardboard scale on the back of the tubing and hold it in place with Scotch tape. You can calibrate the scale later by comparing your readings with that of a regular thermometer. As the water in the bulb warms up to room temperature, it warms the air in the bulb. The air molecules move faster and push harder on the water. This pushes the water up the tube.

A second type of air thermometer has an inverted bulb. Heat the bulb so that some of the air will leave the bulb. Quickly insert the rubber stopper in the bulb or bottle, and stick the other end of the tubing in a pan or jar containing colored water. As the air in the bulb cools, the molecules don't move as fast, and the water pushes part way up the tube. Fix a stand to hold the bulb or bottle and again attach cardboard for a scale.

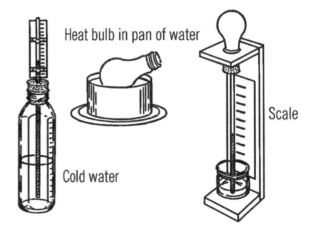

Heat bulb in pan of water

Cold water

Scale

Operation of Equipment: After the air or water has reached room temperature, make a mark on the scale for the height of the liquid. Use another thermometer to give you the correct reading and mark it on the scale. Try your thermometer with different room and outdoor temperatures. Again calibrate with readings from a standard thermometer.

---

17 This apparatus is useful to build for the experience and understanding, but it is not very useful as a thermometer. (Why?) Use a modern thermometer (glass or digital) when you need to monitor temperatures.

Can You Work Like a Scientist?

1. Does air pressure affect your readings? Which thermometer would be the most accurate?[18]

2. Try your thermometer directly in the sun. Is it accurate under direct sunlight? Check with a standard thermometer.

3. How high up a tube can you make the water climb by heating the air?

Air

Cloud

Alcohol or water

Cloud Jar

Purpose: The cloud jar is used to produce artificial clouds in the classroom in order to study the causes of clouds and do experimentation with clouds.

Materials: Gallon jug, one-hole rubber stopper (#6½),[19] tire or hand pump, alcohol[20] or ditto fluid.

What to Do: Pour a little alcohol into the bottom of the jug (you may use water if alcohol is not available). Insert the stopper into the jar. Pump air into the jar. Remove the stopper quickly.[21]

Operation of Equipment: When you remove the stopper quickly, a cloud should form.[22]

Can You Work Like a Scientist?

1. Is the alcohol just in the bottom of the jar? After you pour the alcohol in the jug, smell the opening to the bottle.[23]

Modern Safety Practice

Wear safety glasses, just in case the glass should break. If using alcohol, follow procedures for working with chemicals, Note 17 in Appendix E, and keep away from sparks or flames.

Where does this smell come from? Can you see any other alcohol besides that in the bottom of the jar? How long does it take for the alcohol to evaporate?

2. If you plug the hole in the bottle, will all the alcohol completely evaporate? Why?

3. When you pump air into the bottle, do you increase the air pressure? When you compress air, do you increase the temperature? Will warm air hold more vapor (bits of water or alcohol) than cold air?

---

18 Hint: Compare this apparatus with the one in "Air Barometer" on page 191.

19 A 2-liter soda bottle with a #3 stopper is a good substitute.

20 See Note 6 in Appendix E about alcohols.

21 You will likely need to hold the stopper in place by hand; it will want to pop out on its own. If your pump has a gauge, aim for about 20 PSI of pressure.

22 How is this different from the "Diffusion Cloud Chamber" on page 103?

23 This is improper procedure. Waft the air near the opening to the bottle towards your face, using your hand. *Never smell chemicals directly.*

---

4. When you put extra air into the bottle, does it hold more vapor?

5. When you pull out the cork, what happens to the extra air in the bottle? Does this raise or lower the temperature in the bottle? When air expands, does the temperature increase or decrease?

6. Does cooler air hold as much moisture?

7. What happens if air is suddenly cooled?

8. After you have formed your cloud, insert the rubber stopper and pump air into the jug. What happens to the cloud? Why?

9. If you don't have a pump, try blowing into the bottle. Can you think of a way to make a permanent cloud jar that will make clouds and destroy them as you wish?

10. Put some water in a bucket. Drop in some pieces of dry ice.[24] Does a cloud form? Can you explain what happens?

11. Put some dry ice in a can. Pour some alcohol on the ice. Let the can set for 15 minutes. Does snow form on the can? What causes snow? *(Be careful-don't touch dry ice or a mixture of dry ice and alcohol.)*

Air Barometer

Purpose: The air barometer is used as a tool to measure the pressure of the air and thus predict future weather. The barometer is affected somewhat by temperature, so keep it in a place where the temperature doesn't change very much.

Materials: Small 15-watt light bulb, #1 rubber stopper (one-hole), glass tubing about 9" long (5 mm.), a jar, and wood for a stand.[25]

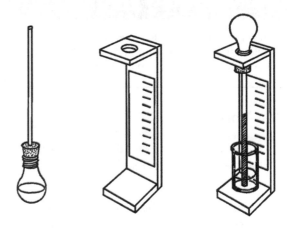

What to Do: You can blow a small ball on the end of your glass tubing (see "Large Pipette-Glass Blowing" on page 12), or you can use the light bulb and stopper. If you use the light bulb, remove the inside (see "Light Bulb Chemistry Flask" on page 5). Insert a #1 one-hole stopper as shown. Fill the bulb about one-fourth full of water. You can color the water with food coloring or ink. Insert the glass tubing into the stopper. Use soapy water on the glass tubing so the tubing will slip easily into the stopper. Turn the bulb and tube over so the small end is in a glass full of water. Attach a small strip of cardboard to the tubing with Scotch tape. Mark your scale by using a regular barometer to take sample readings. If one is not available, make division marks on the card and number from 1 to 10.

Operation of Equipment: When the air pressure increases, it pushes down on the water in the glass and forces the water up the tube. When the air pressure decreases, the pressure

---

24 See Note 14 in Appendix E about working with dry ice.

25 For an alternative design, see the "weather glass," Note 42 in Appendix E.

in the bulb is greater and forces the water down the tube.[26]

Can You Work Like a Scientist?

1. Place your hands on the bulb. What happens? Is this caused by a pressure change?

2. Keep a chart of the readings on your barometer over a period of time. Check your barometer every three hours. Also keep a record of the weather. Does your barometer help to forecast the weather? Remember, your readings forecast changes to occur in the weather one or two days ahead.

3. Place your barometer in a room containing lots of water vapor in the air. What happens? Why?

4. How can air be lighter when it contains water vapor? The bits of air (molecules) are pushed so closely together when the air pressure is high that there is little room for water vapor. When the pressure decreases, the bits of air move farther apart and can hold water vapor. Water vapor molecules weigh about half as much as the average air molecule. Can you demonstrate this using a sponge? When the sponge is squeezed together, this is high pressure. When the sponge is spread out we have low pressure. Which holds the most water?

Convection Current Box

Purpose: This box is used to study air currents caused by convection. When gases such as air are heated, the molecules move more rapidly. They spread out over a larger area and are not as tightly bunched together (dense). This warm air then contains fewer molecules and is lighter then the surrounding cooler air. The warm air, being lighter, rises, and the cooler air moves in to take the place of the rising warm air. Heat being moved or transferred because of unequal heating is known as convection.

Materials: Plywood or cardboard box, two quart jars, candle, and a roll of paper.

What to Do: Make the box as shown. Cut two holes just smaller than the diameter of the two mason jars. Cover the open side of the box with Saran Wrap or a piece of glass. Place a candle in one of the holes as shown.

Cut off the bottom of the two jars with your bottle cutter (see "Bottle Cutter" on page 18). Place the jars on the box as shown.

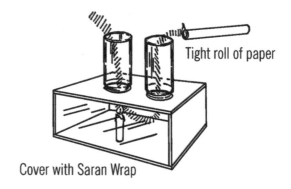

Tight roll of paper

Cover with Saran Wrap

Operation of Equipment: Light the candle. Set the jars in place. Roll up some newspaper into a tight roll. Light the newspaper. The smoke should be sucked down the one jar chimney and up out the other one.

---

26 How is this apparatus different from that of the "Air Thermometer" on page 189? What does that tell you about how the two instruments behave?

Modern Safety Practice

When lighting things on fire, use care, safety glasses, a non-flammable surface, and an area with good ventilation. And, have a fire extinguisher handy, just in case.

Read Note 3 in Appendix E for additional guidance on safety around fire.

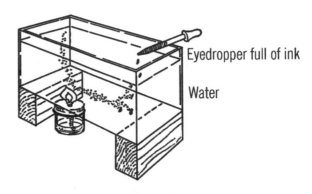

Eyedropper full of ink

Water

Can You Work Like a Scientist?

1. Place a thermometer on the top of each jar. What is the difference in temperature of the two columns of air?

2. What is the effect on the strength of the convection current if the amount of heat given off by the candle is increased?

3. What effect do convection currents have on our weather?

4. What could cause the convection currents that affect our weather? What causes the unequal heating on the surface of the Earth?

5. Can you prove there are convection currents in your house or classroom?

6. Do convection currents occur in liquids as well as gases? Set up an experiment in an aquarium or a large pan of water. Use ink to trace the path of the water currents.[27]

Sunshine Recorder

Purpose: A sunshine recorder[28] is used to measure and record periods of bright sunshine throughout the day.

Materials: Large burned-out light bulb, tin can, cardboard (2" × 10"), wood for base and support, and a rubber stopper to fit the light bulb.

What to Do: Remove the insides of a large light globe (see "Light Bulb Chemistry Flask" on page 5) such as is used in an auditorium or gymnasium. The clear light globe is ideal.[29] Fill the light globe with water and insert a rubber stopper or cork.

Drill a hole in the wooden base so that the neck of the light globe fits in the hole. This sphere focuses the rays of the sun onto the cardboard scale.

Examine a globe of the Earth or a map to find out at what degree of latitude your city is located.[30] Cut a triangular support board so that the

---

27 Small aquariums are normally made from soda lime (window) glass, which does not handle changes in heat very gracefully. (By contrast, Pyrex, or borosilicate glass, is frequently made into glass stovetop cookware.) Thus, putting an open flame directly under an aquarium could crack the glass, resulting in the release of its water. Can you think of a way to heat part of the bottom, but that would be safe for the glass?

28 This particular type of sunshine recorder is known as a *Campbell–Stokes recorder*. Commercial versions are typically made with a glass sphere and metal support.

29 That is, a "globe" style (near-spherical) incandescent light bulb with a clear—not frosted or silvered—glass envelope. Globe bulbs are sometimes called "vanity" bulbs. The "G25" size bulb is a good choice.

30 Alternately, use an online *geocoder* service and enter your address.

sharp angle near the base is equal to the degree of latitude.

Cut a tin strip from a can that is large enough to hold the cardboard strip. The tin strip should be carefully formed so that it follows the curve of the light bulb. A cardboard is fastened on the tin strip so it can be easily removed and replaced by another piece of cardboard.

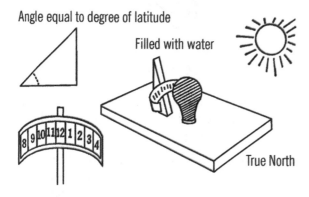

Angle equal to degree of latitude

Filled with water

True North

Operation of Equipment: On a sunny day place your sunshine recorder so it faces true north. Move your support and tin strip until the sun's rays focus on the cardboard. The support is then screwed to the base. The tin strip is then adjusted and fastened to the support.

You can mark the position of the sun's rays each hour on the piece of tin. The longer the tin and cardboard the longer the period of time you can record the duration of sunshine. Usually you will find a period of six to eight hours adequate and easy to check.

The rays of the sun focus on the cardboard strip and burn a line on the cardboard. The longer the period of sunshine, the wider the line going across the cardboard. You can measure somewhat the intensity of the sunlight by the degree of burning. Each mark on the cardboard can be subdivided so that the sunlight can be measured in tenths and even hundredths of an hour.

Each day you must change the cardboard or move it so that a fresh path can be burned across your chart.

Modern Safety Practice

When used correctly, this instrument will generally darken the cardboard, burning the amount of sunlight into the cardboard without causing a flame that spreads. Nonetheless, take precautions as though the cardboard could catch fire. It may be wise to use a fireproof base and locate it on a safe concrete area that can't burn. You may also want to keep a close eye on it during its first use, just in case.

Can You Work Like a Scientist?

1. Why do you use a round lens such as the large light bulb instead of a regular lens?

2. What month of the year do you have the most days of sunshine?

3. How does the amount of sunshine and the duration of sunshine each day affect the growth rate of various plants, seeds, and trees?

Radiometer

Purpose: A radiometer is an instrument that can be used to measure indirectly the amount and intensity (strength) of sunlight. It has many other uses as a measuring device.

Materials: You can purchase a ready-made radiometer for about ten dollars from various sources including scientific supply companies. If you wish to make one, the materials needed are: pint mason jar, cork, long needle, glass eyedropper, 2" square of very thin cardboard, jar lid adapter, and a one-hole rubber stopper.

What to Do: Seal the small end of the eyedropper in the flame of an alcohol burner. After the dropper cools, cut a one-inch piece off the end with a file. Punch a hole in the center of the white cardboard square so that the cardboard will fit snugly over the tip of the eyedropper. Remove the cardboard and make the cuts shown with the dotted lines. Bend the cardboard tips down along the solid lines to form vanes. Blacken the back side of each vane by holding the cardboard carefully over a candle. Mark one white side with an "x."

Stick the large end of the needle into the cork and place the cork in the mason jar. Set the eyedropper tip and cardboard vanes over the needle. Fasten the jar lid adapter (see "Tripod and Adjustable Rings" on page 14) to the mason jar with a jar ring. Connect a vacuum pump hose to the rubber stopper and remove as much air from the jar as possible. Seal the opening to the stopper.

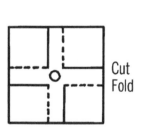

Cut
Fold

Operation of Equipment: Since most of the air has been removed from the inside of the radiometer, the paddle wheel or vanes are free to move. Normally the paddles would strike air molecules and be slowed up or stopped. With most air molecules removed, the slightest energy will start the paddles turning.

Can You Work Like a Scientist?

1. Will the energy from a light cause the paddles to turn?

2. Will heat energy (a match) cause the paddles to turn?

3. Does the paddle wheel always turn in the same direction? Does the energy seem to push the black or the white sides? Light and heat energy seems to travel in bundles somewhat like bullets. This energy striking the black side is absorbed or taken in. The energy striking the white side is repelled or reflected. Can you see why there is more push on the black side?

4. Do the paddles seem to turn faster when the light increases? Could you use this paddle wheel to measure the amount and strength of sunlight?

5. If you change electrical energy to light, can you measure the electrical energy by using the radiometer?

6. Can you measure the strength of heat or light by the rate the paddle wheel turns? How does the strength decrease as you increase the distance from the radiometer to the light or heat source?

Weather Balloon

Purpose: The weather balloon is used to carry various weather instruments high up into the atmosphere in order to record weather conditions at various altitudes.

Materials: Plastic cleaner bag.[31]

What to Do: The plastic bag should be filled with a light gas. Helium is non-explosive and

---

31 A lightweight, clear plastic trash can liner bag is an excellent choice. Would there be any advantage or disadvantage to using a black bag?

can be secured at industries which store and distribute bottled gas.[32] Hydrogen is easier to secure in large amounts but is somewhat explosive, especially if mixed with oxygen.

You can generate your own hydrogen, as mentioned in the chemistry section under hydrogen gas generator. Aluminum foil and hydrochloric acid produce the greatest amount of gas in the shortest time. Care should be taken that the bottle used to mix the acid and the foil can stand a great amount of heat, since much heat is given off during the reaction.

After the bag is filled with gas, the bag is tied. Light-weight instruments such as maximum and minimum thermometers can be attached to the plastic bag.[33]

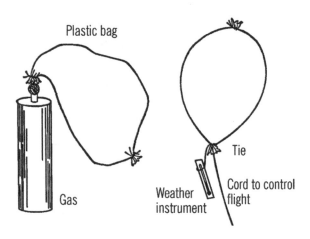

Plastic bag

Gas

Tie

Weather instrument

Cord to control flight

Operation of Equipment: A strong line or cord can be attached to the balloon. When the balloon is released, the cord should be eased through the fingers. Every ten feet can be marked on the cord so altitude can be determined. The plastic balloon can be recovered by reeling in the cord.[34]

Modern Safety Practice

1. A bag filled with hydrogen can make a very impressive explosion, which could cause injuries. If you use hydrogen, take extreme care to keep it away from possible sources of heat or flame.

2. Helium gas presents no explosion hazard, but does present a danger of asphyxiation. Never fill balloons or otherwise open a tank of helium without excellent ventilation. Helium can displace air in the room. Humans (and other living things) require the oxygen in air in order to breathe.

Can You Work Like a Scientist?

1. Can you measure the lifting force of your balloon by adding weights to the balloon until it is just held to the ground?

2. What effect does altitude have on the temperature?

3. What are the problems in trying to record weather data from a balloon?

4. How does the weather bureau secure its information?

5. How high can you get your balloon to rise?

6. Does your plastic balloon increase or decrease in size as it rises up into the atmosphere?

7. What effect does altitude have on air pressure? Can you use your balloon to prove this?

---

32 Helium can now be obtained easily from party stores, to fill balloons.

33 A maximum and minimum thermometer is a special type of thermometer (mercury or digital) that records the maximum and minimum temperatures experienced.

34 See Note 44 in Appendix E about cords and tethered balloons.

---

8. Can you devise a way of parachuting instruments or materials from your weather balloon?

9. Will your balloon rise if you fill it with hot air instead of with the gases mentioned?

10. What can you tell about upper air currents from the movement of your weather balloon?

Problems to Investigate in the Study of Meteorology (Weather)

(P)-Primary

(I)-Intermediate

(U)-Upper

1. Are there different kinds of clouds in the air? What are the names of these various clouds? *(P)*

2. What are clouds made of? Can you make a cloud? *(P)*

3. Where does rain come from? What causes rain? *(P)*

4. How are clouds formed? Are all clouds the same? *(P)*

5. What causes snow? Can you produce snow? *(I)*

6. What happens to snow when it melts? How much snow is necessary to make an inch of rainfall? *(I)*

7. Why is the sky different colors at different times? *(I)*

8. What is fog? Can you make fog? Blow across dry ice. *(P)*

9. Is soil always the same temperature? How about water and air? *(P)*

10. How often do weather conditions change? What causes these changes? *(P)*

11. Do winds warm or cool the earth? Why? *(P)*

12. From which direction does the wind blow most frequently where you live? *(I)*

13. What percentage or part of the time does the wind blow from the different directions around your home or school? *(I)*

14. How do the seasons of the year vary around the world? Why? Can you duplicate these conditions? *(U)*

15. Do all clouds form at the same heights in the sky? What kinds of clouds form at the different levels? Can you measure these heights? *(U)*

16. What is above the clouds? Can you produce this condition in a vacuum jar? *(U)*

17. What happens to the water that forms puddles on the ground? Where does the water go? Can you measure the rate of evaporation of different mud puddles? *(P)*

18. How does the water get into lakes, rivers, and ponds? Can you make a study of local bodies of water? *(I)*

19. Does the level of lakes, rivers, and ponds remain the same throughout the year? Can you measure the level of a body of water for a period of time? *(I)*

20. Is any water ever lost during the water cycle? Can you set up a water cycle? *(P)*

21. What conditions affect the speed of the water cycle? *(I)*

22. Why are some things damp in the early morning? Where does the water come from? *(I)*

23. Does dew make all objects equally damp? *(I)*

24. How much dew is formed on one square foot of surface throughout the year? *(I)*

25. Does the amount of dew vary with the season? *(I)*

26. What causes the windshield of your car to fog up on the inside on certain mornings? *(I)*

27. What causes the windshield of your car to fog up on the outside on certain mornings? *(I)*

28. What causes ice to form on the outside of your windshield? On the inside? *(I)*

29. Why is the outside of a milk carton sometimes wet? *(P)*

30. What causes drawers to stick at certain times of the year? *(I)*

31. Does the wind high up in the atmosphere blow in the same direction as the wind close to the ground? Make observations over a period of time. *(I)*

32. Does the wind far above the surface of the ground blow faster or slower than the wind close to the surface? *(I)*

33. What causes wind? Can you produce wind current by natural means? *(I)*

34. How fast can the wind blow? Measure with an anemometer over a period of time. *(P)*

35. What effect does the speed of the wind have on plant growth? *(I)*

36. What causes seasons? Can you demonstrate this in the classroom? *(I)*

37. What caused the Earth's axis to tilt? *(I)*

38. What effect would the straightening of the Earth's axis have on the Earth? *(U)*

39. Are both poles of the Earth illuminated at the same time? When does this occur? *(U)*

40. Can you record the length of the days and the length of the nights for a period of time? When is a day over and a night begun? *(P)*

41. Are the length of days and nights the same all over the world? Are they the same in your state? In your city? *(I)*

42. What must the temperature of the air be in order for frost to form? *(I)*

43. Does the amount of humidity have any effect on the formation of frost? *(I)*

44. Does the amount of humidity have any effect on the rusting of metals? *(I)*

45. Are frost and dew made of the same percentages of hydrogen and oxygen as regular water? *(I)*

46. Do frost and dew contain minerals? *(I)*

47. What periods of the year have days and nights of equal length? Is this true all over the world? Use a universal sundial to determine this. *(I)*

48. What effect does pressure have on the amount of rainfall in your area? *(U)*

49. How does air pressure vary from day to day in your area? *(I)*

50. Does air pressure vary in different parts of the city? *(I)*

51. Is warm air heavier or lighter than cold air? How much? *(I)*

52. Is the air warmer nearer the floor or the ceiling of your home? *(P)*

53. Does the temperature difference between the floor and the ceiling of your house remain the same throughout the year? Is this difference the same in all parts of your house? *(U)*

54. What conditions are necessary for thunder and lightning in your area? *(I)*

55. What kinds of clouds are present when thunder and lightning occur? *(I)*

56. Can you cause precipitation (rain) using any temperature of air? *(U)*

57. Is the rate of evaporation of water affected by temperature? *(I)*

58. Is this rate affected by humidity? By wind speed? *(I)*

59. What affects the rate of evaporation most: temperature, humidity, or wind speed? *(U)*

60. Does rapid evaporation cause heating or cooling? *(P)*

61. What effect does this rate of evaporation have on the temperature change? *(P)*

62. Does weather have an effect on the amount of curl in your hair? *(I)*

63. Does the amount of humidity have an effect on how fast your hair dries out? *(I)*

64. Can you measure the force of raindrops? Does this force vary? *(U)*

65. How fast does a raindrop travel? *(U)*

66. What is the shape of a raindrop? *(P)*

67. Into what layers is the atmosphere usually divided? Can you produce conditions similar to each layer? *(U)*

68. Does the temperature of the atmosphere increase or decrease as altitude increases? Is there any exception to this? *(I)*

69. What causes meteors? Can you find material on Earth similar to meteors? *(U)*

70. What is a meteorite? Why do meteors burn up? Can you produce similar friction conditions with a grinding stone? *(I)*

71. How hot do meteors get when they enter our atmosphere? Is there any kind of life that can live at this temperature? *(U)*

72. How can we predict meteor showers? *(I)*

73. What velocity must a meteor be traveling when it strikes the Earth? *(I)*

74. Do all solar bodies have atmospheres? Why? *(I)*

75. Do land surfaces or water surfaces retain heat the longest? *(I)*

76. Does salt water retain heat for a longer time than an equal amount of fresh water? *(I)*

77. Does the temperature of the Earth vary with depth? *(I)*

78. Does the temperature of water vary with depth? *(I)*

79. During what part of the day does the Earth receive heat? During what part of the day does the Earth start to lose heat? *(I)*

80. Does air pressure increase or decrease with an increase in altitude? *(I)*

81. Does air density increase or decrease with an increase in altitude? *(I)*

82. What effect does increased temperature have on air density? *(U)*

83. Do water vapor molecules weigh more or less than dry air molecules? *(U)*

84. How much moisture can the air hold at different temperatures? How does increased temperature affect the amount of moisture necessary to saturate the air? *(U)*

85. At what temperature can the air retain the most moisture? *(U)*

86. Can you separate a sample of the atmosphere into its parts? What is the percentage of each part in the total sample? *(I)*

87. Is the make-up of the atmosphere constant, or does it vary from place to place? *(U)*

88. Does the composition of the atmosphere vary with increased altitude? *(U)*

89. What holds the atmosphere to the Earth? What kind of atmosphere could the moon retain? *(I)*

90. Is heat given off by land and water? How does this heat energy travel? *(I)*

91. Does the sun heat the air directly? How? *(U)*

92. What is heat radiation? What factors affect the amount of heat radiated? *(I)*

93. What is conduction? What factors affect the amount of heat conducted? *(I)*

94. What is convectional mixing in the atmosphere? How can heat travel by convection? *(U)*

95. How does pressure rise in the atmosphere increase the temperature of the atmosphere? *(U)*

96. How does the atmosphere retard the heat loss? *(U)*

97. How can impurities in the air affect the amount of radiant energy from the sun that is transmitted to the Earth? *(U)*

98. Would changes in the composition of the atmosphere affect the amount of radiant energy transmitted to the Earth from the sun? *(U)*

99. How accurate are snow forecasts? *(I)*

100. How accurate are rain forecasts in your area? How about sun forecasts? *(P)*

101. How accurate are studies of cloud formations as a basis for predicting weather? *(I)*

102. What parts of the world could benefit from cloud seeding? What conditions are necessary for cloud seeding? *(U)*

103. Would weather change if the amount of heat were constant through the Earth? *(I)*

104. Can you compute what fraction of the sun's energy is reflected by snow? What part is absorbed? *(U)*

105. Can you compute what fraction of the sun's energy reaches the Earth? *(U)*

106. Does a cloud reflect sunlight? *(I)*

107. How do different surfaces affect the amount of sunlight reflected and absorbed? *(I)*

108. What kind of rays pass through glass? Why is it warmer inside a greenhouse? *(I)*

109. Does the Earth cool faster on cloudy or clear evenings? *(I)*

110. What land and water surfaces cause local winds in your area? Can you produce a wind in your classroom? *(I)*

111. Can you collect the amounts of water in the air at different temperatures? *(I)*

112. How much water is given off by the leaves of various trees and plants? What type of plant gives off the most water per square inch of leaf surface? *(I)*

113. What type of plant gives off the least amount per square inch? *(I)*

114. What is the highest temperature at which your breath can form a miniature cloud? *(I)*

115. What causes early-morning sea or lake breezes? *(I)*

116. Why do clouds often form near a mountain peak? *(I)*

117. What is the difference between the temperature in direct sunlight and the temperature in complete shade? Is this difference constant or does it vary? *(P)*

118. What do different cloud formations mean in your area? *(I)*

119. How does the amount of moisture in the air affect a spider's web? *(I)*

120. Why doesn't rain or snow fall constantly from all clouds? *(U)*

121. How does the size of a cloud droplet compare with the size of a raindrop? *(U)*

122. How does a cloud droplet grow? *(I)*

123. How cold must a cloud be before snow crystals will form? *(U)*

124. What causes a halo around the moon? Does it help you to predict weather? *(I)*

125. What causes freezing rain? What causes ice pellets? *(I)*

126. How does a hailstone grow? *(U)*

127. What causes snow crystals? Are snow crystals all alike? *(U)*

128. What causes freezing rain? Can you produce this condition? *(I)*

129. Does dew form on clear or cloudy nights? *(P)*

130. How do frost crystals compare with snow crystals? *(I)*

131. How do different kinds of soil affect the amount of run-off water that results from a thunderstorm or cloudburst? *(I)*

132. What is the height of the water table on your property during the year? *(I)*

133. What is the weight of snow as compared to an equal amount of water? *(P)*

134. What effect has temperature on the rate at which snow melts? *(P)*

135. What causes water to seep through the ground? What affects this seepage? *(I)*

136. How does dry ice seeding cause rain? *(U)*

137. What gases found in air will sustain life? *(I)*

138. What effect does ozone have on the radiation from the sun? *(U)*

139. What is the composition of the atmosphere 500 or more miles above the surface of the Earth? *(I)*

140. What effect does the amount of carbon dioxide in the atmosphere have on the heat energy from the sun? *(U)*

141. How high up in the sky can *you* send a hydrogen balloon? *(I)*

142. Can you study upper wind currents with hydrogen balloons? *(I)*

143. Which balloon will rise quicker and higher, one filled with hydrogen or one filled with helium? Can you measure the difference? *(U)*

144. What effect has ionized air on radio waves? *(U)*

145. Does the temperature increase or decrease as a gas expands? *(I)*

146. Does the temperature increase or decrease as a gas contracts or is compressed? *(I)*

147. Can you make an air conditioner? *(I)*

148. Why does air cool as it rises? *(I)*

149. Why is it warmer on the leeward side of a mountain? Set up a class experiment. *(I)*

150. How do Chinook winds warm temperature? *(I)*

151. What causes the jet streams? *(U)*

152. What causes vapor trails after high-flying jets? *(I)*

153. What causes the temperature range in the ionosphere? *(U)*

154. What causes auroras? *(U)*

155. What is the summer solstice? The winter solstice? What is the equinox? *(I)*

156. Why is the summer warmer than the winter? *(I)*

157. Why do direct light rays of sunlight give off more energy than rays coming in at a slant? *(I)*

158. If the longest day of the year in the Northern Hemisphere is in June, why is the hottest month of the year usually August? *(I)*

159. Why is the warmest part of the day in the afternoon? *(I)*

160. How does the temperature change during the day? *(P)*

161. What time of day is usually the warmest? *(P)*

162. How fast does the Earth rotate at the various latitudes? *(I)*

163. How does the rotation of the Earth affect the pattern of winds in the Northern and Southern Hemispheres? (Use a phonograph for the experiment.) *(U)*

164. What causes the "trade winds" and the "prevailing westerlies?" *(I)*

165. What causes the polar easterlies? *(I)*

166. Are the wind patterns the same in the Southern Hemisphere? *(I)*

167. What causes the "equatorial doldrums" and the "horse latitudes?" *(I)*

168. What causes high- and low-pressure areas? *(I)*

169. Does wind flow from low-pressure regions to high-pressure regions? *(I)*

170. Why do the winds near the ground blow in a different direction from those high above the Earth? *(I)*

171. How are air masses formed? How do air masses affect our weather? *(U)*

172. What is the difference in winter and summer temperatures of the ocean? *(I)*

173. What causes warm and cold fronts? *(I)*

174. Does the speed of cold fronts vary from winter to summer? *(U)*

175. What causes squall lines? *(I)*

176. Can you predict the sequences of a cold front? How accurately? *(I)*

177. Can you predict the sequences of a warm front? *(I)*

178. What causes thunderstorms? Can you tell how far away such a storm is by the sound of the thunder? *(I)*

179. Why does cold air precede a thunderstorm? *(U)*

180. What causes thunder? *(P)*

181. What causes lightning? *(P)*

182. Does lightning originate from the cloud or from the ground? *(U)*

183. How do lightning rods protect people? *(I)*

184. What causes tornadoes? *(I)*

185. How can a tornado cause a house or barn to explode? *(I)*

186. Why do most hurricanes which affect the U.S. occur in the summer? *(I)*

187. Can you forecast weather from reports and maps issued by the weather bureau? *(U)*

188. How does an anemometer measure the speed of the wind? *(P)*

189. How does a rain gauge work? Can you measure the rainfall over a period of time and compare the amount with that of the weather bureau? *(P)*

190. Can you design an instrument to measure the duration of sunshine each day? *(I)*

191. Can you design a new method for making a wind vane? How does the direction of the wind in your area affect the type of weather you will have in the near future? *(I)*

192. Can you compare the accuracy of a mercury barometer with that of an aneroid barometer? Can you construct these two types? *(I)*

193. Can you design and build an automatic recording weather station? Be sure to record your findings over a period of time. *(U)*

194. What is the freezing temperature of mercury? Of alcohol? Of salt water? *(I)*

195. Can you construct an accurate thermometer? *(P)*

196. How does a maximum thermometer work? A minimum thermometer? *(I)*

197. Is a sling psychrometer more accurate than a wet- and dry-bulb thermometer for measuring humidity? Can you construct a sling psychrometer? *(I)*

198. How much does a human hair contract and expand when exposed to changing humidity? *(U)*

199. How does the human hair compare with the hair from other animals as far as contracting and expanding during changing humidity? *(U)*

200. Can you experiment with other materials and compare their expansion and contraction with changes in humidity and also temperature? *(I)*

201. Can you make an automatic recording rain gauge? How does it compare with your regular rain gauge? *(U)*

202. Can you make an aerovane wind speed and direction indicator? Can you make one so it will automatically record both speed and direction of the wind? Be sure to use your instrument for some project. *(U)*

203. Can you check the accuracy of the Beaufort wind scale? *(I)*

204. Can you devise ways of measuring the ceiling height of clouds? *(U)*

205. What is an inclinometer? Can you make a similar instrument and use it? *(U)*

206. Can you make a theodolite and then use a pilot balloon to keep track of winds aloft? Be sure to record your findings. *(U)*

207. Can you devise a radiosonde receiver and transmitter that will work at a limited height? (Be sure you attach a cord so you won't lose your instruments.) *(U)*

208. Does it rain more in the city or in the nearby country? *(I)*

209. What are local problems in forecasting weather in your area? *(I)*

210. Can you determine the average temperature in your area for a period of time? *(P)*

211. What kinds of crops will grow in the temperature ranges you have determined? *(I)*

212. Can you determine the average hours of sunshine during different periods of the year? *(I)*

213. What effect will this average have on different crops? *(I)*

214. How many killing frosts do you have in your area during a year? *(I)*

215. How much snowfall do you have in a year? Compare this with your readings in other localities. *(I)*

216. How many clear days do you have in a year? *(P)*

217. How do animals and insects adjust to temperature changes in your area? *(P)*

218. Can you predict weather by observing the movements or activities of various animals and insects? *(I)*

219. What is the maximum amount of water a cubic foot of air can hold at various temperatures? *(U)*

220. What effect does the relative humidity reading have on bodily health and comfort? *(I)*

221. What effect does relative humidity have on fire danger? Can you experiment with the kindling temperature necessary for fire at different humidity readings? *(U)*

222. Can you determine the dew point of the air over a period of time? What effect does temperature have on the dew point? *(I)*

223. How accurate are old-time weather signs and sayings? *(P)*

224. How accurate is the almanac in predicting weather in your area? *(I)*

225. What is the background count due to cosmic rays in your area? Use a Geiger counter to determine this. *(U)*

226. Can you determine the fallout rate of radiation in your area due to the testing of nuclear weapons?[35] *(U)*

227. Can you determine from weather maps the percentage of times wind currents would carry radiation from a target area to various locations in your state? *(U)*

228. Why would a nuclear bomb suck up dirt into its mushroom? *(I)*

229. Does the temperature of the air have any effect on the rate radiation travels? *(U)*

230. Does radiation affect both living and non-living materials? *(U)*

231. Are students more restless during cloudy weather? *(I)*

232. What effect do changing temperatures have on mental ability? *(I)*

233. What effect does temperature have on the lubricating qualities of various oils? *(U)*

234. What is the air pollution rate in your area? How does this rate vary with industrial activity? *(U)*

235. What effect has a polluted atmosphere on the health of various animals, such as white rats? *(U)*

236. What industries around your area contribute to the pollution problems? What materials are released? *(U)*

237. What causes "smog?" *(I)*

238. What effect does ionized air have on behavior? *(U)*

---

35 Low levels *are* plausibly still present from historical nuclear tests, in some areas.

239. Do negative ions relieve hay fever and other air-borne allergies? *(U)*

240. What is the value of various insulating materials? *(I)*

241. Why is it possible for Eskimos to keep warm in ice huts? *(I)*

242. Can you make an estimate of the heat loss through the windows and walls of your house? *(U)*

243. How can a mercury barometer be used to determine altitude? *(I)*

244. Why do breezes at the shore blow in one direction during the day and in the opposite direction at night? *(I)*

PART III

Biology

General Biology Equipment

Light Source for a Microscope

Purpose: In order to view objects satisfactorily under a microscope, a light source is needed which is neither too bright nor too weak. The light is usually reflected off a mirror at the base of the microscope and up through the lenses.[1]

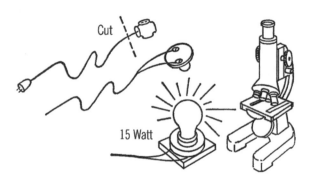

Cut

15 Watt

Materials: Six-foot extension cord, light base ($1-3 porcelain fixture from the hardware store), and a small 15- or 25-watt blue bulb.

What to Do: Cut off the socket end of the extension cord. Divide the cord into two wires. Strip the end of the wire so that about ¾ of an inch is exposed on each wire. Connect wires to

a screw on the bottom of the porcelain fixture. Screw in your light bulb. The porcelain fixture can be mounted on a block of wood so that the screws are not exposed.

Modern Safety Practice

This is an elementary circuit, but great care is required anytime that you design a circuit that directly uses mains (wall) voltage. A shock from mains voltage is potentially lethal. Make sure never to create a situation where a live wire is exposed such that it could be touched. Read Note 10 in Appendix E for safety practice around exposed wiring.

Given all of that, there's no harm in buying a desk lamp or building a low-voltage LED-based alternative instead.

Operation of Equipment: Set the bulb near the base of the microscope. Turn the mirror on the microscope so the greatest amount of light is seen through the eyepiece. Now move the light closer or farther away in order to increase or decrease the amount of light. This light base can be used for your low-powered microscope

---

1 If you purchase a microscope, it *may* already have a plug-in or battery-powered light source. This light source is for microscopes that only have a mirror below the stage. Can you design a modern LED-based apparatus that will work the same way?

or for any light needed in your laboratory. You can improve the illumination by cutting a hole in a black cardboard box and then placing the box over the bulb.

Water Drop Magnifying Glass

Materials: Nail, soft copper wire.

What to Do: Wrap the wire around the nail and make a loop.

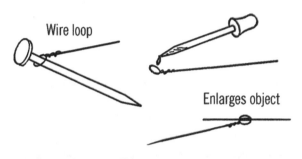

Wire loop

Enlarges object

Operation of Equipment: Place a drop of water on the loop. The water is heaped up, or rounded. This acts as a lens and magnifies objects. It has a very short focal length. Place your eye close to the drop. Try this lens out on newspaper print, bits of salt, etc. Try different size loops. You might try other liquids besides water.

Eyedropper

Materials: Glass tubing, one-inch piece of rubber tubing, material for a small plug for the tubing.

What to Do: Heat the glass tubing with your alcohol burner. When the glass is soft, pull on both ends. Break the tubing at the desired length. Slip the rubber tubing over the large end of the dropper and seal the open end of the rubber to make a bulb.

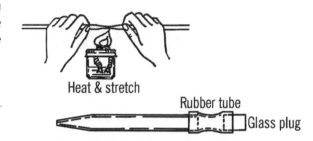

Heat & stretch

Rubber tube

Glass plug

Modern Safety Practice

For this and the other glass-working projects in this section, re-read the safety tips for working with glass. See "**Modern Safety Practice**" on page 6.

Low-Power Microscope

Purpose: The microscope is used to magnify objects that are too small to be carefully studied by the eye alone. Most viewing is done under very low power. The microscope you will make will magnify up to 25 times, so it is said to be 25 power.

Materials: Two or more double convex lenses (five are ideal) with a focal length of about an inch, wood or cardboard box to make the frame, and a small-size light bulb (7½ or 15 watt).

What to Do: Screw the light bulb into a socket. Build a frame to go over the light bulb which will support the slide. The bulb should be three or four inches under the slide. Place the blank microscope slide on the support. Sprinkle some salt on the slide. Now hold a lens as shown and move it up and down until the salt comes into focus. In order to increase the magnification, hold two lenses together and note how the size of the object seems to increase. Now try three, four, or five lenses. Be sure to hold them so one is resting on the next one.

Operation of Equipment: You will notice that when you use only one lens, your eye may be

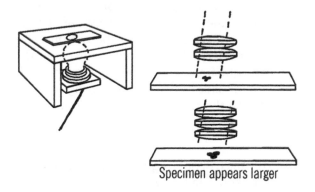

Specimen appears larger

six or more inches away from the lens and the object is still in focus. As you increase the number of lenses, you seem to shorten the focal length, and your eye must be much closer to the lenses. Any substance that is thicker in the middle and will let light through will serve as a lens. The greater the curve the more powerful the lens. A small round glass bead has a lot of curve and is very powerful. A drop of water even makes a good magnifying lens. The light rays hitting the middle of the lens are slowed up. The rays going through the outside edges are thus going faster than those in the center. The rays bend, and it is this bending that causes magnification.

Can You Work Like a Scientist?

1. Place a drop of water on a blank slide (or any piece of glass). Set the leg of a fly in the center of the drop and cover it with a cover glass or another slide. Can you see hairs on the fly's leg? How does the fly hang onto the ceiling?

2. Try the same with the leg of a bee. You may not need to use water and a cover glass, but they are helpful. Can you see pouches on the bee's leg? What does the bee use these for?

3. Look at bits of sugar and salt. Is there a difference between the two?

4. Make a set of slides. Hold the parts of insects down on the glass slide with Scotch tape. To make a permanent slide, glue the cover glass to the slide with clear Karo Syrup or gelatin.

5. Place a whole fly on the slide. Look carefully at the eye of the fly. Can you see many little checks or holes? Each one of these is an eye.

6. Put a drop of silver nitrate on a strand of copper wire. You will see a beautiful crystal tree grow on your slide. Silver nitrate is available at the drugstore.[2] Be careful. It stains skin or clothes.

7. Can you use two lenses and make a compound microscope?

Leeuwenhoek-type Microscope

Purpose: To make a microscope very similar to the first microscope ever invented: a water drop microscope used by Leeuwenhoek about 300 years ago. Leeuwenhoek used a drop of water for a lens. Your microscope will use a glass lens which will be superior. It will magnify up to 160 power.

Materials: Glass rod or tubing, piece of spring brass or iron, three machine screws and nuts about an inch long, and some quick-drying cement.[3]

What to Do: Heat about a one-inch area of the glass tubing over your alcohol burner. When the glass becomes soft, remove the tubing from the flame and pull both ends of the tubing to stretch the glass (see Part I). Break the thin glass filament about six inches long. Then

---

2 *Possibly.* Silver nitrate is still available at *some* drugstores. See Appendix A for additional material sources.

3 Model cement, Duco cement, or five-minute epoxy.

slowly feed this filament into the flame from above. A thin bead will form. The bead should be not more than 1/16" in diameter. Experiment with different sizes. Break the bead so that it has a tail about an inch long. This is your lens. Drill a hole just smaller than the bead in the center of the spring brass. Cement the bead to the bottom of the brass. Build the rest of the microscope as shown. Your major focusing is done by moving the stage. Tighten the stage by tightening the nuts. Your fine focusing is done with the focusing screw attached to the spring brass.

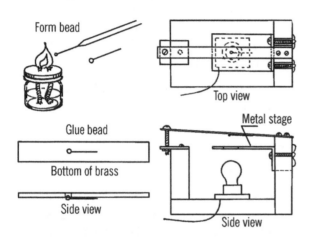

Safety Tips

Read Note 4 in Appendix E about working with hot glass.

Be very careful about touching the heated glass tubing. Glass keeps its heat for a long period of time and can burn your fingers badly.

Operation of Equipment: Place a glass slide on your stage and adjust the stage so you can see the object on your slide. The lens has a very short focal length so you must bring your eye close to the lens. Try salt crystals for your first attempt as these are easy to view. After you sight your object, slowly turn the nut on your fine adjustment. For a light source, either bounce light from a bulb off your mirror and through the slide or place your viewing light directly under the stage for sub-stage lighting.

Can You Work Like a Scientist?

1. This microscope requires care in making. You may have to make many beads before you find one that satisfies you. Patience is the mark of a scientist. Can you think of a way to figure the magnification power of your lens? The power of a lens generally can be figured by dividing the diameter of the lens (in fractions of an inch) into the number ten. If the diameter of your lens is 1/16 inch, how powerful is your microscope?

2. How powerful would your microscope be if the diameter of the lens is ⅛"?

Bacteria Garden

Purpose: A bacteria garden is used to grow or culture various strains of bacteria and molds.

Materials: Ideally, you should have about six glass Petri dishes with lids. Glass castors (the type used under legs of furniture) can be substituted. Another substitute for a Petri dish is a regular bowl.[4] The bowl or castor can then be covered with a glass plate or Saran Wrap. You will also need several half-pint bottles or light bulbs with their insides removed. Some agar-agar is desirable. It can be purchased at many drugstores.[5]

What to Do: Get a can of beef broth from the grocery store. Thin the beef broth with about

---

4 Small glass kitchen prep bowls or ramekins will work well.

5 More commonly at grocery stores, especially Asian or health food stores.

one quart of water. Add about three table-spoons of agar-agar to the broth. Pour the mixture into the small bottles or light bulbs. Do not fill the bottles more than two-thirds full. Plug the bottles with a wad of cotton and then cover these cotton plugs with pieces of paper toweling and fasten with a rubber band. In order to sterilize the mixture, steam the bottles for about 30 minutes in a pressure cooker.

Plain jello or gelatin can be substituted for the agar-agar and beef broth mixture.

In order to sterilize the Petri dishes or castors, scrub and dry the dishes. Wrap each dish in a piece of paper towel and then wrap about three dishes together with aluminum foil. Place the dishes in an oven that has been heated to a temperature of about 400 °F. Leave the wrapped dishes in the oven for about one hour. Let the dishes cool completely before opening the oven door.

When you are ready to use the dishes and the agar-agar, melt the agar-agar mixture slowly by placing the bottle in hot, but not boiling water. Care has to be taken when you pour the culture mixture into the Petri dishes. The neck of the bottle probably has bacteria on it. Therefore, you should pass the neck of the bottle through the flame of an alcohol burner after you remove the plug. Then remove the lid to the Petri dish and pour the mixture into the dish. If you are filling several dishes follow a certain pattern. Remove the stopper, place the neck of the bottle in the flame, lift the lid, pour the mixture, and cover the dish. Be sure to plug the bottle as soon as you are finished.

Operation of Equipment: Bacteria exist almost everywhere. In order to start a culture, remove the lid of the dish and touch the suspected object to the gelatin. Replace the lid immediately.

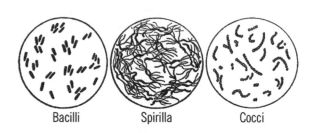

Bacilli   Spirilla   Cocci

Safety Tips

1. Most cultures are harmless. However, some bacteria can be harmful. Therefore, when you wish to clean out an old culture from a dish, place the dish in the pressure cooker for about 30 minutes. This should kill any bacteria. Remove the lid only after you have killed the bacteria.

2. Don't touch hot Petri dishes. Glass holds heat for quite a period of time.

3. Be careful when you are working with the flame from the alcohol burner.

Modern Safety Practice

Review Note 3 in Appendix E about working with fire.

Can You Work Like a Scientist?

1. Do bacteria grow better in daylight or darkness?

2. What effect does temperature have on bacteria growth?

3. Do all forms of bacteria grow at the same rate?

4. What effect does humidity have on the growth of bacteria?

5. Try bacteria from fingernails, air, chewing gum, mouth, comb, shoes, etc.

Micrograph

Purpose: The micrograph is used for group viewing or making drawings of objects viewed through a microscope. The micrograph uses the principle of the microprojector and therefore the sketches made are quite accurate in all details.

Materials: Microscope with a sub-stage light, wooden box and stand, and a piece of frosted glass. Type B requires only a chemistry flask of the proper size to fit over the eyepiece of the microscope, a piece of translucent plastic (such as the sheet that covers a stencil), and a rubber band.

What to Do: Type A is simply a wooden box mounted on legs. A hole is cut in the bottom of the box so that the box will fit over the eyepiece of the microscope. A hole is cut in the top of the box and a piece of frosted glass is used to cover the hole. This is the viewing screen.

Type B is much easier to make. It requires a 250 ml flask. The neck of the flask should just fit over the eyepiece of the microscope.

A piece of frosted (translucent) plastic covers the flat bottom of the flask. This plastic is held in place by a rubber band. The bottom of the flask forms the drawing table.

Operation of Equipment: Place your micrograph over the eyepiece of the microscope. Plug in your sub-stage light. (If you do not have a sub-stage light, use the light from a slide projector. This light can be reflected off the mirror and through the lens system.) Place the slide containing the object to be drawn on the stage. The light shining up through the slide and the lens system strikes the frosted glass plate or

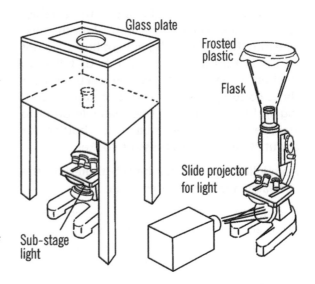

plastic. This image can be focused by turning the focusing knob on the microscope. Place a sheet of thin white paper over this viewing screen. You should be able to see the object to be drawn through the paper. The drawings can be made with pencil, charcoal, water colors, crayons, or other art medium. You should use the micrograph in a darkened room.[6]

Can You Work Like a Scientist?

1. Can you make drawings and compare the legs of various insects such as the bee, fly, grasshopper, and butterfly? Your drawings make a good permanent record.

2. What insects have compound eyes?

3. Can you compare the mouth parts of various insects? How do insects and spiders bite?

4. Can you make drawings of live insects? You can hold the insects down on the slide by using Scotch tape.

---

6 Question: What advantages and disadvantages does this micrograph have over a modern USB microscope?

Polarized Light Filters

Purpose: Polarizing filters on a microscope are used to strain out certain light rays. Substances can then be examined under this polarized, or directed, light for their optical properties.

Materials: Lenses from a pair of polaroid sunglasses or a small 2 × 2" polaroid filter,[7] and any kind of compound microscope.

What to Do: Remove the lowest power eyepiece on the scope. Place the barrel of the eyepiece on the polaroid filter or lens. Trace around the outside of the barrel and then cut the circle out of the filter. The circle should be cut so that it will fit tightly in the barrel at the bottom of the eyepiece.

A second filter can be installed on the microscope or can just be held in position while the microscope is in use (see "Operation of Equipment," below). If you wish to install a second filter permanently, cut out a small circle from the polaroid material and glue over the large hole in the rotating light aperture disc below the stage of the microscope.

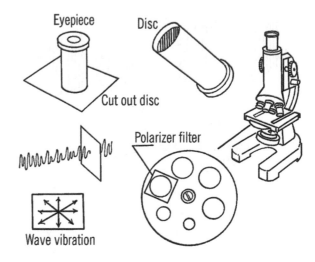

Operation of Equipment: In order to understand how to use your polarizing filters with your microscope, it is necessary to understand how polarizing filters work. There are many theories about how light travels.[8] One of the most widely accepted is that light travels in a wave motion similar to waves made in water by the dropping of a stone. These waves travel[9] in all directions, some of which are vertical and some horizontal. Most of the waves are neither vertical nor horizontal, but travel at many different angles through the air.

The polarizing filter consists of a thin layer of tiny needlelike crystals of herapathite (iodoquinine sulfate). These needlelike crystals are lined up parallel to each other so they all face in the same direction. The crystals are then enclosed for protection between two transparent plates. Only light rays that vibrate in the same direction as the crystals face on the filter can pass through the filter. All other light rays are screened out.

---

7 Most often, this material is called "Linear polarizing film."

8 It would be more correct to say that there is *a lot to say* about how light travels; our understanding of how light travels is *not* controversial.

9 Oscillate.

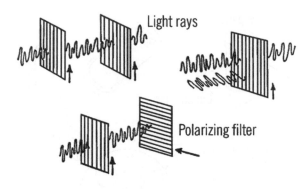

Light rays

Polarizing filter

If two such filters are used, light rays will pass through both if both are turned so the crystals are pointed in the same direction. If we turn or rotate one of the filters, we screen out more and more of the light rays. When the two filters are crossed, all the rays are screened out.

A transparent object placed between the polarizers causes an unusual effect. Instead of being able to screen out all the light rays completely, light seems to be transmitted through the material. Some materials, such as crystals, allow colored light rays to come through. The colors are not a property of the crystal but depend on the thickness of the crystal and the material the crystal is made of as well as the position to which the rotating polaroid filter is turned.

Can You Work Like a Scientist?

1. Look through one piece of a polaroid filter toward the light. Is the light as bright? Part of the light has been polarized.

2. Turn one filter while you hold another filter steady. Can you explain what happens?

3. Place the two filters in the path of light from a flashlight or slide projector. Can

you explain what happens to the light cast on the wall as you turn one of the filters?

4. Cross several strips of Scotch tape on a blank microscope slide and place the slide under your microscope. Use either the polarizer you have glued on the disc under the stage, or hold a filter between the sub-stage light and the stage. As you look through it, rotate the eyepiece containing a second filter. The light should polarize the slide containing the Scotch tape strips into beautiful colors.

5. Remove the bottom polarizer (if this is fastened to the disc, turn the disc). Catch a goldfish in your hand and place the head of the goldfish in a wet handkerchief.[10] Lay the tail of the goldfish on a blank microscope slide. Examine the tail of the goldfish under low power. You should see the blood circulating through the tail. Add the bottom polarizer. Now rotate the eyepiece. Does the tail of the goldfish polarize?

6. You can grow and polarize crystals under the microscope. Place a few (two or three) acetaminophen crystals on a blank slide.[11] Heat the crystals with a match. When the crystals melt, cover them with a cover glass. This spreads the acetaminophen out into a very thin solution. If any carbon from the match gets on the bottom of the slide, wipe this off with a cloth. As the solution starts to cool, it crystallizes. You should see crystals suddenly form and shoot across the slide.

7. Melt the crystal on this slide again with a match. Place the slide under low

---

10 See Note 1 in Appendix E about working with animals.

11 Originally "acetamine." Pure acetaminophen may come as powder or crystals. Acetaminophen tablets (and other forms for human consumption) are not suitable for this project.

power and use both polarizers. As the crystal starts to cool again, needlelike crystals will move across your field. Turn your eyepiece as you watch the crystals grow. Can you explain the unbelievable color patterns? Could you melt the crystals again and get the same pattern?

8. What effect does pressure have on the speed the crystals form and move across the screen? Press on the cover glass if the crystals haven't started to grow. Does this added pressure start the reaction?

Microbiology Cultures

Purpose: Cultures are started and maintained in order to have a continuous supply of one-celled and multi-celled plants and animals.[12] Observations and experimentations with these simple living things enables us to learn much that will also apply to the higher forms of plants and animals.

Materials: Eyedropper, jars with lids, and food material (hay, lettuce, oat or wheat kernels).

What to Do: Cultures are very easy to start and maintain providing you think of this microscopic life in much the same manner as a civilization of human beings. Human beings cannot stand chlorine in their atmosphere. Micro-organisms cannot stand chlorine in their atmosphere and their atmosphere is the water in which they live.

In order to start a culture, get several quarts of water that do not contain chlorine. Pond water, rain water, river water, or water found in a mud puddle will all support a microscopic civilization. Pour the water into several bottles.

Life needs food. The simplest food material is bits of lettuce. Break bits of lettuce into several of the bottles.

If you are going to try hay, wheat, or oats as a food material, boil these food materials first. This softens the material and makes it easier for micro-organisms to start. You may use any of these food materials in the same way as you used the lettuce. Cover the jars and let the jars stand for about a week.

Operation of Equipment: After about a week examine your cultures by placing a drop of the culture on a blank microscope slide. Examine the drop under the low power of your microscope. You should see small bits of life swimming around. This civilization will continue to grow. As these small animals[13] mature and reproduce, the numbers will increase to a point where the food material and oxygen in the water will not support all of the life. To avoid this, divide your culture every week or ten days. Place a few eyedroppers of life into a jar containing new water and food material. In this way you will have a continuous supply of micro-organisms on which to experiment.

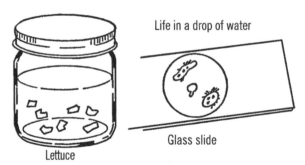

Lettuce   Life in a drop of water   Glass slide

Figure 8-10   Culturing microbes

Can You Work Like a Scientist?

1. Can you separate certain micro-animals from the others and try to start a

---

12 In modern scientific usage, the terms "plant" and "animal" refer strictly to multicellular organisms.

13 And other micro-organisms.

pure culture from just one or two organisms of the same kind?

2. What effect does light have on the growing of a culture?

3. Will protozoa (one-celled animals)[14] grow in liquids other than water?

4. What effect does an electrical current in the water have on protozoa such as paramecium?

5. What effect does temperature have on the rate of growth of a culture? Does each kind of micro-organism have its favorite temperature?

6. What micro-organisms feed on other micro-organisms?

Growing Brine Shrimp

Purpose: Brine shrimp are large-size members of the microscopic world. They are easy to hatch and raise and make ideal subjects for experimentation. The eggs can adapt to extreme temperature changes.

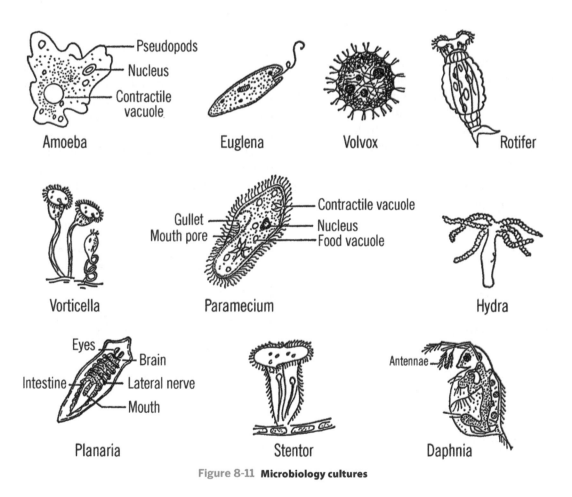

Figure 8-11 **Microbiology cultures**

---

14 Protozoa is a historical classification, no longer in use, that refers to "animal-like" single-cell organisms.

Materials: Sauce dish or half-pint jar, non-iodized table salt, and brine shrimp eggs. These eggs may be purchased at some pet stores.[15]

What to Do: Pour one cup of water into the sauce dish or jar. Let the water warm up to room temperature. Mix one teaspoonful of non-iodized table salt into the water. Draw a circle about the size of a dime on a paper. Fill the circle with brine shrimp eggs and then place the eggs in the salt water. They should start hatching after 24 hours. The shrimp should be fed a very small pinch of oatmeal flakes once a week, or a grain or two of dried baking yeast each day. Be careful not to overfeed, or your colony will die off.

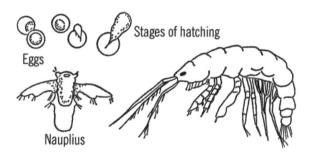

Can You Work Like a Scientist?

1. Place a drop of water containing the brine shrimp on a blank microscope slide. Observe under the microscope, microprojector, or magnifying glass. If the brine shrimp grow too large to go through the opening of the eyedropper, remove the bulb from the eyedropper and place it on the other end of the dropper so that you can use the large opening.

2. When some of the brine shrimp grow larger, select pairs and place them in separate containers (medicine vials). If you selected a male and female, eggs should be laid in the vial. Can you keep a record of the life cycle of a pair of shrimp?

3. How fast does a brine shrimp grow? Hatch an egg inside an eyedropper using a very small amount of water. Place the brine shrimp on a glass slide each day and observe with a microscope or microprojector. In order to avoid losing the brine shrimp, use a well slide and then suck the water up with an eyedropper when you have finished. You can make a well slide by gluing a round paper reinforcement ring to the slide.[16]

4. What effect does temperature have on the hatching rate of the eggs? Try some water at 100°, 90°, 80°, 70°, 60°, 50°, 40°, and at almost freezing temperature.

5. What effect does chlorine in water have on brine shrimp? Try hatching shrimp in tap water containing chlorine. Place a drop of Clorox in a drop of water containing brine shrimp. Observe the effect under the microscope.

6. Try different concentrations of salt. What effect does the salinity of the water have on the hatching rate of brine shrimp? Use a hydrometer to measure the salt concentration in the water.

7. Can a brine shrimp be conditioned gradually to fresh water?

8. Will brine shrimp hatch in fresh water?

9. What percentage of the eggs hatch normally? What conditions affect the hatching of the eggs?

---

15 Brine shrimp eggs come in a jar of dry granules. They are about the size of sand grains.

16 See "Microscope Slide Making" on page 221 for more about well slides. Take care not to leave your living specimens on a well-lit microscope stage for too long; they may not be able to handle the heat.

10. How hardy are brine shrimp eggs? Place some eggs in the oven for a short period of time. Keep a graph of the number that hatch after exposure to such unusual temperature conditions.

11. What percentage of the eggs will hatch if the eggs are frozen and then thawed? Find the temperature range the eggs can stand.

12. What effect has light on the percentage of shrimp that hatch? Place some in a darkened area and another in full light.

13. Will brine shrimp regenerate body parts? Remove a leg and then observe if the shrimp can regenerate the leg.

14. What effect has electricity on brine shrimp? Attach two wires to a battery. Place the end of the wire into a small container of brine shrimp. When the electricity passes through the salt water, are the brine shrimp attracted to either wire (pole)?

15. What is the effect of different foods on the growth of brine shrimp? Try other grains as well as foods you think the shrimp might eat. Compare the size and number of shrimp in each container by observing the life under magnification. The number of shrimp observed in one minute moving through the field of the microscope or microprojector will give you an average number for that container. You can compare the containers with their different foods in this way.

16. What do brine shrimp breathe? Fill one bottle and seal the top so no oxygen from the air can enter. As a control, remove the lid from another bottle.

17. What effect do various protozoa have on brine shrimp? Place a brine shrimp into a drop of protozoa culture and observe.

18. What effect does a hydra have on a brine shrimp? Place a shrimp into a large drop of water containing a hydra. Hydra are similar to salt water jellyfish and contain stinging cells. They will sting a shrimp to death and then eat it. This sometimes can be observed through the microscope or microprojector.

19. Can brine shrimp see?

20. What water pressure do brine shrimp prefer? Use a piece of glass tubing or a test tube. If you use glass tubing, plug up the bottom end or seal it in the flame of an alcohol burner. Place the shrimp in the tubing or test tube. Observe at what depth most of the shrimp collect. Glass tubing filled with water is ideal for this experiment as the pressure from the top of the tubing increases tremendously with the depth of the water.

21. What effect does wave action have on the hatching of brine shrimp eggs? Stir one container regularly. Use as a control a container in which the eggs are not stirred.

22. Will brine shrimp hatch if you use iodized salt instead of non-iodized salt?

23. Will brine shrimp continue to live if placed in a solution containing iodized salt after they have hatched in a solution containing non-iodized salt?

24. Do brine shrimp molt or shed their skin?

25. If the water in your brine shrimp dish evaporates, does the salt evaporate too? If not, does the salinity (salt content) of the water change? If you wish to control this change in salinity, mark

the level of the water in your container. Replace the water that evaporates with rain or tap water.

26. Can you observe and record the embryology of the brine shrimp? Place the eggs in a shallow glass dish under the microscope or microprojector. Observe the progress of the eggs every hour. Here are some things to look for:

    a. Are the eggs completely round, or do they have a dent on one side?

    b. Do the eggs seem to swell after the first hour?

    c. How long is it before the eggs start to split open?

    d. How do the shrimp get out of their eggs? Are they regular adults at first or are they covered with a thin membrane? What is a nauplius?

    e. Can you locate a red spot? Is this the eye?

    f. How do the shrimp get out of their transparent case? Can you observe this?

    g. How many legs do the brine shrimp have? What are the other body parts? Can you compare the body parts with insects and spiders?

Microscope Slide Making

1. **Water Drop Slide:** Place the object such as a leg of a bee on a blank slide. Place about one drop of water on the object. Lower a cover glass over the water drop and the object. The drop spreads out under the glass and forms a strong seal. Water molecules have a strong attraction for glass molecules. When you wish to remove the cover glass from the slide, soak the slide in warm water. Life cultures and many temporary slides can be made in this way.

2. **Vaseline Slide:** Insert a toothpick into a jar of vaseline. Draw the edge of the toothpick across the slide to form two vaseline walls. Insert the toothpick in vaseline again and then draw the toothpick along both edges of the slide to finish forming a pen of vaseline on the slide. The pen should be about the same size as the cover glass you use. Place a drop of the culture in the center of the vaseline "swimming pool." Carefully lower a cover glass down on the vaseline pool. The vaseline should form a seal around the edges of the cover glass and completely trap any microscopic life. An advantage to this kind of slide is that the water cannot evaporate. You can trap rapidly moving micro-organisms by placing a few cotton fibers in the pool before you put the drop of water on the slide. The fibers will trap many small creatures after the cover glass is lowered down on the vaseline ring.

3. **Scotch Tape Slide:** Small insects and spiders such as mites, aphids, and garden spiders can be held in place and examined "Live" under the microscope. The tape is lowered down on the insect or spider. The tape is then fastened to a blank microscope slide. The small animal is free to kick or otherwise move even though part of the animal is stuck to the tape.[17]

---

17 It is kinder to use a small transparent pillbox or petri dish with a lid on the microscope so that the animal can be set free later.

4. **Gelatin Slide:** A semi-permanent slide can be made by using gelatin as a glue. A small amount of clear jello is mixed up. The object such as an insect leg is placed on the slide. A drop of liquid gelatin is placed on the object. A similar drop is placed on the blank slide. The blank slide is then lowered down on the cover glass containing the object. The two drops flow together forming a perfect seal between the slide and the cover glass. The slide is left in this position in order to give the gelatin time to harden. Excess of gelatin can be cleaned off with warm water.

5. **Well Slide:** A well slide is a slide with a depression ground into the center of the slide. This well or concavity forms a "swimming pool" for larger microscopic animals such as brine shrimp, daphnia, and planaria. A well slide can be made from a blank slide by punching a hole in the center of a piece of cardboard with a paper punch. The cardboard should be about the same size as a cover glass. The cardboard square is soaked in shellac and then held in position on a blank slide. The shellac not only sticks the cardboard to the slide, it also waterproofs the cardboard. You can make well slides of any thickness by adding extra layers of cardboard. After the drop of water has been placed in the center of such a well slide, a cover glass is used to flatten out and also hold the water still.

A quick substitute for a well slide can be made by sticking two pieces of Scotch tape about a half inch apart across a blank slide. The space between the two strips serves as a well. The well again can be deepened by adding additional layers of tape.

A wick for a well slide can be made by running a piece of string from a medi-

cine vial containing water to the well on the slide. The water will travel from the vial to the well through the string. The vial can be glued to part of the slide. The water supplied through the string wick replaces the water evaporating from the well.

Permanent Microscope Slides

1. **Mounting Small Dry Objects:** Objects which are thin and do not contain water can be easily mounted without much preparation. The object such as an insect wing is placed in the center of the cover glass. A drop of Canada balsam is placed on the object. Next, a drop of balsam is placed in the center of a blank slide. The slide is turned over and the hanging drop is lowered into the drop on the cover glass. The two drops flow together and remove any air bubbles. The slide is left in place several days. Then xylene is used to remove the excess balsam around the cover glass.

2. **Mounting Objects Containing Water:** Objects containing water cannot be mounted in the manner just described. The water in the cells evaporates and the object dries up or withers away. Such a specimen needs to have the water gradually removed from the cells (dehydration). This gradual dehydration is done by soaking the object in several mixtures of alcohol and water. Each succeeding mixture should have a higher proportion of alcohol.

The first mixture should contain about 30% alcohol. Mix one ounce of alcohol with each two ounces of water. You can use a baby bottle to accurately measure out the correct amount. The specimen should be left in this weak mixture for about five minutes.

The second mixture should be about 50% alcohol. Mix equal amounts of water and alcohol. Leave the specimen in this solution for five minutes and then transfer the specimen to a third solution containing 70% alcohol (7 parts alcohol and 3 parts water). Again the specimen should soak for about five minutes.

The fourth solution is 90% alcohol. After five minutes, transfer the specimen to a fifth solution which is 100% alcohol. After five minutes in this solution, place the object in turpentine and let the object soak for several hours. This increases the object's transparency and makes it easier for light to pass through.

The specimen is finally mounted on a slide by using Canada balsam as has already been described.

The small bottles of alcohol can be saved and used over again. Since alcohol evaporates, the bottles should be capped after every use. The alcohol will eventually be weakened by the addition of water removed from various specimens.

3. **Mounting Cross-sections:** Microscopes have a very limited depth of field. By this we mean that the microscopes will focus on only one level or spot at a time. Thus, an object that is very thick such as a piece of cork or a stem of a plant cannot be viewed satisfactorily because only a small part of the object will come into focus at any one time. Also, light will not travel through thick sections. Thin sections of specimens are sliced on a microtome. The sections are then dehydrated as

mentioned above and mounted by using Canada balsam.

4. **Stains:** Many times it is helpful to stain or dye parts of a specimen. Since some parts of a specimen react differently to a dye than other parts, a stained slide reveals much detail that cannot be detected without a stain.

Common dye such as methylene blue or Congo red as well as such improvised dye as India ink, merthiolate, and iodine can all be used as stains. Usually the specimen is placed in the dye for a few minutes and then the specimen is washed off in water. The specimen is then mounted as previously described.

Microtome

Purpose: Many objects are too thick to be viewed under the microscope. The microtome enables you to slice thin sections or cross-sections to be mounted on a slide.

Materials: Either a ¼" or ½" bolt and nut (can be any length), paraffin wax, and a single-edge razor blade.

What to Do: Screw the bolt about one fourth of the way into the nut. Prop the bolt up so the opening of the nut is facing up. Place the object to be viewed in the opening in the nut. Heat some paraffin with a candle and drip the melted paraffin into a spoon.[18] Then pour the melted paraffin into the hole around your object.

Operation of Equipment: After the paraffin has completely hardened, screw the bolt into the nut. This pushes some of the paraffin out. Try to slice as thin a section as possible. Move the razor blade down along the face of the nut so as to trim off a very thin section. Trim off several such sections until you have a good cross-

---

18 Alternative method: Melt a small amount of paraffin, in a small beaker in a microwave.

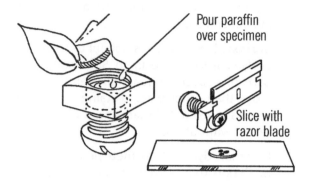

Pour paraffin over specimen

Slice with razor blade

section of the object to be viewed. Place this thin paraffin section on a blank slide. Heat the slide slightly and the paraffin will melt and stick to the slide. You can try dropping the sections of paraffin into warm water and then bringing the slide up from underneath the floating sections and lifting the sections out of the water onto the slide. The sections will cool and can then be viewed.

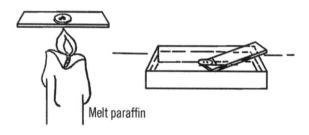

Melt paraffin

Modern Safety Practice

1. Razor blades can cut your skin very easily, so handle them with extreme care.

2. You will be working with open flames; review Note 3 in Appendix E.

Can You Work Like a Scientist?

1. Use the microtome with leaves of lettuce.

2. Can you see the cells in an onion?

3. Try other vegetables. Which ones work the best?

4. Try stems of various plants.

5. Take cross-sections of tissue of animals. Try different meats.

Precision Microtome

Purpose: A microtome is used to slice very thin cross-sections of objects so they may be mounted and viewed under the microscope. The microtome described here is capable of producing very thin cross-sections for both viewing and slide-making.

Materials: Two short pieces of copper or steel tubing about a half inch in diameter and about two inches in length, a three-eighths-inch bolt and nut about three inches in length, two small pieces of plastic or two microscope slides, wood for the frame, and a sharp razor blade.

What to Do: Screw the base and upright together. A hole should be drilled in the upright large enough for one of the pieces of metal tubing to slide through while resting on the base. Place the tubing in position. Cut two side support pieces and notch them so that the nut can fit in the slot. Nail the side support pieces alongside the tubing. The nut is placed in the slot and the bolt is screwed through the nut and the tubing. Plastic strips or microscope slides are glued on the back side of the upright as shown.

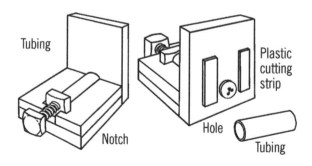

Tubing

Notch

Plastic cutting strip

Hole

Tubing

Operation of Equipment: Place the object to be sliced (such as a stem of a plant) in the extra metal tubing. Pour melted paraffin around the material. Allow the wax to cool completely. After the wax has cooled, push the wax plug out of the tubing and insert the plug into the microtome cylinder. Then, when you wish to slice off a cross-section, screw the bolt through the nut and against the wax plug. The plug can be gently pushed forward until a thin piece sticks out beyond the plastic or glass-cutting surfaces. The wax can be sliced off neatly with a razor blade. The slice should be allowed to drop into a pan of warm water. The slice will float on the surface. As the wax starts to melt, bring the slide up underneath the section and lift up carefully.

If you wish to mount the specimen, place a small drop of balsam on the blank slide. Place a drop on the cover glass as well. The object is held to the slide by the balsam. Turn the slide over and slowly lower the slide and object down on the cover glass. When the two surfaces of balsam meet, they will flow together because of the attraction the like molecules have for each other. You should not have any air bubbles trapped between the slide and cover glass. Allow the slide to dry for several days. Excess balsam can be cleaned off by using xylene, which is a thinner for balsam.

Can You Work Like a Scientist?

1. Can you mark the head of the bolt so that you can accurately measure the thickness of the cross-sections you cut? If the bolt moved a sixteenth of an inch during one complete turn, how far would it move if you turned the bolt only a quarter turn?

Animal Maze

Purpose: An animal maze is used to test the intelligence and learning rate of small animals, such as rats, hamsters, etc.

Materials: Two pieces of 1″ × 8″ wood about 8 feet long, other short wood pieces to make partitions as shown.

What to Do: Cut the boards in half. Make a 4′ × 4′ fence as shown. Nail various short pieces of wood together to form the inside chambers. You can design your own pattern.

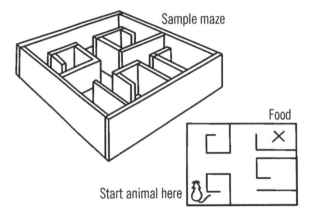

Sample maze

Food

Start animal here

Operation of Equipment: You can use a table or the floor for the base or bottom of the maze. Set down the fence and then place the inside chambers into the position you wish.

Rats, hamsters, etc. need a reason for learning. Don't feed your test animal for about two days. Then place the food at one end of your maze and the animal at the other end. Time the progress of the rat to the food. Also record the number of wrong turns. Try this test twice a day. Reward the animal with food or water. Give the animal no other food or water.[19]

---

19 If you were to follow these directions as written—and *please do not*—you might observe that what begins as a simple "harmless" experiment (to see how rats navigate mazes) quickly turns into what could be called an experiment in animal cruelty. This is an excellent illustration of why even "informal" research with animals needs guidelines and oversight. In this and other cases where animals are involved, please carefully read and understand Note 1 in Appendix E.

Can You Work Like a Scientist?

1. Is a rat or a hamster more intelligent?

2. Does the female learn more quickly than the male?

3. Is the desire to care for her young a stronger motivation for a female than food? Than water?[20]

4. Is learning passed on to the young?

5. Do animals learn faster with a reward or a punishment as a stimulus? Is fear contagious?[21]

6. Can odor or scent help the animal find his reward?

7. Can the animal relearn a changed pattern more easily? (Move the chambers.)

8. How long does an animal retain learning?

9. What effect has light upon the learning rate?

10. You can make a turtle maze by making a raised platform. The walkways should be about three inches wide and about a foot above the table. A bowl of water serves as a good reward. If you make the walkways out of metal, you can heat the bottom of the metal to get the turtle to move.[22]

Insect Net

Purpose: The insect net is used to collect insects and trap them until they can be placed in small bottles to be studied.

Materials: Round handle (broom handle or dowel), heavy wire (coat hanger wire could be used), and cheesecloth or lightweight curtain material for netting.[23]

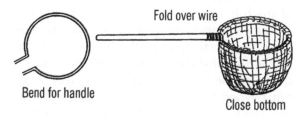

Fold over wire

Bend for handle

Close bottom

What to Do: Bend the wire into a circle, leaving about three or four inches of wire sticking out so the wire can be attached to the handle. Fold the end of the cloth over the wire and stitch with a needle and thread. Attach the wire to the handle with tape. Gather the net together at the bottom.

Operation of Equipment: Insect nets are used to catch more than just butterflies. In fact, most insects caught are not flying and are not even seen. With the net make several passes at the bushes. You are bound to collect, along with a few leaves, many insects you do not normally see. Close the net off with your hand and then slowly reach in with a small vial and collect the insects of particular interest. You may place the insects in a killing jar if you wish to dissect them for study under your microscope or to mount them for your collection. The insects can be studied with a hand lens through the vial and released if not wanted.

---

20 Can you find a way to test this that does not involve starving the animal?

21 Tread carefully here (if at all) and work within animal research guidelines.

22 Under no circumstance go hotter than an electronic "heat stone" from a pet store (the kind designed for reptiles to bask upon).

23 Most fabric stores sell lightweight mosquito netting, which would be perfect for this project.

Killing Jar

Purpose: The gas chamber is used on insects in order to preserve the insects in a life-like position.

Materials: Bottle with lid, cotton, liquid that evaporates rapidly and gives off strong gas fumes (fingernail polish remover, or carbon tetrachloride[24]).

What to Do: Dip the cotton into the liquid. Drop the cotton into the jar. Tighten the lid so the fumes can't escape. In order to use the jar, remove the lid and drop the insect or insects into the jar. Tighten the lid, and observe the results.

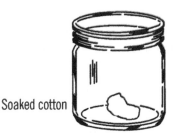

Soaked cotton

Can You Work Like a Scientist?

1. Are all kinds of insects affected by this gas chamber?

Modern Safety Practice

1. Read through Note 17 in Appendix E about safety around chemicals.

2. As you will be storing (a small amount of) toxic material in this jar, it is your obligation to label it clearly: "KILLING JAR (POISON)." Do you understand why you should never smell chemicals to identify them?

2. What gases kill insects? Will the odor of perfume kill?

Insect Mounting Box

Purpose: A mounting box is used to preserve and display insects.

Materials: Cigar box, paraffin, and Saran Wrap.

What to Do: For the mounting material in the bottom of the box use melted wax (paraffin) or a piece of styrofoam. If you use paraffin, melt the wax in a can over a burner and pour it into the box to a depth of about half an inch. Mount the insects with pins pushed into the wax or styrofoam. After you fill a box with a particular type of insect (such as moths), cover the box with Saran Wrap. This will allow you to study your collection, and yet the insects will be protected. If you wish to keep it intact over a period of time, drop in several mothballs before covering the box.

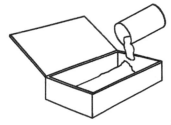

Pour paraffin in box

Cover with Saran Wrap

Quick Freeze Chamber

Purpose: This chamber produces extremely cold temperatures. The chamber is used for carrying out experiments to determine the effects of extreme cold on plants and animals.

---

24 Ethyl acetate is also widely used. Avoid carbon tetrachloride; see Note 21 in Appendix E about it.

Materials: Coffee can or pyrex beaker, rubbing alcohol or duplicator fluid, and two pounds of dry ice. Dry ice is frozen carbon dioxide gas. It costs about $1-2 a pound and is available at ice cream stores, refrigeration plants, and sometimes in school cafeterias.[25]

What to Do: Put dry ice into the container. Don't touch the dry ice directly, but instead use a cloth to pick up the pieces. Dry ice can burn the skin.[26] Pour the alcohol over the dry ice. *Do not place fingers inside the container. Do not touch the outside of the container.* The extreme cold will freeze your fingers in a few seconds. Use tweezers to put materials in and out of the can.

Modern Safety Practice

1. Dry ice can cause severe frostbite injuries. Handle only briefly, using oven mitts or a towel to protect your hands. See Note 14 in Appendix E for more information. Alcohol at the temperature of dry ice is also very hazardous. If it is spilled on someone's skin or clothing, they could be severely injured.

2. Alcohol vapor is flammable. Keep sources of spark and flame away, and have a type ABC fire extinguisher handy.

Can You Work Like a Scientist?

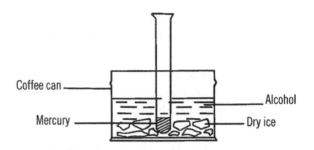

Coffee can — Alcohol
Mercury — Dry ice

1. Place a few drops of mercury in a test tube and lower into the mixture. What happens to the mercury? How low does the temperature get? Mercury freezes at -38 °F.[27]

2. Place a frankfurter in the mixture. Try plant leaves, rubber, etc. Try hitting the rubber with a hammer. Then throw the frankfurter to the floor.[28]

3. Will cultures stay alive at these temperatures?

4. If you placed a goldfish in the tank, could it be revived?

5. What effect does extreme cold have on insects? Will seeds germinate if exposed to extreme cold?

6. What liquids will freeze in the chamber? Use test tubes.

7. Will a turtle survive a quick freeze?[29]

---

25 Many major grocery stores and ice cream shops carry dry ice; you may need to ask for it.

26 Obviously, it cannot *burn* but instead *freezes* skin causing frostbite, which is a serious injury.

27 The safest way to perform this experiment is to use a glass-encapsulated mercury switch.

28 You may be able to predict the results. But also think about the cleanup afterwards!

29 Once again, please read Note 1 in Appendix E.

Chest Cavity

Purpose: The chest cavity shows how your stomach muscles aid you in breathing. It also shows an effect of air pressure.

Materials: Gallon jug, glass or copper tubing, one- or two-hole rubber stopper, two balloons for lungs, and a large piece of rubber to stretch over the bottom of the jug.

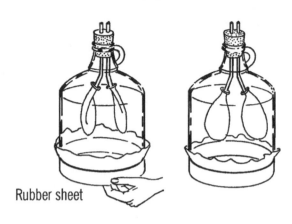

Rubber sheet

What to Do: Cut off the bottom of the jug with the bottle cutter ("Bottle Cutter" on page 18). Be sure to smooth the bottom of the jug with emery paper under water. The easiest way to make the chest cavity is to bend two pieces of glass tubing ("Bending Glass Tubing" on page 6). Fasten a balloon on the end of each piece with tape or a rubber band. Stick the ends of the glass tubing into the holes of a two-hole rubber stopper. Put the stopper in place in the jug and then twist the tubes so the balloons (lungs) are in their proper places. Now cut open a large balloon and stretch the rubber across the bottom of the jug. Fasten the rubber with tape or rubber bands. Sheet rubber is stronger and is easier to use for the diaphragm.

Operation of Equipment: The rubber across the bottom of the jug represents your diaphragm (muscles of your abdomen). As you push in and out on the rubber, the balloon lungs inside your chest cavity seem to breathe. You can feel the air move in and out of the rubber stopper. Your breathing is controlled in much the same way by the movement of your diaphragm muscle.

Safety Tips

1. Be sure to read all the directions before you use the bottle cutter.

2. Be sure to smooth your jug under water with emery paper.

3. Don't stretch the balloon too tight on the bottom of the jug or it will split.

Can You Work Like a Scientist?

1. Can you use compressed air and a partial vacuum to explain why the lungs seem to move as the diaphragm moves?

2. Would this chest cavity work on the moon?

3. Can you make the chest cavity snore?

4. What causes snoring? How can you stop a person from snoring?

5. Do fish breathe in this same way?

6. How do birds breathe? Do birds snore?

7. How do animals breathe when they hibernate? Why do animals hibernate?

8. Can you measure the changes in pressure by running a rubber tube from the chest cavity to the vacuum gauge?

9. Can you measure the changes in pressure as your chest cavity moves in and out?

Metabolism Chamber

Purpose: The purpose of this chamber is to measure the oxygen consumption of small animals and thus indirectly the metabolism of the animal.

Materials: Milk bottle or wide-mouth quart jar, glass tubing, mason jar adapter lid, one-hole rubber stopper, a wire mesh to enclose the animal, and soda lime or some other material that will absorb carbon dioxide. A tablespoonful is needed for a one-quart container.

What to Do: Place the bottle on its side. Insert soda lime or similar material. Enclose the small animal, such as a white rat, in a wire screen. Tape the screen so the animal can't move about.[30] Measure how much water the glass tubing will hold by using an eyedropper that is marked in cc. Divide the glass tubing into convenient marks for measurement. Place the rat or other small animal into the bottle and insert the stopper and glass tubing. Form a soap bubble at the end of the tubing outside the bottle.

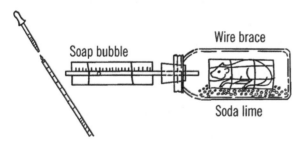

Soap bubble
Wire brace
Soda lime

Operation of Equipment: As the rat breathes, he uses oxygen. The rat exhales carbon dioxide but the carbon dioxide is absorbed by the soda lime. Therefore, as the oxygen is used up, the soap bubble travels up the glass tubing. The volume of the space traveled by the soap bubble is equal to the oxygen used. Time this for a one minute period. Multiply this volume (for example, 3/100th of a liter) in liters (about a quart) times 4.8 calories to find the metabolism or rate of burning or using food material.

Can You Work Like a Scientist?

1. If you grow plants in the same bottle, can you produce enough oxygen so that none needs to come from outside the bottle? Would the soap bubble move?

2. Is the metabolism rate always the same? See the effect of fear, different foods, alcohol, tobacco, excitement, etc. on the rate of metabolism.[31]

3. Is the metabolism rate of all animals alike? Try frogs, etc.

4. Start small plants growing and place some water and a plant in a test tube. Insert a snail and see how long you can keep it alive in a balanced aquarium.

5. Try to make a balanced aquarium with goldfish, water plants, a gallon jug, and a rubber stopper.

6. You can measure the metabolism of very small animals by placing a number of them in the jar. If you use ten animals, you divide the oxygen consumption by ten.

Aquarium

Purpose: The aquarium is used to carry out experiments in the science laboratory, which requires a large container of water. It is also, of course, a place to keep fish and other aquatic life so that a study can be made of their habits, food, and reaction to stimuli.

---

30 Is this a reasonable thing to do to a small animal? One might argue either way. See Note 1 in Appendix E.

31 As we have noted repeatedly, much more care must be taken with animal experiments than in the era when this book was originally published. See Note 1 in Appendix E.

Materials: Large-mouth gallon jar or gallon jug, nichrome wire bottle cutter.

What to Do: You may use a large gallon jar, just as it is, for your aquarium. A better aquarium, however, can be made by cutting the top off a gallon jug with your nichrome wire bottle cutter. Be sure to smooth the glass edge with wet or dry emery paper.

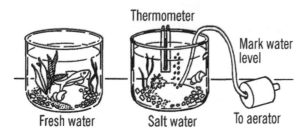

Thermometer

Mark water level

To aerator

Fresh water   Salt water

Operation of Equipment: Fill the bottom of your fish aquarium with about one inch of gravel. Be sure the rocks contain no limestone or acid-forming metals. Slope the gravel so you can place the plants near the back. Plants are necessary to help provide the necessary oxygen. Your pet or feed store will recommend the right types. Fill the aquarium and let the plants root for about a week before adding fish. Since tap water contains chlorine, you must let it stand for at least 48 hours before using it for the aquarium. When you add fish to the tank, set the box from the pet store in the water. The fish then get used to the new temperature. After 15 minutes you can place the fish directly in the tank. Don't overfeed fish. Never give them more than they can eat in ten minutes. Clean the tank once a week unless it is well balanced with plants.

Can You Work Like a Scientist?

1. Why do you need plants in the water?

2. You can add more oxygen to the tank by pouring the water over and over with a cup. How could you keep the oxygen supply coming in at night?

3. Why is the wider bottom of a gallon jug better than a widemouth gallon jar for an aquarium?

4. What effect does chlorine have on organisms? If you have a culture of paramecium growing, try a drop of chlorine in a spoonful of culture. Examine under the microscope.

5. You can start a saltwater aquarium. Collect salt water from the ocean. This water contains the right amount of minerals. Add your sea life, but make sure you have only a few small specimens. The temperature should be about 65 degrees, so keep the aquarium in a basement or other cool place. Make a mark as to the height of the water. As the water evaporates, the remaining water gets saltier, a condition which will soon kill off the life. In order to keep the salt content the same, add distilled water to fill the aquarium to the mark. Rain water will serve as distilled water. Bits of hamburger or liver serve as food for sea anemones and other life. Hermit crabs help keep the tank clean. You will probably need an aerator.[32]

Terrarium

Purpose: The classroom or home terrarium allows a student to study animal life in its natural environment.

---

32 Saltwater aquariums are now common. Pet stores that sell them can show you how to build a sustainable environment. What (other than the salt) is different in a saltwater aquarium?

Materials: Gallon jug, wire screen or glass plate. A second type utilizes a gallon jar with lid, and a wooden stand as shown.[33]

What to Do: Cut off the top of a gallon jug with your nichrome wire bottle cutter. Smooth with wet or dry emery paper in a bucket. Use a wire screen or a piece of glass plate for a cover. If you want to make the gallon jar terrarium, cut part of the lid off and cover with a wire screen.

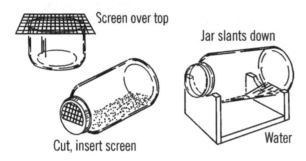

Screen over top

Jar slants down

Cut, insert screen

Water

Operation of Equipment: Wash some coarse sand and fill the jar to a depth of two inches. Put an equal amount of rich soil on top of the sand. Add some rocks and dead wood. Plant small ferns and bushes. Have some in the shade and some exposed. Water with a squeeze bottle sprinkler. Make this by punching holes in the lid of a squeeze bottle. This woodland terrarium makes a good home for toads, small snakes, young box turtles, some salamanders and lizards.

A semi-aquatic terrarium is suitable for frogs, turtles, and most salamanders. The woodland terrarium will work if the water and land areas are separated. The gallon jar aquarium is best for this. The jar can be tipped at a slight slant, and the water is confined to the bottom of the aquarium. Moisture in both aquariums should be kept quite high.

A desert terrarium is suitable for cactus and desert plants as well as lizards. The sand should be washed and mixed with charcoal and a small amount of lime. Ask your florist about this and about plants.

Turtles eat flies, turtle food, ground meat, bits of fish, lettuce, and other leafy vegetables.

Snakes eat live mice, frogs, earthworms, insects, and they can be coaxed into eating eggs and raw meat. If a snake refuses to eat after a week, it should be released.

Tadpoles eat small bits of lettuce and finely chopped meat. Frogs eat all kinds of insects and worms. Also big frogs eat little frogs. Frogs eat only moving objects, so dangle the worm or insect in front of the frog with tweezers or a string.

Lizards eat all types of insects, live flies, beetles, grasshoppers, and crickets. In the wintertime mealworms will serve as a food supply. Ask a pet store employee how to raise mealworms.

Salamanders in the larval stage eat bits of liver and the yolks of hard-boiled eggs. As adults they eat insects, worms, and bits of raw meat. Most salamanders are active only at night.

Dissecting Needle

Purpose: The dissecting needle is a tool of many purposes. It can be used for making a vaseline ring on a microscope slide. The main purpose of the needle is to help hold tissue during dissections or to remove small parts of insects for further study.

Materials: Common pin, short piece of ¼" wooden dowel (available at any lumber store).

What to Do: If you use a pin, cut off the head. File this tip and then force it into the end of the wood dowel. The end of the pin can then be bent to form any kind of hook needed. You can make a dissecting needle out of a regular nee-

---

33 This is only enough room for a *very* small animal. A terrarium should provide ample moving space for the animals within. Consider starting with a 10-gallon aquarium, for all but the very smallest animals.

dle in the same way described. However, you cannot bend the point, or the needle will snap.

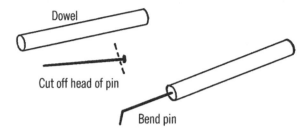

Dowel

Cut off head of pin

Bend pin

Dissecting Pins

Purpose: The pins are used to hold the skin or tissue in place during a dissection. A pine board is soft and makes a good dissecting board.

Materials: Common pins will serve for dissecting purposes. However, a better substitute is the large round-headed pin used by dressmakers. These are available at most sewing supply counters.

What to Do: Try dissecting an earthworm or a small snake. Remember, do not torture animals. Have a purpose before doing dissecting work.[34] An excellent book to consult is *The Living Laboratory* by James Donald Witherspoon and Rebecca Hutto Witherspoon. This book is published by Doubleday & Company, Inc.

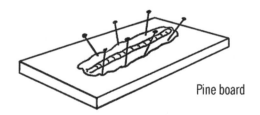

Pine board

Dissecting Knife

Purpose: The dissecting knife is used in work on animals, insects, etc. It must be very sharp and capable of making a fine incision.

Materials: Single-edge razor blade, ½" wood dowel, small pin or brad.[35]

What to Do: Cut a slit at the end of the dowel. Insert the razor blade as shown. Drive a pin or brad through the side of the dowel to hold the razor blade in place.

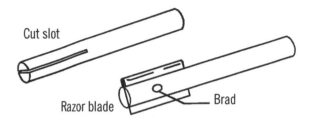

Cut slot

Razor blade

Brad

Animal Cage

Purpose: An animal cage is used to house animals during periods of observation or experimentation. The animal cage described is easy to make, clean, and store.

Materials: Round candy or cookie tin, half-inch mesh hardware cloth, and materials for a water bottle (baby bottle, one-hole stopper, glass tubing).

What to Do: Bend the hardware cloth into a cylinder. Slip the cylinder of wire into the bottom of the candy tin. The lid of the candy tin fits over the top of the wire cylinder. This lid can be held in place with rubber bands or wire hooks attached to the lid.

A water bottle can be made by bending a piece of glass tubing and inserting it in a rubber stopper. The stopper is then inserted in a bottle

---

34 See Note 45 in Appendix E about dissections.

35 A common hobby knife (X-Acto, for example) will also work well.

containing water. The "drinking fountain" is held in place with rubber bands. The water does not leak out because air is pushing against the glass tubing. The water does fit the glass tubing and forms a bubble of water at the end of the tubing. Animals such as white rats, hamsters, and others lick the water off the end of the tubing.

The animal cage pictured can be stored by removing the top and bottom of the candy container and straightening out the wire cylinder. The wire can be stored flat and thus does not take up needed shelf space.

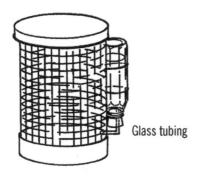

Glass tubing

Problems to Investigate in the Study of Biology

Note: Many of the problems here involve working with animals. Please use appropriate research guidelines. see Note 1 in Appendix E.

(P)-Primary

(I)-Intermediate

(U)-Upper

Conditions Necessary for Life

1. Are there any places on earth where no living things exist? *(P)*

2. How far down into the ground would you have to go to find no life? *(I)*

3. What are the temperature limits for life of different plants and animals? (What are the coldest and warmest temperatures at which they will live?) *(I)*

4. Is water necessary for all forms of life? *(P)*

5. What is the minimum water requirement for various plants and animals? *(I)*

6. Does water contain air? *(P)*

7. Do fish need air in water in order to live? *(P)*

8. How much oxygen is used by various forms of plants and animals? *(U)*

9. Can various plants and animals use pure oxygen? *(U)*

10. How can we get nitrogen to dissolve in water? *(U)*

11. How do plants get nitrogen? *(I)*

12. Can plants live without nitrogen? *(I)*

13. How much nitrogen is needed by different forms of animals? *(U)*

14. How much carbon dioxide is used by different plants? *(I)*

15. Can plants live in pure carbon dioxide? *(U)*

16. How much food is required by different forms of animal life? How does this compare with the oxygen consumption? *(U)*

17. What plants need light in order to grow? *(P)*

18. How much light a day must different forms of plant life have? *(I)*

19. Will some plants grow in different forms of artificial light? *(I)*

20. What effect does colored light have on the growth of various plants? *(I)*

21. How much light is necessary for mushrooms to grow? *(I)*

22. What animals seek light? *(I)*

23. What animals avoid light? *(I)*

24. How are the eyes of various living things alike and unalike? *(I)*

25. What plants and animals can adjust to changing environments? What plants and animals are unable to adjust? *(P)*

26. How do the different animals find food? *(P)*

27. How do the different animals avoid being eaten? *(P)*

28. How intelligent are various animals? Can you set up a test for them? *(I)*

29. What effect does the changing of the coloration of an animal have on its ability to live in its environment? (Paint an insect or other animal.) *(I)*

30. How are some animals able to change their coloration in order to blend in with their environment? *(U)*

31. What insects cannot be eaten? Experiment with insect-eating animals. *(I)*

32. Why are some animals colored with warning colors? (Examples: skunk, bumblebee) *(P)*

33. What animals are protected because of their general appearance? *(I)*

34. What animals (particularly insects) depend on being mistaken for other animals? *(I)*

35. How far do various animals move in a day? Can you calculate the speed? *(I)*

36. Do all animals move? *(P)*

37. How many different kinds of motions can you observe among animals? *(P)*

38. Do plants move? *(P)*

39. What causes movement in plants? *(P)*

40. Can some plants eat living things? How? *(I)*

41. What kind of digestive juices are used by various animals to digest their food? *(U)*

42. How are various animals equipped to get air? *(P)*

43. Do plants breathe? *(P)*

44. How much of various food materials is turned into energy? How much food material is not used but excreted? *(U)*

45. What are the processes of excretion (getting rid of waste materials) in various forms of animals? *(U)*

46. Can you time the rate of response to various stimuli? (Some common stimuli are heat, pain, electrical current, odor, fear, hunger, light, darkness, noise, moisture, and gravity.) *(I)*

47. Do the roots of all plants grow toward moisture? *(P)*

48. Do the leaves of all plants grow toward light? *(P)*

49. Are plants intelligent? *(I)*

50. How do various plants and animals reproduce? *(P)*

51. How closely do the young plants and animals resemble their parents? *(P)*

52. What plants and animals are not sensitive to stimuli? *(I)*

53. What do various objects look like under the microscope? *(I)*

54. How do the cells of various plants and animals compare? (Use a microscope.) *(I)*

55. Do all living things have cells? *(P)*

56. Are all the cells of a living thing alike? *(I)*

57. What do the cells of an onion membrane look like? (Place in a drop of water and cover with a cover glass.) *(I)*

58. How does staining a material help in locating the parts of a cell? (Use iodine, methylene blue, or ink as a stain on the onion membrane.) *(I)*

59. What is the cell structure of a geranium or a rhubarb leaf? *(I)*

60. Do all plants have cell walls? *(P)*

61. Is the protoplasm of all plant cells alike? *(U)*

62. Is the nucleus of all plant cells alike? *(U)*

63. Do all cells have a nucleus? *(I)*

64. How are the vacuoles in various cells alike and different? What is the purpose of vacuoles? *(U)*

65. Does dandruff contain cells? *(I)*

66. What does the protoplasm inside your cheek look like? (Scrape your cheek with a toothpick and place the material in a drop of water on a glass slide.) *(I)*

67. How do the size of cells vary in various plants and animals? *(I)*

68. How big are the largest cells? (Try the yolks of various birds' eggs.) *(P)*

69. To what lengths can cells grow? (Try various nerve cells.) *(U)*

70. How big are the smallest cells? *(U)*

71. Are bacteria living things? Are they plants or animals? *(I)*

72. Are viruses living things? *(U)*

73. What is the difference between cells and tissues? *(I)*

74. How do different kinds of tissues in various animals compare (covering tissue, muscle tissue, and connective and supporting tissue, such as blood, nerve, fat, bone, and fibrous connective tissue)? *(U)*

75. What is the difference between tissues and organs? *(I)*

76. How do the tissues from various heart organs compare? *(U)*

77. Do plants have organs? Can you compare the organs of various plants? *(U)*

78. Is protoplasm a mixture or a compound? *(U)*

79. What percent of various plants and animals is water? (Try different fruits and vegetables as well as various insects.) *(I)*

80. What materials make up protoplasm? Can you make up artificial protoplasm? *(U)*

81. Are plants as sensitive to stimuli as animals? *(P)*

82. How do various living things grow? *(P)*

83. How does a bone grow? *(U)*

84. Grow a mimosa plant. Try various experiments with this unusual plant. *(I)*

85. What are the conditions necessary for the life of a brine shrimp? *(I)*

86. How are living things classified? *(I)*

87. Collect various plants. Can you divide the plants up into their correct phyla? (Remember the four basic parts: roots, stems, leaves, and flowers.) *(U)*

88. How are animals classified? Can you classify various animals? *(I)*

89. Collect various forms of thallophytes (plants without roots, stems, leaves, or flowers). Can you divide these into their correct subphyla: algae or fungi? *(U)*

90. Are algae found in both fresh and salt water? *(I)*

91. How many different kinds of algae can you locate? What is the difference in the cells of these algae? *(P)*

92. What is the appearance of algae? Scrape some algae off the side of a tree (pleurococcus) and compare under the microscope with pond scum algae (spirogyra). *(P)*

93. Can you divide various algae into their color classes: green, blue-green, brown, and red? *(U)*

94. Will fish eat algae? *(P)*

95. Can algae be used as fertilizer? *(I)*

96. Can you extract iodine from kelp algae? *(U)*

97. Can you use algae as a food material? *(I)*

98. What is the rate of growth of different algaes? *(P)*

99. Will algae grow without sunlight? *(P)*

100. Under what conditions will algae grow the best? *(P)*

101. Does algae seek light? (phototropism) *(I)*

102. How does algae reproduce? *(U)*

103. Is some saltwater algae adaptable to fresh water? *(U)*

104. What animals will eat food made of algae? *(U)*

105. Is algae nourishing? *(I)*

106. Can you collect different kinds of fungi and divide them into parasites and saprophytes? *(U)*

107. Which fungi are helpful and which are harmful? *(U)*

108. How small are various forms of bacteria? Can you measure bacteria? *(I)*

109. Can you find bacteria in the air? *(P)*

110. Can you find bacteria in water? *(I)*

111. Can you find bacteria in soil? What conditions produce the most bacteria? *(P)*

112. Which bacteria live on dead material? *(U)*

113. Which bacteria live on living material? *(U)*

114. What do bacteria look like? Can you divide bacteria into their groups (rod-shaped, spherical, curved, or spiral)? *(U)*

115. Where do bacteria grow? *(P)*

116. What kinds of bacteria do you find around your home? *(P)*

117. What kinds of bacteria do you find in your school? *(I)*

118. What kinds of bacteria do you find on your body? *(U)*

119. How fast do bacteria grow under certain conditions? *(P)*

120. Will bacteria grow in darkness? *(P)*

121. Will bacteria grow in freezing temperatures? *(P)*

122. What effect does humidity have on the growth of different kinds of bacteria? *(I)*

123. Can you experiment with beneficial uses of bacteria? *(U)*

124. Can you experiment with harmful uses of bacteria? *(U)*

125. How does bacteria help the food industry? Be sure to experiment with these processes. *(U)*

126. Can you repeat the work and discoveries of Anton van Leeuwenhoek and his work with bacteria? *(U)*

127. Can you repeat the work of Louis Pasteur and his experiments with bacteria? *(U)*

128. What does yeast look like under the microscope? How large are yeast cells? *(P)*

129. What bacteria cause disease in animals? *(U)*

130. Are all yeast cells alike? *(I)*

131. What effect do yeast plants have on sugar? Can you identify the products given off? *(U)*

132. Under what conditions does yeast grow the best? Vary the temperature, solution, light, etc. *(I)*

133. Are all molds alike? *(P)*

134. How do molds reproduce? What conditions are favorable for mold reproduction? *(I)*

135. How large do molds grow? *(P)*

136. Is light necessary for the growth of molds? *(I)*

137. Are the tubes of mold plants attracted by the pull of gravity? *(I)*

138. Can you produce mutations with bacteria, yeast, and molds? Try exposing them to ultraviolet, and other forms of light rays and radiation. *(U)*

139. What mushrooms are edible? Which are poisonous?[36] *(U)*

140. Will mushrooms grow in the dark? In the light? *(I)*

141. What temperature extremes can mushrooms stand? *(I)*

142. How do mushrooms reproduce? What conditions are the most favorable? *(U)*

143. Under what conditions will lichens exist? *(P)*

144. Will lichens grow without moisture? Sunlight? *(I)*

145. On what material will lichens grow? *(P)*

146. Will lichens continue to grow under freezing conditions? *(U)*

147. What do the cells of lichens look like under the microscope? *(I)*

148. What is the effect of radiation on lichens? *(U)*

149. What is the altitude range for various forms of lichens? *(U)*

150. What environments are favorable for the growth of mosses? *(I)*

151. Why do mosses grow so low to the ground? *(P)*

152. Where are liverworts found? What conditions are favorable for their growth? *(I)*

153. Will liverworts live in water? *(I)*

154. What are the beneficial uses of liverworts? *(I)*

155. Can you experiment with the environment of various ferns found in the area? *(U)*

156. Why do ferns grow higher than mosses? *(P)*

157. Can ferns adapt to a saltwater environment? *(I)*

158. Under what conditions will the horsetail fern continue to survive? *(I)*

---

36 Many mushrooms—even some that look very similar to edible mushrooms—can sicken or kill you. Learn safe mushroom collecting practice from guides or a mushroom club before starting. *Never* test your results while collecting mushrooms without a 100% positive identification, preferably from more than one source. Consult with an expert (a researcher or professional who works in the field of mycology) when necessary.

159. Can you produce mutations in various seed plants? *(U)*

160. How are the seeds of various plants transferred for reproduction? *(P)*

161. How much water is really needed by the various seed-growing plants? *(P)*

162. What kind of conifers are in the region around you? What percentage of each do you find? *(I)*

163. How do the seeds of various conifers compare? What is their weight and area? Under what conditions will the seeds start growing? *(I)*

164. Can you experiment with different kinds of seeds and then test whether they will grow? *(P)*

165. Can you experiment with the reproductive parts of various flowering plants? *(U)*

166. Can you divide seeds from many different plants into their classes: monocots (one-piece) and dicots (two-piece)? *(U)*

167. Are Euglena animals or plants? *(U)*

168. Do the Euglena need light in order to live? *(I)*

169. How do different protozoa (amoeba, paramecium, vorticella, etc.) react to changing environment, such as heat, light, electrical field, magnetic field, various gases, and other protozoa? *(U)*

170. What do various forms of protozoa eat? *(I)*

171. Where are various forms of protozoa found naturally? *(I)*

172. What are the problems of culturing various forms of protozoa? *(I)*

173. Can you cause mutations among the different protozoa? *(U)*

174. How do the different protozoa reproduce? Study only one at a time. *(U)*

175. Can protozoa think? Experiment with different stimuli. *(U)*

176. Can sponges adjust to water containing less salt than ocean water? *(U)*

177. How much water can various types of sponges hold? What determines this capacity? *(U)*

178. Can you start sponges growing in a different environment? *(U)*

179. How do sponges reproduce? *(U)*

180. What do coral eat? Can coral adjust to changing temperature of water? *(U)*

181. How are coral islands formed? Can you determine the rate of growth of such a coral deposit? *(U)*

182. Where can hydra be found around your area? *(I)*

183. What will hydra eat? How do hydra feed? *(I)*

184. Are hydra affected by electrical stimuli? *(U)*

185. How do hydra reproduce? *(I)*

186. How do jellyfish compare with hydra? *(I)*

187. How long does it take the nerve cells of a jellyfish to react? *(U)*

188. What will jellyfish eat? How do jellyfish locate their food? *(U)*

189. What is the power of the stinging cells of the jellyfish? *(U)*

190. Are jellyfish adaptable to changing environment? *(U)*

191. Will jellyfish regenerate parts? *(I)*

192. What parasitic worms live in or on different animals? *(U)*

193. Can parasitic flatworms regenerate parts? *(I)*

194. Can Planaria regenerate parts? Try different experiments with Planaria. *(I)*

195. What hostile conditions can Planaria survive? *(I)*

196. Can you experiment with the life cycle of the tapeworm? *(U)*

197. Can tapeworms regenerate parts? *(U)*

198. What temperature is necessary to destroy various types of roundworms and flatworms? *(I)*

199. Does apple cider contain roundworms? Are these worms harmful? *(I)*

200. Can you conduct various experiments with vinegar eel worms? *(U)*

201. What effect do earthworms have on the soil? *(P)*

202. Do earthworms prefer an acid or alkaline soil? *(I)*

203. How do earthworms breathe? *(P)*

204. How do earthworms reproduce? Can you raise worms? *(P)*

205. How does light stimuli affect earthworms? *(P)*

206. What temperature variations can earthworms stand? *(I)*

207. What parts of an earthworm can regenerate? *(P)*

208. Are earthworms attracted by electricity? *(I)*

209. How do earthworms react in a magnetic field? *(U)*

210. Can earthworms learn? *(I)*

211. What is the metabolism rate of the earthworm? *(U)*

212. Are earthworms ever frightened? *(U)*

213. How far under the surface of the ground can earthworms live? *(U)*

214. Can you experiment with the internal structure of the earthworm? *(U)*

215. Are fish attracted to the color or odor of the earthworm? *(I)*

216. What kind of diet is best for earthworms? *(I)*

217. Can you cause mutations in earthworms? *(U)*

218. How strong is the grip of a starfish? *(I)*

219. What will starfish eat? *(I)*

220. How far will a starfish move in a day? *(I)*

221. Will all starfish atomize (throw off an arm)? *(P)*

222. Will one arm of a starfish regenerate a whole body? Experiment with various forms and conditions of regeneration among starfish. *(I)*

223. Can starfish tolerate some fresh water? *(I)*

224. Can starfish stand great temperature variations? *(I)*

225. Do starfish exhibit a positive or negative trophism toward light? Is this true for all wave lengths of light? Try colored light. *(I)*

226. How strong is the resistance of the bivalve mollusk against having its shell opened? *(I)*

227. How fast can a mollusk move? (Pick a certain member, such as a clam.) *(I)*

228. Can you grow a pearl in an oyster? *(U)*

229. How does a sea snail keep water in his shell and air out of his shell when he is exposed to the air between tides? *(I)*

230. How watertight is the operculum of different members of the ocean snail

family? What materials and liquids can these snails resist and for what periods of time? *(U)*

231. How fast can different members of the land snail family move? *(P)*

232. How fast is a slug? Compare the land slug with the sea slug. *(I)*

233. Can members of the octopus family re-generate parts of their body? *(U)*

234. What is the composition of the poison of the octopus? *(U)*

235. Can lobsters see? *(P)*

236. Can a lobster or crayfish swim? How do they move? *(P)*

Spiders and Insects

237. How do different spiders spin their webs? *(P)*

238. How strong are the webs of different spiders? *(I)*

239. Do all spiders bite? How? *(P)*

240. What do spiders eat? *(P)*

241. Do all spiders spin webs? *(P)*

242. Will one spider get caught in the web of another kind of spider? *(P)*

243. How does a spider avoid being caught in his own web? *(I)*

244. How fast can the different members of the spider family move? *(I)*

245. Can spiders see? How well can they see? *(I)*

246. Can spiders detect odors? *(U)*

247. What environmental changes can the spider withstand? *(I)*

248. What environmental changes (gases, temperature, etc.) can the different mites withstand? *(U)*

249. Does each spider have but one web? Do more than one spider share the same web? *(I)*

250. How many legs do centipedes have? How about millipedes? Do they vary? *(I)*

251. What do centipedes eat? *(I)*

252. Do millipedes exhibit a positive or neg-ative tropism toward light? *(I)*

253. What do millipedes eat? *(I)*

254. What kind of insects live in your area? Pick out one small area and keep a re-cord throughout the year. *(P)*

255. What insects are helpful to man? How? Can you find examples? *(P)*

256. What insects are harmful to man? How? Can you find examples? *(P)*

257. What are the characteristics of an in-sect? (Examine a grasshopper.) *(I)*

258. How do insects smell and feel? *(P)*

259. Can you investigate the vision of vari-ous insects? *(I)*

260. Can you measure the number of vibra-tions per second that various insects move their wings? *(I)*

261. How do insects hear? Can you compare the hearing ability in various insects? *(U)*

262. How do various insects breathe? Can you figure the metabolism rate? *(U)*

263. How do various insects reproduce? How do you tell the males from the fe-males? *(U)*

264. Can you experiment with some of the body processes of insects? *(I)*

265. Can you experiment with the eggs of various insects to determine hatching conditions? *(U)*

266. Can you determine the amount of plant material a caterpillar or other insect will eat in the larval stage? *(I)*

267. Can you collect various insects and divide them into their correct order according to their wings or other determining factors? *(I)*

268. Can you collect and study several types of moths and butterflies in order to determine the difference between the two? *(I)*

269. Can you breed houseflies and determine their life cycle? *(I)*

270. Can you experiment with the environment of the housefly, with an idea of exterminating the fly? *(U)*

271. Can you experiment with the raising of mosquitoes? *(U)*

272. Can you experiment with the environment of various types of mosquitoes? *(U)*

273. Can you experiment with different types of odors and their effects on mosquitoes? *(U)*

274. How does man control the mosquito? *(I)*

275. Can you measure the strength of various insects? *(I)*

276. Can you measure the flying speed of various insects? *(U)*

277. What insects will destroy others? Can you experiment to determine this? *(I)*

278. How strong are the wings of various insects? *(U)*

279. Can you experiment with the diet of various insects? *(I)*

280. Can you determine the weight of various types of insects? Can you take other physical measurements? *(U)*

281. Can you compare the wings of various insects as to construction, size, etc.? *(P)*

282. What insects are attracted to light? What insects have a negative tropism? *(P)*

283. How much light is given off by a firefly? *(U)*

284. What insects do damage in your locality? *(P)*

285. Can you compare the pollination of certain plants by insects as compared to similar plants that have been screened to prevent insect pollination? *(U)*

286. How do various insects spend the winter? Where can you find them during the winter? *(P)*

287. Why are insects so successful? *(U)*

288. Can insects think? How intelligent are the various forms? *(U)*

289. What insects must be exposed to a constant source of moisture in order to avoid drying up? Can you test various insects, including some found in the soil? *(I)*

290. What is the effect of changing temperature on the rate and duration of light from the firefly? *(U)*

291. What are the mineral requirements of the fruit fly, Drosophila? *(U)*

292. What do fruit flies eat? *(P)*

293. What is the effect of high and low temperatures on the housefly? *(I)*

294. What is the effect of different sound waves on bees, particularly drones? *(U)*

295. Can insects such as grasshoppers communicate? Try separating the males and females and use the mating sounds to test your idea. *(U)*

296. How do different insects react to the stimuli of loud sounds? *(P)*

297. Can water beetles hear under water? What frequencies are the easiest to hear? *(U)*

298. When you try to catch a fly, is it the shadows or the air current which give the fly a warning? *(I)*

299. Can flies learn or become conditioned? Try conditioning them to escape from some trap. *(I)*

300. What insects exert a positive tropism or attraction toward a breeze or small wind? Which insects will face into the wind? *(I)*

301. Will the human breath activate insects? Try ants, stick insects, or butterflies on cold mornings or in cool temperatures. *(I)*

302. What is the effect of nicotine on various insects? Can certain insects stand cigarette smoke? *(U)*

303. How long can certain insects live on the following diets: sugar water; raw meat? Be sure to try blowflies as well as other insects. *(I)*

304. Does an insect's diet affect reproduction? *(U)*

305. How long does it take mosquitoes to digest blood? Remember, you can see the red coloration of the blood through the swollen abdomen. *(I)*

306. What is the effect of vision on an insect's movement? Blindfold the insect by painting the eyes black. *(I)*

307. Can you hypnotize stick insects? Try rolling them between your fingers. *(I)*

308. What other animals can become immobilized? (chicken, crayfish?) *(U)*

309. Why do some people react more violently to mosquito bites than others? *(U)*

310. How does the length of the trachea affect the size of an insect? *(I)*

311. What effect does temperature have on the frequency and amplitude of the respiratory movement of insects? Use stick insects, bees, moths, or beetles. The insect can be magnified by being placed in the beam of a projector. Watch the shadow. *(I)*

312. What is the effect of added carbon dioxide in the air on the respiratory rate of insects? *(I)*

313. What conditions are necessary for a rich oxygen supply in water? *(P)*

314. What is the effect of stagnant or warm water on the amount of oxygen dissolved in the water? *(I)*

315. What effect does water with little dissolved oxygen have on the respiration of such water insects as water beetles and water insect larvae? *(U)*

316. What periods of the day are certain insects the most active? *(P)*

317. Will the pattern of activity set by insects be upset by periods of continuous darkness and periods of continuous illumination? *(I)*

318. What is the walking order of the legs of various types of insects? How does the insect modify his walking pattern with the removal of one or more legs? *(I)*

319. What insects will atomize their legs easily (shed one or more legs)? *(P)*

320. How long does it take various forms of insects to make the necessary adjustments in coordination after the removal of one or more legs? *(U)*

321. Can you calculate the wingbeat of various insects? What is the effect of added load to the wings (wax added) on the number of vibrations? *(U)*

322. What is the effect of temperature on the time it takes a moth to flutter before it can fly? (Keep one hot and one cold.) *(I)*

323. How long can a fly stay submerged in water or alcohol and still live? Submerge an insect in a mixture of alcohol and paraffin and then observe under a lens. *(I)*

324. Why don't insects dry up from evaporation of the water in their bodies? Remove the wax from the cuticle of the insect by dusting the cuticle with charcoal powder. *(U)*

325. What insects may be charged by static electricity? *(I)*

326. What insects can change their body color? How is this change possible? *(U)*

327. How do caterpillars react to a physical stimulus such as a slight touch? Experiment on various parts of the insect. Are all reactions alike? *(I)*

328. What are the reactions of the caterpillar to stimuli such as heat and cold? *(P)*

329. How do various insects, such as caterpillars, react to vibrations? *(P)*

330. Do water beetles react to red light? *(I)*

331. How do water beetles avoid contact in a pond or aquarium? *(U)*

332. What effect has contact of the tarsi of an insect with a solid object have to do with the flight of the insect? *(U)*

333. How do different insects regain their normal posture when turned over? Can you inhibit this? *(U)*

334. Do stick insects exhibit a falling reflex? How does this compare with a cat? *(I)*

335. What effect does vision have on the movement of the dragonfly? Paint the eyes. *(U)*

336. Can bees be trained to go to a water container? *(U)*

337. How does the smelling ability of various insects differ? Have a test tube containing food and another without food. Make an insect maze. *(I)*

338. How does the touching of the tarsi of the front legs of various insects (bees, flies, etc.) with a sugar solution affect the movement of the mouth parts? *(U)*

339. Can bees distinguish between saccharin and sugar? Can you experiment by adding different substances to syrup and then testing the bees' preference? *(U)*

340. Can bees discriminate or detect higher and lower concentrations of sugar? *(U)*

341. Can bees learn? Use the results from the previous question. Use dishes of different concentrations of sugar. Later test the memory of the bee. *(I)*

342. How do ants follow the trail of other ants? Try making an artificial trail with a weak solution of formic acid. *(I)*

343. How does a bee communicate to other members of a hive the direction and location of a rich source of honey? *(U)*

344. Which insects have a negative tropism toward the pull of gravity? Try water bugs and other insects. Use a seesaw that tilts under the weight of the insect. *(I)*

345. Why does an insect crawl up to the end of a branch or leaf and then reverse direction and go downward? Try flies, ants, etc. in a sealed bottle. Can you

change the direction the insect wants to go by turning the bottle end for end? *(P)*

346. Is the pig louse attracted by temperature and/or smell? What is the effect of different kinds of surfaces upon its movement? *(U)*

347. What is the preferred temperature of various insects, such as flies, beetles, fruit flies, etc.? *(I)*

348. Can you experiment with the effect of vision on an insect's ability to move? Paint the right eye on some, the left on others, and both on some. *(U)*

349. Can you experiment with the effect of different colored light and the absence of light on the flight of the moth and other insects? *(P)*

350. What larvae, pupae, and marine life react to the stimuli of a moving shadow? Is the reaction due to the movement or the change in light intensity? *(U)*

351. When you chase flies, why do they head for the window? Is the reaction due to the light? *(I)*

352. Do social insects such as ants, bees, and termites use the position of the sun in order to help find their directions? *(U)*

353. Will certain insects, such as beetles, caterpillars, and stick insects change their direction of movement when the direction of light (use a flashlight bulb) is changed? Try using two different light sources that can be turned on and off. *(U)*

354. Can you investigate the mating instincts of the housefly? *(U)*

355. What are the problems involved in training bees? Can you devise a bee IQ test? *(U)*

356. What forms or designs can bees distinguish? *(U)*

357. Can bees tell time? Feed bees a sugar solution at a certain time each day. Will bees show up only at this time after a few days of conditioning? *(I)*

358. What changes occur in insects (such as stick insects) after molting? Be sure to check all body parts and weigh the insect. *(U)*

359. What insects have the power of breaking off their body parts? Which ones can regenerate these parts? Try stick insects and grasshoppers. *(I)*

360. Does regeneration of legs occur just in the young insects, or can adults do it too? *(I)*

361. Do bees visit more than one type of flower, or does each individual bee stick to one particular type? *(U)*

362. Upon what conditions does the number and size of larvae in a given culture medium depend? *(I)*

363. What is the difference between a butterfly and a moth? Can you observe both through a life cycle? *(P)*

364. What insects are bugs? What is the difference between bugs and other insects? *(P)*

365. How do the wings of various insects differ? *(P)*

Fish, Reptiles, and Amphibians

366. Are fish cold- or warm-blooded? Can you measure the temperature of a fish? *(I)*

367. How do scales help fish? How do the scales of various fish compare? *(I)*

368. How do different fish move their fins? What determines the rate of movement of fins? *(U)*

369. How do the gills in a fish remove oxygen from the water? *(I)*

370. What fish can adapt to changing amounts of salt in the water? *(I)*

371. What is the difference between toads and frogs? Can you discover this by observation? *(P)*

372. Can you keep a record of the changes in the life cycle of a frog from the egg to the adult stage? *(P)*

373. What effect does temperature have on the hatching of frog eggs? *(I)*

374. What animals will eat frog eggs? *(U)*

375. What effect does temperature have on the mating instinct of frogs? *(U)*

376. Do people get warts from touching toads? *(P)*

377. What is the difference between salamanders, newts, and lizards? Can you discover this by observation? *(P)*

378. How far can frogs jump? *(P)*

379. What effect does temperature have on the rate of movement of different kinds of amphibians? *(I)*

380. How and what does an amphibian eat? *(P)*

381. What effect does temperature have on the rate of motion of various lizards? Try snakes, turtles, horned toads, etc. *(P)*

382. What changes in environment can turtles withstand? Can turtles be frozen in a block of ice and still live? *(I)*

383. Can you analyze the poison from different reptiles? *(U)*

384. How are poisonous animals useful? *(U)*

385. Will turtles eat frogs? *(I)*

386. How strong are turtles? What is one turtle power? *(P)*

387. How strong is the turtle's jaw? *(I)*

388. What effect has the direction of light on a turtle's motion? *(I)*

389. How does a chameleon change color? *(I)*

390. What happens to reptiles and amphibians in the winter? Can you use a refrigerator to duplicate these conditions? *(I)*

391. What is the difference between crocodiles and alligators? *(P)*

392. What kinds of snakes are found in your locality? What can you observe about them? *(P)*

393. Are snakes intelligent? Can snakes remember? *(I)*

394. What percentage of the population is afraid of snakes? How does this change with age? *(P)*

395. Why do many people dislike snakes? Is this based on personal experience? *(P)*

396. What are the food habits of various snakes? *(I)*

397. What squeezing force can be exerted by different members of the constricting family? *(U)*

398. What snakes are nocturnal? Can some snakes see in the dark? *(I)*

399. How does a snake smell? *(I)*

400. Can snakes see? Do they depend on vision or smell in searching for food? *(U)*

401. How do snakes react to different stimuli, such as sound, different colored lights, etc.? *(P)*

402. How do snakes get moisture? Can snakes live without water? *(P)*

403. How does a snake move? Do all snakes move in the same way? *(P)*

404. Can lizards regenerate parts? How long does regeneration take? *(I)*

405. What effect does age have on a lizard's ability to regenerate? *(U)*

Birds and Mammals

406. What birds do you observe around your locality? Do you see different types of birds at different times of the year? *(P)*

407. Do birds have backbones? How do the bones of a bird compare with those of a mammal? *(I)*

408. Are birds cold-blooded? What is the temperature of the body of different kinds of birds? Does this temperature vary? *(P)*

409. What can you tell about a bird from its feet? *(I)*

410. What do the different varieties of birds eat? How much will they eat? *(I)*

411. Does the food consumption of birds depend upon the temperature of the air? *(U)*

412. How do hawks and owls locate their food? *(U)*

413. Can hawks and owls see in the dark? *(U)*

414. What birds eat seeds of weed plants? *(U)*

415. Are scavenger birds helpful? How do they locate their food? *(U)*

416. What is the body temperature of different kinds of mammals? Does this temperature vary? *(P)*

417. How does the milk of different kinds of mammals compare? Which is the healthiest? *(U)*

418. Do all mammals have four-chamber hearts? Compare heart size with body weight. *(U)*

419. What temperature do the various types of mammals prefer? Try white rats, guinea pigs, kittens, etc. *(I)*

420. What can you tell about an animal from its teeth? How do the teeth of various mammals compare? *(I)*

421. Compare the intelligence of various types of mammals with your own IQ test. *(U)*

422. How do various mammals make sounds? How are these sounds helpful? *(I)*

423. Can you compare the blood of various mammals, reptiles, amphibians, and birds? How do the cells compare? *(U)*

424. What animals are easy to condition? *(I)*

425. What is the temperature range of various mammals? *(I)*

426. How do nests of various birds compare? Which is the strongest? *(P)*

427. What is the safety factor of bird nests? How much weight can various nests hold as compared with the number and weight of the eggs of different birds? *(P)*

428. How do birds and mammals prepare for the winter? Can you observe some? *(P)*

429. Can birds smell? How do birds hear? *(U)*

430. Can birds and mammals recognize their own young? Does odor or appearance help? *(I)*

431. Do birds live in one particular area, or do they travel great distances in a day? *(U)*

432. Can birds be trained by artificial calls? *(U)*

433. Can you compare the strength of various animals? How does this strength compare with the size of the animal? *(I)*

434. Can you measure the metabolism rate of various mammals and other forms of life? *(U)*

435. What causes fear in different animals? Can you measure fear by using the metabolism rate? *(U)*

436. Are white rats instinctively afraid of snakes, or is this learned behavior? *(U)*

437. What effect do various stimuli have on the metabolism rate of certain birds and mammals? *(U)*

438. What effect do different diets have on the growth rate of mammals? *(U)*

439. How does the temperature affect the hatching rate of various fowl? *(U)*

440. Can chickens be hypnotized? *(U)*

441. What effect do different kinds of music have upon the egg production of chickens? *(U)*

Plants and How They Grow

442. How do leaves vary in size, shape, and structure? *(P)*

443. Which seeds are monocots (single seed leaf), and which seeds are dicots (seeds with two halves)? Try corn, beans, peas, rice, and oats. *(P)*

444. How do the leaves of monocots and dicots differ? Examine corn and beans as two examples. *(P)*

445. Where do the colors of the leaves in autumn come from? *(P)*

446. How does a tree lose a leaf? *(U)*

447. Do evergreen trees ever shed their leaves or needles during a year? Mark certain needles and examine during the year. *(I)*

448. What are the internal parts of a leaf? *(U)*

449. Is the green substance in leaves (chlorophyll) distributed evenly throughout the leaf cell? Cut a cross-section of a leaf (elodea) and examine under the microscope. Try other plants. *(U)*

450. What plants secrete a waxy substance (cuticle) that covers the outside of the leaves? What is the purpose of this waxy covering? *(P)*

451. What purpose is served by the veins in leaves? *(P)*

452. Will a plant live in pure oxygen? *(I)*

453. Will a plant live and grow in pure carbon dioxide? Make some carbon dioxide and find out. *(I)*

454. What percentage of carbon dioxide can plants tolerate? *(I)*

455. Is the epidermis or outside covering of a leaf transparent? If so, why? *(I)*

456. What is the function of the guard cells and the stomates in leaves? On which side of the leaf are most of these found? Why? *(U)*

457. What are the sizes of stomates in various plants, such as corn, apple tree leaf, and lily? Can you measure these and determine the number per square inch? *(U)*

458. What product or gas is given off as a result of photosynthesis? Can you collect the gas by covering a water plant with a funnel and collecting the gas with a

test tube? Be sure the water contains carbon dioxide. *(P)*

459. What effect does light have on the rate of photosynthesis? You can use the same equipment as you used in the previous problem. *(P)*

460. Do plants require light to manufacture food? Place one plant in the dark and another plant in sunlight. Check the leaves for sugar. Grind up the leaves to get the sap out of them. Add Benedict's solution to each. The solution should turn brick red if glucose (sugar) is in the sap. *(U)*

461. Will a plant carry on photosynthesis if the stomates are covered in the bottom part of the leaf? Cover the bottom part of the leaf with Scotch tape or a thin wax. *(I)*

462. What kind of gas is given off by animals as a waste product of breathing? *(I)*

463. Can you measure the amount or rate of carbon dioxide production by different animals? *(U)*

464. Can plants use the carbon dioxide given off by animals? Place plants and animals in a box. Note how long the plant remains healthy. Try the experiment again, but this time without the animal. *(I)*

465. Can animals use the gas given off by plants? Try the experiment given before, but this time remove the plant. *(I)*

466. Is the rate of photosynthesis influenced by the temperature? *(I)*

467. Is the rate of photosynthesis influenced by the intensity of the sunlight? *(U)*

468. Is the rate of photosynthesis influenced by the amount of carbon dioxide in the air? *(U)*

469. Is the rate of photosynthesis influenced by the amount of water present? *(I)*

470. At what temperature extremes can algae carry on the processes of photosynthesis? *(U)*

471. What are the light requirements of various plants? *(I)*

472. How do plants adapt themselves in order to receive the exact amount of light necessary for growing? *(P)*

473. What color rays in sunlight are used in the process of photosynthesis? Experiment on plants by using colored lights or a colored filter with sunlight. *(I)*

474. What color ray from sunlight is not absorbed at all by the leaf? Remember the color of the leaf. *(I)*

475. How much water is used by different plants? Can you think of a way to measure the water consumption? *(I)*

476. Do plants manufacture proteins? How can you prove this? Do you know a test for proteins? *(U)*

477. How strongly do the root systems of various plants anchor the plants to the soil? Can you measure the pull required to remove the plant? *(I)*

478. What effect do different kinds of soil have on the ability of roots to anchor plants in the ground? *(I)*

479. How do the roots of various plants compare? *(P)*

480. How does the amount of moisture available affect the length of the root system in various plants? *(U)*

481. Do all roots grow in the soil? How about air and water roots? *(U)*

482. How does the size of the root system affect the amount of water evaporated

from the leaves of various plants (transpiration)? *(U)*

483. What is the structure or make-up of roots and root hairs? Examine the roots of young bean or wheat plants. You may start the plants by placing the seeds on damp blotting paper in a dish and then covering the dish. *(P)*

484. How does soil water enter a root? What is osmosis? Use a carrot for your root. Make a carrot osmometer and experiment with sugar solutions. *(I)*

485. Why does grass die when a strong solution of salt water is poured on it? *(U)*

486. Can you experiment and determine the osmotic pressure of various solutions through different membranes? *(U)*.

487. Can you analyze the soil around you as to composition and percentage of composition? Can you make an artificial soil with chemicals? *(U)*

488. How are soils kept fertile? Can you experiment with the fertility of different soils? *(I)*

489. How is nitrogen supplied to the soil naturally? Can you experiment with the bacteria found in legume plants (clover, peas, beans, and alfalfa)? *(U)*

490. Can you extract nitrogen from the air and make your own fertilizer? *(U)*

491. Can you experiment by making your own fertilizer and using it on plants? *(I)*

492. Will fish serve as a good fertilizer? Plant some beans or corn over a buried fish. *(P)*

493. What foods, animals, or plants make good fertilizer? *(I)*

494. Can you experiment with crop rotation? *(I)*

495. Can you grow plants in water (hydroponics)? *(P)*

496. Can you experiment with different kinds of chemicals in the growing of plants in water? *(I)*

497. What effect has temperature on the growth rate of plants grown in water? *(U)*

498. What plants can be grown successfully by the hydroponic method? *(U)*

499. What is the effect of different amounts of sunlight on plants grown by the hydroponic method? *(I)*

500. How do stems serve plants? Can you compare the stems of different plants? *(I)*

501. How do the roots of plants get food material? Where does the food come from? *(I)*

502. Can stems carry on the process of photosynthesis? *(U)*

503. How does a cactus plant carry on the process of photosynthesis? *(U)*

504. Can the cactus plant adapt to a changed environment? *(I)*

505. How does a cactus plant retain its moisture? *(I)*

506. How do the stems of dicots and monocots differ? Use a cross-section of a corn stem and a cross-section of a sunflower stem. *(U)*

507. What plants (dicots) in your region are killed to the ground by frost in the fall? What plants survive frost and are able to grow again in the spring? *(P)*

508. What plants (herbaceous) in your region live only one year? What plants live two years? What plants live more than two years? What determines the length of life of a plant? *(I)*

509. How does a tree or shrub grow? Can you observe and record the results over a period of time? *(I)*

510. Can you take cross-sections of stems of a plant to show the growth and change in the cells? *(U)*

511. What time of the year do plants (trees, shrubs, flowering plants) show the most growth? Can you measure the growth at different times of the year? *(P)*

512. How does a tree grow? Does the area of the xylem (water-conducting tubes) increase faster or slower than the area of the phloem (food-conducting tubes)? *(U)*

513. Why do the cells of the xylem vary in size? Can you compare xylem cells in different plants and trees? *(U)*

514. Can you determine the age of trees around you from examining the growth rings on stumps? Can you determine what years the tree grew the most and what years it grew the least? How does this compare with the weather cycles? *(P)*

515. Can you determine the year in which various trees started to grow by comparing growth rings on stumps with the growth rings of a tree whose age is known? *(I)*

516. Will a tree die if the bark is removed in a ring all around the tree? Try different places on a tree. Try this experiment on branches. *(P)*

517. What use is the bark to a tree? How much bark is necessary for the tree to keep alive? *(I)*

518. What effect has the removal of some bark of a tree on the growth rate of the tree? Can you find examples in nature of bark removed from a tree? *(U)*

519. How do twigs of different kinds of trees differ? *(P)*

520. Can you determine the growth rate of growing trees by measuring the distance between the bud scale scars? How does the growth rate compare with the climate? *(U)*

521. Do all trees grow at the same rate? Use the method above to determine this. *(I)*

522. How does water move upward in plants and trees? Is this movement of water due to osmotic pressure of water moving into the roots and forcing the water up? Is it due to capillary action of water being lifted up by the hairlike tubes in the plant? Is it due to the evaporation of the water off the leaves? *(U)*

523. How does temperature affect the rate the water rises through the capillary tubes of a plant? Use a stalk of celery and colored water. *(P)*

524. Will plants lift up liquids other than water? How does the rate compare with that of water? *(I)*

525. How much water is given off by leaves of various plants and trees (transpiration)? *(P)*

526. How does the rate of transpiration vary with the season? *(I)*

527. How does the rate of transpiration vary with the humidity of the air? With the temperature? *(U)*

528. How do various insectivorous plants (Venus flytrap, sundew, and pitcher plant) attract insects? *(P)*

529. What effect has temperature on the sensitivity of the Venus flytrap? *(I)*

530. How do insectivorous plants digest and use their food? *(U)*

531. Do insectivorous plants need sunlight? *(P)*

532. What kind of insects will various insectivorous plants eat? *(I)*

533. Can you compare the growth rate of insectivorous plants with the amount of food supply? *(U)*

534. What effects have various fertilizers on the growth of insectivorous plants? *(I)*

535. Can you analyze the digestive juices of insectivorous plants? *(U)*

536. How do various insectivorous plants capture insects? *(P)*

How Plants and Animals Use Food

537. What foods contain starches? Test by using a few drops of dilute iodine. A resulting color of blue-black indicates starch.[37] *(P)*

538. What foods contain sugars? Test by putting a sample of the food material in a test tube with an equal amount of water. Add a few drops of Benedict's or Fehling's solution. The color indicating sugar may be red, orange, or yellow depending on the concentration of sugar. *(P)*

539. Can you determine the concentration of sugar in various foods? *(I)*

540. Can you measure the amount of energy produced by various foods? Use a calorimeter, an instrument that consists of a chamber for burning food and a jacket for water surrounding the chamber. Be sure to measure accurately the amount of food and the amount of water used in the test. *(U)*

541. What type of food (fats, carbohydrates, proteins) contains the largest number of calories? *(I)*

542. How much heat is a calorie? Does it take the same amount of heat to raise all liquids one degree Celsius? *(I)*

543. What are fats composed of? Can you analyze various fats for their composition? *(U)*

544. How much energy do fats produce per gram of weight? *(U)*

545. What foods contain fats? You can test peanuts by chopping them into small pieces and covering them with carbon tetrachloride in order to dissolve the fatty oil. Remove the nuts and allow the carbon tetrachloride and oil mixture to sit. The carbon tetrachloride will evaporate and leave just the oil. Compare the weight of the oil to the weight of the dried peanuts to determine the composition. Try this method on other food.[38] *(I)*

546. Where is fat stored in the body? Examine and dissect some animal. *(U)*

547. Can you test various foods for protein? Try the white of an egg. Place some in a test tube and add one drop of a 1% solution of copper sulfate and five drops of a 10% solution of potassium hydroxide. The presence of protein is indicated by a purple color. If you are testing solid foods, crush the food and cover with water in the test tube before adding the chemicals. *(U)*

548. What element is found in proteins and not in fats and carbohydrates? Can you test to find out? *(U)*

---

37 It should go without saying, but once you begin testing a piece of food with chemicals, you should no longer consider it to be food.

38 Don't use carbon tetrachloride. See Note 21 in Appendix E.

---

549. What effect does a lack of proteins have on the body? Run an experiment on white rats. *(I)*

550. Can you identify the various amino acids in proteins by paper chromatography? *(U)*

551. Which proteins contain the ten essential amino acids? *(U)*

552. How much energy does one gram of protein produce? Do all proteins give off the same amount of energy? *(U)*

553. Can you dehydrate various foods and small animals to determine the amount and percentage of water compared to the total weight? *(I)*

554. How much water is given off by various animals each day? What factors determine this? *(I)*

555. How much water do different animals drink in a day? How does this amount compare with their body weight? *(P)*

556. What factors affect the amount of water an animal drinks in a day? *(P)*

557. Why must the body have water? *(I)*

558. Why is iron needed in the body? What is the effect of a lack of iron in the body? Can you carry on this experiment on a white rat? *(I)*

559. What foods are good sources of iron? Can you test this food for the presence of iron? *(U)*

560. Why is phosphorus needed in the body? Carry out an experiment using a diet which lacks phosphorus. *(U)*

561. Why is calcium needed in the body? Carry out an experiment to determine the effect of a lack of calcium in the body. *(U)*

562. What effect does a lack of iodine have on the body? Does your soil contain iodine? Can you test to find out? *(U)*

563. What effect does a lack of sodium have on the body? Can you experiment using a salt-free diet? *(I)*

564. Is salt given off by the body when we perspire? *(P)*

565. What other minerals are needed in the body? *(I)*

566. How can you detect minerals in foods? Burn bread or sugar by heating them in a jar lid. The ashes left are minerals. Can you determine the proportion of minerals to the total weight of various kinds of foods? *(I)*

567. What is the effect of a lack of vitamin C in our diet? Can you carry on a controlled experiment with a white rat? *(P)*

568. What is the effect of a lack of vitamin A in our diet? *(I)*

569. What foods contain vitamin A? Can you test for vitamin A? What does this vitamin look like? *(U)*

570. How do fats in the diet help prevent vitamin A from being wasted? *(U)*

571. What symptoms result from a lack of the vitamin B complex? *(U)*

572. What do the different vitamins look like under the microscope? How can you locate vitamins? *(U)*

573. How can you detect vitamin C in food? What foods contain vitamin C? Boil about a fifth of a quart of water. Add a teaspoonful of starch. Use about ten drops of this mixture with one drop of iodine. Add this to one-fifth of a quart of water. Then slowly add lemon or orange juice until the blue color disappears. Try apples, pineapples, and other fruits. *(U)*

574. Are vitamin D deficiencies found more among dark- or light-skinned people? Does sunlight contain vitamin D? *(U)*

575. What effect has vitamin E on the ability of rats to reproduce? Is this true of other animals? *(I)*

576. How is vitamin K produced? What help is vitamin K? *(U)*

577. What effect has temperature on vitamin loss? *(I)*

578. What vitamins are destroyed by water? *(I)*

579. Can you determine the basal metabolism (chemical changes going on while the body is at rest) of yourself? Of other animals? Measure the amount of oxygen used and the amount of carbon dioxide given off since chemical action in your body uses up oxygen in its production of energy. *(I)*

580. How many calories of energy are required to perform various activities? Remember, again figure the amount of oxygen used up. *(I)*

581. How does age affect the calorie requirement of people? *(U)*

582. How does the size of the individual affect the calorie requirement? *(U)*

583. What is a balanced diet? *(P)*

584. What is the most effective way to lose weight, by exercising or eating less food? *(U)*

585. Is milk a perfect food? *(I)*

586. Is fish brain food? Can you get smarter by eating lots of fish? *(I)*

587. Are tea or coffee harmful to children? *(I)*

588. How true are advertisements about certain foods on the market? *(I)*

589. What foods lose their vitamins by canning? *(I)*

590. Is soda pop harmful to the human body? *(I)*

591. Is fluorine in drinking water harmful to white rats? *(I)*

592. Can humans detect fluorine in drinking water? *(I)*

593. Are you taller in the morning or at night? *(P)*

594. What effect has good posture on health? *(I)*

595. Can you keep track of your growth rate during the year? Do you grow more in the winter or the summer? *(P)*

596. Does the season of the year affect your intelligence? Are you smarter in the winter? *(I)*

597. Is alcohol harmful to the body? Experiment on rats with an alcoholic diet. *(U)*

598. What is digestion? What happens to food during digestion? *(I)*

599. What effect does saliva have on starchy foods? Make a starch solution test for sugar. Add equal amounts of saliva to starch solution. Wait for ten minutes and then test again for sugar. *(I)*

600. What effect does temperature have on the rate of digestion of starch to sugar by saliva? Carry on the experiment as above, but this time hold one test tube in your hand while leaving a control in a stand. *(I)*

601. Can you analyze your saliva for the chemical composition? *(U)*

602. Can you compare human saliva with saliva from other animals? *(U)*

603. Is the food within seeds stored as a starch? Use the iodine test. *(P)*

604. Do seeds digest starch? Crush bean seeds and then test for starch and sugar. Place other bean seeds on a wet blotter until they germinate. Test again for starch and sugar. What happens to the starch food in the seed? *(I)*

605. What effect has temperature on the rate of digestion of starch into sugar by seeds? *(I)*

606. What effect has light on the rate of digestion of starch into sugar by seeds? *(I)*

607. How does mold digest its food? Wet a piece of bread. Sprinkle some dust on it. Place the bread in a jar and screw the lid down. What do you observe? *(P)*

608. What is the effect of temperature on the digestive action of the bread mold? *(P)*

609. What is the effect of light on the digestive action of the bread mold? *(P)*

610. Can you trace the digestion of various food elements, such as protein, through the various stages into the end product of digestion? *(U)*

611. Are the end products of digestion the same in all plants and animals? *(U)*

612. What is an enzyme? Can you separate various enzymes found or secreted by plants and animals? Can you test these enzymes on various food materials? *(U)*

613. How long is the alimentary canal in various animals? What has the length of the alimentary canal to do with the rate of passage of food from mouth to the anus? *(U)*

614. Why do you chew food? Put a whole piece of bread in a bottle. Add a measured amount of dilute hydrochloric acid. Next break up the bread into smaller pieces. Repeat the experiment. Which took longer? Why? *(P)*

615. How can enzymes digest solids? Pour Coke on a piece of raw meat. *(P)*

616. How does reducing a particle's size affect the total surface area of the material? *(I)*

617. How do the teeth of various animals compare with those of a human? What can you learn about an animal's habits from its teeth? *(I)*

618. Do toothpastes reduce the bacteria content of the mouth? Compare various toothpastes. *(I)*

619. What effect does fear, excitement, or other stimuli have on the digestive rate of an animal? Try this on rats. *(U)*

620. Is the saliva in your mouth acid? Test with litmus paper. *(P)*

621. Does hydrochloric acid help the enzyme pepsin to digest proteins? Coagulate egg white in a piece of glass tubing by heating in boiling water. Break the glass tubing into sections. Use a .4% solution of pepsin on one part. Use a .5% solution on another part. Use both on a third piece. In which tube does the greatest amount of digestion take place? *(U)*

622. How does bile in the liver help to prepare fats for digestion? Study the process of homogenizing milk. Compare cream and milk fat droplets under the microscope. *(U)*

623. Is acid formed in the digestion of fat? Add pancreatin to homogenized milk. Test with litmus paper. *(U)*

624. Can an animal live without a stomach? Use white rats for this experiment. Be sure to try to determine why.[39] *(U)*

625. What is the function of the small intestine? The large intestine? *(U)*

626. How is the small intestine adapted so that digested food can be absorbed into the blood and lymph stream? Examine some tissue under the microscope. *(U)*

627. How does the glucose content of the blood affect the sensation known as hunger? Take a sample of blood before and after meals. *(U)*

628. How do earthworms digest their food? Dissect earthworms to determine how they compare with humans. *(I)*

629. What foods do earthworms eat? Observations are necessary. *(P)*

630. How often and how much do earthworms eat? *(P)*

631. What effect on food material does the saliva have from the mouth of the grasshopper? This saliva is a dark brown juice. *(U)*

632. What food will the grasshopper eat? How does the grasshopper chew its food? *(P)*

633. How often must grasshoppers eat? *(P)*

634. Can grasshoppers do without water or food for the longer period of time? *(P)*

635. Do fish chew their food? What use are teeth to a fish? *(I)*

636. Can you compare the teeth of various fish? *(U)*

637. Can you compare the digestive systems of fish, frogs, and chickens? *(I)*

638. Why do birds need bits of gravel or sand in their diet? *(P)*

639. How can you preserve dissected animals? What is the use of formaldehyde? *(I)*

640. Will alcohol serve to preserve specimens? *(I)*

641. What radio or television commercials about digestive problems are correct? Which ones are false advertising? *(I)*

Cell Structure and Development

642. What part of blood is made up of plasma? What part is made up of cells? Add sodium oxalate to the blood to keep it from clotting. Let the cells settle to the bottom of the test tube or bottle. *(I)*

643. How large are red blood cells? Can you measure the size under the high power of the microscope? How do human red blood cells compare with those of animals? *(U)*

644. What is hemoglobin? What effect does it have on the red blood cells? *(I)*

645. What effect do carbon dioxide and oxygen have on the color of blood? Get some blood from a slaughterhouse.[40] Find out how much sodium oxalate is necessary to keep the blood from clotting. Check with a doctor. Bubble air or oxygen through the

---

39 There are several experiments here that seem very hard to justify (this among them). Can you learn what happens in this case by looking up other people's research on the subject?

40 Many butchers and supermarkets sell blood.

blood. Then bubble carbon dioxide through the blood. *(I)*

646. How can you determine the hemoglobin content of blood? Get a blood color scale (hemoglobinometer)[41] and compare a drop of your blood with the color on scale. *(P)*

647. How does the hemoglobin content of a human compare with that of different animals? *(I)*

648. How can you make a microscope slide of blood and stain it? *(U)*

649. How does blood of cold-blooded animals compare with that of warmblooded animals? *(I)*

650. How does the liver select only the worn-out corpuscles to destroy? *(U)*

651. How long do red corpuscles of various animals live? *(U)*

652. Can you cure anemia in white rats with vitamin B_{12}? *(U)*

653. How do you count red corpuscles? *(U)*

654. What effect does exercise have on the number of red corpuscles in the blood? *(U)*

655. Can you determine the percentage of white corpuscles as compared to red? Make a blood smear on a slide. Stain with Wright's stain. After two minutes add an equal amount of distilled water. Pour off the mixture after two minutes and rinse. The red cells will be pink. The white cells will appear blue with purple nuclei. *(I)*

656. How does the white cell compare with the amoeba? *(I)*

657. What effect do white blood cells have on bacteria in the body? Can you devise a method of observing this? *(U)*

658. Do all animals have blood? *(P)*

659. What is pus composed of? What causes pus? *(I)*

660. How does an infection affect the number of white corpuscles in the blood stream? Can you take blood counts of the white corpuscles? *(U)*

661. Does the white corpuscle count go up during a virus infection? *(U)*

662. How long do white corpuscles live? Where are they produced? *(I)*

663. What are platelets? What is their function in the blood? Can you devise a way of observing platelets? *(U)*

664. What percent of plasma is water? Can you collect animal plasma and boil off the water? *(U)*

665. Is there salt in plasma? After you boil the water off, observe under the microscope. *(U)*

666. Does calcium aid in the clotting of blood? Add calcium to a sample of blood. *(I)*

667. What causes blood to clot? How does vitamin K affect blood clotting? *(I)*

668. How can you determine blood clotting time? Heat a piece of glass tubing. Pull the ends apart so you have a length of fine tubing. Sterilize the tip of your finger with alcohol. Prick with a sterilized needle. Pack the fine glass tubing over the drop of blood. The blood will rush up the tubing. Examine a small bit of tubing each 15 seconds. When you ob-

---

41 This chart is called the *Tallquist Hemoglobin scale*.

serve fine threads being formed, this is the blood clotting time. *(I)*

669. How does the blood clotting time of various animals compare with humans? *(I)*

670. What factors (temperature, humidity, exercise, etc.) affect the blood clotting time? *(U)*

671. What do you observe when blood clots from a cut? Examine under a microscope or a magnifying glass. *(P)*

672. Can you keep blood from clotting? Investigate the substance dicoumarol, a substance made from spoiled sweet clove. *(U)*

673. How is blood typed? Can you type the blood of members of your class?[42] *(I)*

674. What is the most common blood type in your class? Among the boys? The girls? *(U)*

675. What is a universal donor? A universal receiver? What percentages of each do you find? *(U)*

676. Do animals have the same types of blood as humans? Can you type animal blood? *(U)*

677. Can animals (white rats, etc.) receive transfusions? *(U)*

678. How can you determine the Rh factor in blood? *(U)*

679. How long can whole blood be stored? What effect does temperature have on the storage of blood? *(U)*

680. Why is plasma given instead of whole blood? How can plasma be stored? *(I)*

681. What is dextran? How can it replace plasma for transfusions? *(U)*

682. What does the heart really sound like when it is pumping blood? Make or borrow a stethoscope and listen to heart beats of various people. *(P)*

683. What is the average heartbeat of students in your room? *(P)*

684. Does the average heartbeat change with age? *(I)*

685. What effect do different amounts of activity have on the heartbeat? *(P)*

686. What is the average heartbeat of different animals? *(P)*

687. What is the difference between the auricles and ventricles in a heart? Dissect a beef or sheep heart obtained from a butcher. *(I)*

688. Can you design an artificial heart? *(U)*

689. Can you observe blood circulating through capillaries? Place the head of a goldfish in a wet handkerchief. Lay the tail of the fish on a black microscope slide. Can you see the lines of corpuscles moving? *(P)*

690. Can you locate the valves in your veins? Pump your fist open and closed for several minutes in order to make the veins in your arm stand out. Start at your elbow and move your finger along the vein toward the wrist. You will force the blood out of the vein. The vein will be empty from your finger to the valve. *(I)*

691. How does blood circulate through the body? Can you trace the route? How does blood circulate through different animals? *(I)*

---

42 Blood typing a classroom of students was once a common activity. *Do not* attempt to collect biological data or samples from your classmates or colleagues without informed consent (and for minors, parental permission).

692. What material is added to the blood, and what material is removed from the blood at different locations throughout the body? *(U)*

693. Does the heartbeat rate of different animals increase with fear? *(I)*

694. Does lymph contain white blood corpuscles? Examine the fluid from a blister under the microscope. *(U)*

695. Why is the circulation of lymph important to the body? *(U)*

696. What causes swollen ankles if you stand a lot? Why are your hands sometimes swollen in the morning? *(I)*

697. What part do lymph nodes play in getting rid of bacteria? Can you examine lymph nodes of small animals? What do these contain? *(U)*

698. Does grasshopper blood contain corpuscles or just lymph? How do grasshoppers get their oxygen? *(I)*

699. How many hearts does an earthworm have? Are all earthworms constructed the same? *(I)*

700. Does the blood of earthworms contain red corpuscles? Is the blood red? *(U)*

701. Can you observe the flow of blood in an earthworm through a microscope? *(I)*

702. Can you take the pulse rate of a fish? What does the heartbeat sound like? *(U)*

703. Can you observe the circulation of blood in the frog's foot? What conditions affect the circulation of the blood? *(I)*

704. How does a frog's circulation system operate? *(U)*

Appendix

Sourcing Chemicals and Materials

Where to Buy Chemicals and Other Materials

You may also want to look on our website, *http://biyscience.com*, for an online list of sources for materials, complete with purchasing links.

General abbreviations used in this appendix:

- AM—Automotive supply (e.g., Pep Boys, O'Reilly)

- AS—Art supply store (e.g., Michaels, Blick, cheapjoes.com)

- CH—Chemical supply house (e.g., Fisher, Sigma-Aldrich, Cole Parmer)

- CS—Craft store (e.g., Michaels, JoAnn)

- DS—Drugstore or pharmacy (e.g., CVS, Walgreens)

- ES—Electronics store (e.g., Newark, Digi-Key)

- GS—Grocery store (e.g., Safeway, Lucky, Kroger)

- HB—Homebrewing/Winemaking store (e.g. MoreBeer, Northern Brewer)

- HC—Hobby chemical store (e.g., Home Science Tools, United Nuclear)

- HI—Home improvement store (e.g., Lowe's, Home Depot)

- HS—Hardware store (e.g., Ace, True Value, OSH)

- IS—Industrial supply (e.g., McMaster-Carr, Grainger)

- LS—Laboratory supply (e.g., LabPro)

- OS—Office supply store (e.g., Staples, Office Depot)

- SG—Sporting goods store (e.g., REI, Dick's, Sports Authority)

Abbreviations for specific stores:

- AMZ—Amazon.com

- ASC—Arbor Scientific (*arborsci.com*)

- HCS—Hobby Chemical Supply (*hobbychemicalsupply.com*)

- HST—Home Science Tools (*hometrainingtools.com*)

- MCM—McMaster-Carr (*mcmaster.com*)

- TSC—The Science Company (*sciencecompany.com*)

- UNU—United Nuclear (*unitednuclear.com*)

Acetaminophen (pure, as crystals or powder)	*AKA* Paracetamol. CH.
Acetic acid	GS (as distilled white vinegar).
Acetone	DS or GS (as nail polish remover), CH, HI, HS, IS.
Agar agar	GS. AMZ.
Alcohol	See Note 6 in Appendix E about alcohol. CH, DS, GS, HS.
Alum	May refer to *ammonium aluminum sulfate* or *potassium aluminum sulfate*.
Aluminum metal	GS (as aluminum foil), HI, HS, IS. MCM.
Ammonium aluminum sulfate	GS (spice aisle, as *alum*, for pickling).
Ammonium dichromate	CH, HC. UNU HST, TSC.
Ammonium hydroxide ("household ammonia")	GS, HS (cleaning section) HST.
Ammonium sulfate	HI or HS (as a fertilizer) AMZ part number: B000BZ4RJY.
Baby bottle (glass)	AMZ part number: B001F50FFO.
Baking soda (Sodium bicarbonate)	GS, HS.

Bamboo skewers	GS, HI or HS (grilling section), kitchen goods store.
Battery, flashlight	"Heavy duty" type, *not* alkaline, for carbon rods, zinc, and manganese dioxide powder. HS.
Battery, 240-volt photoflash	See Note 29 in Appendix E.
BBs (metal shot)	SG.
Beaker	LS. UNU, AMZ part number: B008VEHLBI.
Bell jar, vacuum	ASC.
Bell wire	HI, HS. Solid conductor insulated wire (as opposed to stranded wire). Commonly found as a bonded pair (zip cord) of two separately insulated solid-core wires, typically 20 to 24 gauge wire. AMZ part number B000UE471O.
Benedict's solution	CH. HST.
Bicycle pump	SG. Bike store, second-hand store. See also: vacuum pump, bicycle tire pump style.
Blotting paper	See *paper, blotting*.
Borax (sodium tetraborate)	HS (laundry or cleaning section). AMZ part number: B000R4LONQ.

Boric acid	CH, GS, HI, HS (as an insecticide). UNU, AMZ part number: B001V9WY0S.
Bottle, baby (glass)	AMZ part number: B001F50FFO.
Bottle, dropper	LS, AMZ part number B0081SRRFO.
Bottle stoppers	See *stoppers*.
Brine shrimp eggs	Pet store.
Broom straw	A filament of plant matter, taken from a household broom that has natural fibers. HI, HS.
Calcium oxide	CH. UNU.
Camphor	Indian grocery stores (as *edible camphor*). AMZ part number: B000UYIQYI.
Canada balsam	AS, CH.
Carbon rods	Found as welding electrodes (welding supply store). MCM, AMZ. Inside "heavy duty" flashlight batteries. UNU. Thick pencil leads may also be used for light duty applications: AS, CS, DS, OS
Carbon tetrachloride	Not recommended. See Note 21 in Appendix E.
Chrome alum	See *chromium potassium sulfate*.

Chromium potassium sulfate	CH, HC. TSC, AMZ part number: B001D7NHEA.
Cigar box	CS. AMZ part number: B001767QPS. Alternately, use an empty shoe box or box from a set of chocolates.
Citric acid	GS, HB, HS (canning section). Health food store. AMZ part number: B00EYFKKZC.
Clips electrical (alligator)	ES, AMZ part number: B0002JJU28, Newark part number: 82R5255.
Clothespins	CS, GS, HS.
Coal, bituminous	Small quantities are frequently available on eBay.
Coal, anthracite	AMZ part number: B00845R1AU.
Cobalt chloride	CH, HC. UNU.
Cobalt nitrate	CH.
Compass (directional)	SG. AMZ.
Congo red	CH. Recommended substitute: Eosin Y.
Copper sheet or strip	HC, HI, or HS (roofing section) UNU. US Penny coin (pre-1982).
Copper tubing	HS, HI, IS.
Copper acetate	CH, HC. UNU.
Copper sulfate	CH, HC, HS (as root killer). AMZ part

	number: B007HU6G22, HST.	Epsom salt (Magnesium sulfate)	DS, HB.
Cotton-covered wire	Use any modern insulated wire as a substitute. (Plastic insulation is much more common than cotton.) HI, HS.	Emery paper (sandpaper)	HI, HS, IS.
Cotton rope	See *rope, cotton*.	Fahnestock clips	AMZ part number: B008C4S31Y Newark part number: 96F9060. Substitute: Shiny (nickel/chrome plated) paper clips or binder clips, OS. Substitute: *Alligator clips*.
Curtain rod, lockseam	HI, HS. AMZ part number: B007ULEZV2.		
Ditto fluid	See Note 3 in Appendix E.		
Dark glasses	See *welding shade*.	Fehling's solution	CH.
Denatured alcohol	See Note 6 in Appendix E about alcohol. CH, DS, GS, HS.	Ferrous sulfate	Garden supply (as a soil additive). AMZ part number: B00KJL5ZL4.
Diffraction grating	Carnival "Rainbow" glasses. HST, AMZ part number: B006ZBISI4.	Filter paper	GS (coffee section), CH, HC. TSC.
		Full wave rectifier	See *rectifier bridge*.
Dicoumarol	CH.	Funnel, plastic	HS, IS, LS, Kitchen Goods store. MCM.
Duco cement	HI, HS, CS, AS. AMZ part number: B0000A605H.	Funnel, metal (steel or aluminum)	HS, LS, IS. MCM.
Duplicator fluid	See Note 3 in Appendix E.	Gallon jug	HB. AMZ part number B006408Z76.
Dropper bottle	See *bottle, dropper*.	Galvanometer	ES.
Dry ice	GS.	Gelatin	GS (use unflavored, unsweetened gelatin), HB.
Ethyl acetate	CH. HST.	Glass pane	HI, HS.
Eosin Y	CH. HST, AMZ part number: B00K33JCL2.	Glass tubing	TSC, AMZ part number: B001949476 UNU.

Glass microscope slide	See *slide, microscope*.
Glasses, dark	See *welding shade*.
Glue, white (e.g., Elmers, wood glue)	Also known as PVA (Polyvinyl acetate) glue. AS, CS, GS, OS.
Graduated cylinder	HC, LS. TSC, HST, ASC.
Grating, diffraction	See *diffraction grating*.
Hacksaw blade	HI, HS, IS.
Hemoglobinometer (Tallquist Hemoglobin scale)	Kemtec.
Helium gas	CS or party stores, for balloons.
Hibiscus flower, dried	GS (in herbal tea or spice section).
Hydrated lime	See *lime*.
Hydrogen peroxide (3%)	DS, GS.
Hydrochloric acid (HCl, muriatic acid)	HB, HI (as cleaner and etchant for concrete), HS (as pH regulator for swimming pools), IS. UNU.
Hydrochloric acid (food grade)	Health food store.
Hypo	See *sodium thiosulfate*.
Infinite loop	See *loop, infinite*.
Iodine (solution)	CH. UNU.
Iron filings	CH. ASC, UNU.
Isolation transformer, fused	Electrical supply store. AMZ part number: B00006HPFH, AMZ part number: B000LDLF3M.
Knitting needles, steel	CS. *Nickel-plated steel is ideal*.
Lath, wood	HI, HS. Lumberyard.
Laundry bluing	GS, HS. Mrs. Stewarts.
Lead metal	SG (as fishing sinkers or lead metal shot). UNU.
LEDs	ES.
Lenses	See Note 24 in Appendix E.
Lime (calcium hydroxide)--not the fruit	HS (canning section, Mexican spices), HI (garden section, as hydrated lime). AMZ part number B0084LZU1Q.
Lime water (saturated solution of calcium hydroxide)	Make lime water by mixing a teaspoon of hydrated lime with one pint of water.
Loop, infinite	See *infinite loop*.
Lye (sodium hydroxide, historically potassium hydroxide)	CH, HC, HS (as a drain opener). Soap making supply companies. Gourmet food store (as Caustic Soda). HST, Bulk Apothecary, AMZ part number: B001EDBEZM.
Magnesium powder	UNU.
Magnesium ribbon	Rotometals, UNU.
Malic acid	HB. AMZ part number: B0064GZB58.

Magnesium sulfate	See *epsom salt*.	Muriatic acid	See *hydrochloric acid*.
Manganese dioxide	CH. Can be found in zinc-carbon batteries (see Note 12 in Appendix E). HST.	Nails, finishing	HI, HS.
		Naphthalene	Traditional mothballs. See Note 20 in Appendix E. UNU.
Manganese sulfate	CH, HI (garden section). AMZ part number: B004RXD1HK.	Neon bulb	ES. Jameco part number: 210315.
Masking tape	CS, HI or HS (painting section), OS.	Nichrome wire, 20 AWG (gauge)	Electrical supply store. CS (as replacement wire for hot-wire foam cutter). MCM, UNU.
Masonite	AS, HI, HS.		
Mason jar	GS, HS.	Nickel sulfate	CH. Available in nickel plating kits. AMZ part number: B00LET73NI.
Measuring spoon set	GS, HS.		
Methylene blue	CH. Pet store (aquarium section). AMZ, HST.	Nitric acid	CH. Hydroponic supply store. AMZ part number: B002LHW8PA, TSC.
Mercurochrome	Neither advised nor available. See Note 18 in Appendix E.	Pancreatin	CH. Veterinary supply. Health food stores (in tablet form with additives).
Mercury	Use not advised. See Note 26 and Note 32 in Appendix E. CH. UNU.		
Merthiolate	Also known as thiomersal or thimerosal. Use not advised; contains mercury. CH, possibly DS (as antiseptic).	Paper, blotting (Most commonly, as watercolor paper)	AS, CS, OS.
		Papier-mâché (paper mache paste/wheat paste)	HI or HS (wallpaper section). HCS.
		Paraffin wax	GS, HS (canning section) MCM.
Mercury switch	AMZ part number: B001QG76B8.	Pepsin	CH. HST.
Milk of magnesia	DS.	Petri dish	LS. HST, TSC, AMZ, UNU.
Mothballs/crystals	See *naphthalene*.	Phenolphthalein solution	HST, UNU.

Phonograph (AKA record player/ turntable)	Second-hand store. AMZ part number: B00AOBHDPA.	Potassium nitrate (saltpeter)	CH. UNU.
Phosphoric acid	CH, HB, HI (as concrete etchant), IS (as cleaner). AMZ	Potassium permanganate	CH. UNU.
pHydrion paper	CH, LS. TSC.	Potassium sodium tartrate	See *Rochelle salt*.
Polarizing film, linear (AKA Polaroid Filter)	ASC, HST.	Potentiometer (50 ohm, 1000 ohm)	ES.
Plaster of Paris	AS, HI, HS.	Propane torch	HI, HS, IS.
Plastic wrap (clear stretch type)	GS as "Saran Wrap," plastic cling film.	Radioactive source	See Note 28 in Appendix E. Thoriated tungsten welding electrode: MCM. Uranium ore available from UNU.
Potassium alum	See *potassium aluminum sulfate*.		
Potassium aluminum sulfate	DS or GS (in health/ beauty, as *alum block* for razor burn), AS or CS (as *alum powder for tanning and dyeing*). AMZ part number: B004NEHR28 for Alum block, AMZ part number: B0009IN1FY for Alum powder.	Radiometer	AMZ part number: B0007YFJI2, ThinkGeek.
		Rectifier bridge (full wave)	Suggested type: GBU8D (see Note 35 in Appendix E for usage). Digi-key part number: GBU8D-ND.
Potassium chromate	CH. HST.	Resistors	ES.
Potassium chromium sulfate	See *chromium potassium sulfate*.	Rochelle salt (potassium sodium tartrate)	CS, AMZ.
Potassium dichromate	CH. UNU.	Rope, cotton	HI, HS.
Potassium ferricyanide	CH. UNU.	Rope, clothesline	See *rope, cotton*.
		Rubber band	OS.
Potassium hydroxide	CH, CS (for soapmaking, as *caustic potash*). HST, TSC, UNU, AMZ.	Rubber cement	AS, OS.
		Rubber tubing	Used on gravity wash bottle. Typically, latex tubing is the best choice when general
Potassium iodide	CH. UNU.		

	purpose "rubber tubing" is specified. Medical supply stores. MCM.	Sodium oxalate	CH. HCS.
		Sodium silicate	CH. AMZ part number: B0019LVJO0, UNU.
Rubbing alcohol	See Note 6 in Appendix E about alcohol. CH, DS, GS, HS.	Sodium sulfate	CH. UNU.
		Sodium thiosulfate ("hypo" to photographers)	CH. HST, AMZ part number: B00HEYUWVW, UNU.
Ruler (plastic or wood)	AS, GS, OS.	Solar filter (eclipse viewing film)	See Note 13 in Appendix E. A #14 or darker welding shade, from a welding supply store. Telescope store. AMZ part number B0071213JA, ASC.
Saran Wrap	(See *plastic wrap*.)		
Sealing wax	AS. Stationery store.		
Shellac	Real shellac may come as flakes that can be dissolved with alcohol. AS, HI.		
		Spectra cord	Kite store. SG (as kite line or fishing line).
Shot, metal	SG (as BBs).	Spectrometer	AMZ part number: B00FGARIAO.
Silver bromide	CH. Photographic supply stores.		
		Splint (as in "glowing splint")	GS, HS (as long fireplace matches or wooden coffee stirrers).
Silver nitrate	CH. TSC, HST, UNU.		
Slide, microscope	LS. Hobby store TSC, HST, UNU.		
		Star charts	See Note 22 in Appendix E.
Soda lime	Medical/veterinary supply. AMZ part number: B00GRPVEL6.		
		Starch	GS (in laundry or in baking as corn, potato, tapioca, or arrowroot starch).
Sodium acetate	CH. UNU.		
Sodium bicarbonate	See *baking soda*.		
Sodium bisulfate	CH. UNU.	Steel wool	HI, HS, IS. AMZ, MCM.
Sodium bromate	CH.	Stirring rod, glass	CH, HC. Gourmet kitchen goods store (bartending section). TSC.
Sodium carbonate	See *washing soda*.		
Sodium chlorate	CH.		
Sodium nitrate	CH. UNU.	Stoppers (bottle stoppers, sizes #1, 3,	LS, IS, HB. HST, TSC, AMZ, MCM.

6½, 7 solid or with 1-hole or 2-hole)	
Sulfur	CH, HC. UNU, HST, HCS, TSC.
Sulfuric acid	AM (as *battery acid* in low concentration), CH, HC, HS (certain drain openers). AMZ part number: B00KAIQI1W, UNU, HST.
Telescope	Note 27 in Appendix E.
Test tubes	CH, HC, LS. TSC, HST, UNU.
Thoriated tungsten	See *radioactive source*.
Tide chart	Online, *http://www.saltwatertides.com* for example.
Tire pump	See *bicycle pump*.
Transformer, filament. Input: 117 V. Output: 6.3 V, at least 0.5 A	ES. eBay. Suggested part: Hammond Mfg part number 166L6 (2 A output). Mouser.com Part number: 546-166L6, Newark part number: 92F1309. Wiring diagram from Hammond.
Turpentine	AS, HS. AMZ part number: B001V9X8ZI.
Uranium glass	UNU, various other internet shops as "Uranium glass marbles." Sometimes labeled as "Vaseline glass."
Vacuum gauge	Included with some vacuum pumps. AM, HS (automotive section). AMZ part number: B006K2RIY8 MCM.
Vacuum grease	Vaseline (original 100% petroleum jelly version): DS. Purpose-built vacuum grease (e.g., Dow Corning brand): Scientific supply. AMZ part number: B001UHMNW0.
Vacuum pump, hand operated	AM (as brake bleeding pump). AMZ part number: B00265M9SS.
Vacuum pump, bicycle tire pump style	Build one: page "Vacuum Pump" on page 80. AM (as oil extractor pump). AMZ part number: B0000BYO97.
Variac with isolation	Electrical supply store. AMZ part number: B006NGC6HU.
Water glass	See *sodium silicate* .
Washing soda (Sodium carbonate)	GS, HS (laundry section). AMZ part number: B0029XNTEU.
Weather stripping	HI, HS.

Welding shade	Welding supply store, AMZ.	Zinc metal	Inner case of "Heavy Duty" flashlight batteries. White metal inside copper coating of post-1982 US penny. HST, UNU.
Wheat paste	See *papier-mâché*.		
Window glass	See *glass pane*.		
Wood glue	See *glue, white*.	Zinc sulfate	CH, HC. HST.
Wright's stain	CH, HC, LS. HST.	Zinc sulfide	CH, HC. Steve Spangler Science, UNU.
Xylene	CH, HI, or HS (painting section, as a solvent).		

Weights and Measures

B

SI base units:

Base quantity	Name	Symbol
length	meter	m
mass	kilogram	kg
time	second	s
electric current	ampere	A
thermodynamic temperature	kelvin	K
amount of substance	mole	mol
luminous intensity	candela	cd

Linear measures:

12 inch	1 foot
3 feet	1 yard
1760 yards	1 mile
1 inch	2.54 cm (exact; definition of inch.)
1 centimeter (cm)	10 millimeters (mm)

1 meter (m)	100 cm
1 kilometer (km)	1000 m
1 astronomical unit	149,597,871 km (mean distance between Earth and the sun)
1 light year	9.4607×10^{12} km (distance travelled by light in a year)

Quantity measures:

1 dozen	12 units
1 score	20 units
1 gross	144 units
1 mole	6.022×10^{23} units

Liquid volume measures:

1 milliliter (ml)	1 cubic centimeter (cc or cm^3)
1 liter	1000 ml
1 cubic inch	1.11 tablespoon

1 tablespoon	3 teaspoons
1 teaspoon	4.93 ml
1 fluid ounce (fl oz)	2 tablespoons
1 fl oz	8 drams
1 cup	8 fl oz
1 pint (pt)	2 cups
1 quart (qt)	2 pints
1 gallon (US)	4 quarts
1 liter	1.057 quart
1 cubic inch	16.4 milliliter

Weight/mass measures:

1 kg	1000 grams (g)
1 kg	2.204 pounds (lb)

1 lb	454 g
1 lb	16 ounces (oz)
1 oz	28.35 g
1 metric ton	1000 kg
1 ton	2000 lb

Time measures:

1 minute (m)	60 s
1 hour	60 m
1 day	24 h
1 day	86400 s
1 week	7 days
1 year	365.25 days
1 century	100 years

Temperature Conversion Table | C

This table gives the approximate relation between temperature in degrees Celsius (°C) (known historically as degrees Centigrade) and degrees Fahrenheit (°F) for a useful range of temperatures. It is worth noting that while the Fahrenheit temperature scale is commonly used in the United States, degrees Celsius or Kelvin are used in essentially all modern scientific work, both in the US and abroad.

°C	°F	°C	°F	°C	°F	°C	°F	°C	°F	°C	°F
-80	-112	-10	14	24	75	52	126	80	176	120	248
-75	-103	-5	23	26	79	54	129	82	180	125	257
-70	-94	0	32	28	82	56	133	84	183	130	266
-65	-85	2	36	30	86	58	136	86	187	135	275
-60	-76	4	39	32	90	60	140	88	190	140	284
-55	-67	6	43	34	93	62	144	90	194	145	293
-50	-58	8	46	36	97	64	147	92	198	150	302
-45	-49	10	50	38	100	66	151	94	201	155	311
-40	-40	12	54	40	104	68	154	96	205	160	320
-35	-31	14	57	42	108	70	158	98	208	165	329
-30	-22	16	61	44	111	72	162	100	212	170	338
-25	-13	18	64	46	115	74	165	105	221	175	347
-20	-4	20	68	48	118	76	169	110	230	180	356
-15	5	22	72	50	122	78	172	115	239	185	365

The exact formulas for conversion between the Fahrenheit and Celsius are:

$$\text{Temperature } [°F] = \frac{9}{5} \times \text{Temperature } [°C] + 32$$

and

$$\text{Temperature } [°C] = \frac{5}{9} \times (\text{Temperature } [°F] - 32)$$

The Kelvin temperature scale, widely used by scientists, has intervals of the same size as the Celsius scale. That is to say, a temperature difference of 1 K is the same as a temperature difference of 1 °C. However, the Kelvin scale has its zero not at the freezing point of water, but at *absolute zero*, the lower limit of any thermodynamic temperature scale. Conversion from degrees Celsius to the Kelvin scale is straightforward:

$$\text{Temperature } [K] = \text{Temperature } [°C] + 273.15$$

The degrees symbol should not be used when referring to temperatures on the Kelvin scale: Room temperature is around 300 K, *not* 300 °K.

Table of the Elements

This periodic table of the elements has been prepared by NIST, the US National Institute of Standards and Technology. You can download a current PDF version of this chart at www.nist.gov/pml/data/.

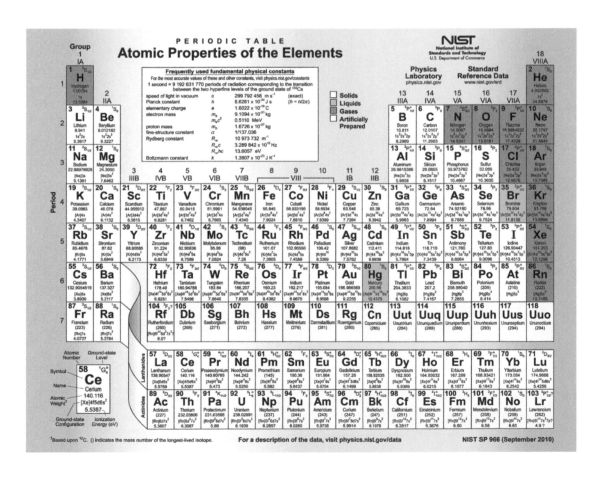

Extended Notes

Note 1. Working with Animals

It is wise to approach the idea of experimenting upon animals — especially living animals — with reservation and skepticism. If you wish to experiment with animals (and there is no blame *if you do not*), we urge you to read and follow the International Rules and Guidelines for Science Fairs, from the Society for Science & the Public, available online at *http:// student.societyforscience.org/forms*.

The guidelines are particularly strict about experiments involving vertebrate animals, and in many cases, will require a committee of qualified individuals to approve and review your work. If you are working independently at home and need a review committee of your own, you can likely find volunteers to help you. Contact the biology teacher or department at your local high school, college, and/or university, as well as your local veterinarian's office. Tell them what you are doing and why you need their help.

For the example case of seeing how vertebrate animals react to changes in air pressure ("Vacuum Pump" on page 80), it is *probably* not worth the hassle of forming a research committee just to investigate that particular question. And that is precisely the sort of reasoning that you should employ: An experiment involving animals is only worth doing *if it is worth doing right*. Doing it right can be a lot of work.

Note 2. Safety: Working with Glass

When cutting, drilling, etching, or shaping cold (i.e., room temperature, not heated) glass, follow these guidelines:

- Always wear safety glasses. Glass can break unexpectedly, and tiny shards can fly when glass shatters.

- Use great care to avoid cutting yourself. Glass has very sharp edges that can easily cut through your skin. In some cases, it is appropriate to wear heavy leather gloves to help protect your skin.

- Work over a tray or cardboard box, that will collect your bits of broken glass. Dispose of any leftover broken glass in such a way that no one (including someone who handles your garbage) could be cut by it. If you have a proper "sharps" container, use it. Otherwise, put your broken glass inside two nested cardboard boxes, each taped shut with packing tape, and label all sides of the outer box clearly: "CAUTION: BRO-KEN GLASS."

Additional precautions (Note 4) are required when working with hot glass.

Note 3. Safety: Working with Open Flames

- *Always* wear safety glasses while working with a torch or flame.

- Have a type ABC fire extinguisher on hand and **know how to use it**.

- Take care to avoid additional fire hazards: Secure any loose hair or clothing, and remove any other flammable materials from the vicinity where you are working. A clean metal, stone, ceramic, or concrete tabletop or board is an excellent surface for using a burner. If you don't have a good surface to work on, you can purchase a large stone or ceramic tile at a home improvement store for several dollars.

- Make sure that your work area is adequately ventilated such that smoke and fumes will not be an issue. If you do not have wide open windows or an exhaust fan that guarantees plenty of fresh air, consider working outside.

- Know the escape route (preferably, *more than one* escape route) from your work area.

- Do not work alone. Have a responsible human on hand who can call the fire department if necessary, even if you are busy with a fire extinguisher.

- Make sure that children and pets do not enter your work area.

- Never leave a burning flame unattended.

Note 4. Safety: Working with Hot Glass

Hot glass is potentially very dangerous, and requires a *consistent focus on safety*. In addition to the guidelines about working with cold glass in Note 2, strictly follow all of the safety guidelines in Note 3 *whether or not* you have an exposed flame.

- Wear only natural fiber clothing (no synthetics or blends), including long pants and a long-sleeved shirt. Wear sturdy, closed-toe shoes. Do not wear jewelry such as necklaces, rings, or watches.

- Wear your safety glasses all the time, even when you don't have an open flame. Glass can still crack and shatter. Accidentally dropped glass is efficient at creating flying shards.

- Remember that *hot glass looks exactly the same as cold glass.* An important aspect of working with glass is that your torches, tools, and glass may remain dangerously hot long after they appear to be safe and cool.

- Keep your face as far away as reasonably possible from hot glass and flames.

- When blowing glass by mouth, *exhale only.* Never risk inhaling hot fumes.

- Understand that hot glass can potentially drip or stick to clothing or other materials, and cause severe burns or start a fire. *Know how you would handle these kinds of situations.*

- Have metal tongs and a tin-can receptacle at hand, to pick up and contain hot glass pieces that end up on your workbench.

- If you are doing any more than a trivial amount of glassblowing, or work with heat sources hotter than an alcohol burner, consider using a welding helmet or special glasses (for example, Rose Didymium or ACE) designed to protect your eyes from harmful IR and UV rays.

- Ideally, seek professional training to learn how to work with glass safely. Many communities have resources for learning to blow glass.

Note 5. Baby Bottles

The bottle referred to here is a glass baby bottle in the 8 oz. (240 mL) size. These bottles are still produced under (for example) the EvenFlo brand as the "Classic Glass Bottle." They are also available in a 4 oz. size. It should be noted that *most* baby bottles today are of a different shape and plastic composition, and thus far less suitable for laboratory work.

Note 6. About Alcohols

Common "rubbing alcohol" sold in drugstores is now *typically* a 70% mixture of isopropyl alcohol with water. For applications like an alcohol burner or cloud chamber, you will have much better luck with a 90%, 99%, or higher grade. Denatured alcohol (odorized ethanol, aka "methylated spirits") is the better choice whenever available: it isn't watered down. *CAUTION: Both rubbing alcohol and denatured alcohol are poisonous and highly flammable*.

Alcohol products available at liquor stores (Everclear et al.) tend to be more expensive and are rarely suitable. While some states do allow sales of relatively pure (95% alcohol by volume/ 190-proof) liquors, it will be more expensive than the alternatives that we have described since it is taxed as liquor. Liquor is also problematic in that if you are below a certain age, it may be illegal to purchase or possess.

Note 7. Ditto Fluid

Spirit duplicating or "ditto" machines were common document duplicating machines in the era before photocopiers and laser printers became inexpensive. They used a fluid (solvent) to operate. That "ditto fluid" is technically still available, but not in common use. If you're curious, the formula is typically a mixture of isopropanol and methanol. Pure (99%+) rubbing alcohol (see the prior note) is generally a good substitute.

Note 8. Tin Cans

"Tin" cans are actually made of tin plated steel, and come in a number of standardized sizes. A "#300" tin can is 3" in diameter by 4 7/16" tall, and is a traditional "14 oz" size used for canned vegetables and the like. A "#303" tin can is 3 3/16" in diameter by 4 3/8" tall, and is a traditional "16 oz" size used for soups and the like.

Note 9. Mason Jars

Rings ("Bands") and lids for traditional glass canning jars (mason jars) by Kerr are still made in standard and wide-mouth sizes. The "small mouth" #63 size is no longer manufactured. While it is possible to purchase them from various vintage sources over the Internet, it may be better to simply purchase a wider variety of tin can sizes and adapt them to the purpose.

Note 10. Safety: Exposed Wiring

Several projects in this book involve line voltage ("mains power" i.e., power from household outlets).

Do not touch (or risk touching) exposed line voltage wiring when connected to power. You will receive a serious shock that carries a risk of injury or death. A good rule is to unplug the project from power—such that you can see the unplugged end of the power cord—before touching the wiring. Do not assume that rubber gloves or shoes will protect you from a shock.

Minimize risk when working with line voltage by using a fused isolation transformer and having an easily accessible power switch (such as a wall switch or power strip) that is upstream and separate from any power switch on your project. *Never operate projects with exposed line voltage (or any apparatus in this book with a power cord!) without easy access to a separate electrical cut-off switch (e.g., wall switch or circuit breaker panel) that can disconnect power if needed.*

If you are a young person, projects with exposed wiring require adult supervision. And *regardless* of your age, good safety practice requires that you have another responsible human being nearby when working on projects that present potentially lethal hazards, such as exposed line voltage.

The projects in this book are designed for use with the power grid in the US, 117 V AC, at 60 Hz. If you live in an area with different wall power, do not assume that a project involving line voltage can be built without accounting for that change.

In addition to these basics:

- *Always* wear safety glasses while working with exposed line voltage.

- Have a type ABC fire extinguisher on hand and **know how to use it**.

- Have a planned escape route (or better, *more than one*) from your work area.

- Make sure that children and pets do not enter your work area.

Note 11. An Alternative Load for the Carbon Arc

For the carbon arc furnace in particular, another alternative that might be worth considering is a "reactance" (inductor) specifically designed for this type of carbon arc furnace, described in Popular Science in 1933. You can find the article online at *http://blog.modernmechanix.com/experimental-arc-furnace-melts-anything/*.

Note 12. Flashlight Batteries

The term "flashlight battery" in this book does not refer to a modern alkaline battery, but instead to a *zinc-carbon battery*. That type of battery, normally sold under the keywords "heavy duty" (as opposed to "alkaline"), has a zinc inner case, a center carbon rod, and a paste filling made with manganese dioxide. Heavy duty batteries are still manufactured and available as low-cost alternatives to alkalines. The C size is likely a good starting point for the experiments within this book. Modern zinc-carbon batteries have a steel outer case around the inner zinc case (which serves as one of the electrodes).

If you only need the carbon rods, you may wish to consider purchasing carbon rods directly, rather than removing them from batteries. See Appendix A for sources.

Wear protective rubber gloves (preferably nitrile) when handling the contents of a battery, and wash your hands thoroughly with soap and water after. Read Note 17 for additional advice about safety while working with chemicals.

Note 13. Safety: Sungazing and Carbon Arcs

Looking directly at the sun or a carbon arc furnace without proper protection causes permanent damage to your eyes *in addition to* any temporary damage (e.g., "seeing spots") that you might notice. Both the part of the sunlight that you can see and the part that you *cannot* see can damage your eyes. *Your eyes are not replaceable; use a proper protective filter between your eyes and the light source.*

Improvised solar filters are generally good enough to protect your eyes. Historically, black-and-white photographic film was common, and a fully exposed piece of it was very dark due to its high silver content, and sometimes used as a solar filter. (Color film, including chromogenic "color-process" or "C-41" black-and-white film, is not suitable whatsoever.) Smoked glass (glass with soot applied) is *potentially* dark enough, but should not be used because it not consistent and wears away easily. Other unsafe filters include sunglasses, neutral density filters, and polarizing filters. Some CDs, DVDs, "space blankets," and other aluminized materials *may* be safe, but are not consistently so, since they are not designed to be.

Purpose-made solar viewing film is inexpensive and easy to obtain, either in sheets or glasses. (See "Solar filter" in Appendix A.) A #14 or darker welding shade — the type specified for use with carbon arc welding — is also safe and inexpensive. Most welding lenses, goggles, and helmets (including auto-darkening helmets) are *not* dark enough for directly viewing the sun, carbon arcs, or burning magnesium. Do not assume a welding shade is safe unless it is labeled with a shade number of 14 or higher.

As a complete alternative to the "shoebox" solar viewer described in the text ("Sunspot Viewer" on page 76), you can make an excellent sunspot and solar eclipse viewer by using a set of binoculars to project sunlight onto a board. See spaceweather.com/sunspots/doityourself.html for instructions.

Note 14. Safety: Dry Ice

Dry ice, which is solid frozen carbon dioxide, should be treated with care. It is very cold: -109.3 °F or -78.5 °C. Direct or indirect contact between dry ice and your skin (whether by holding it or storing it in your pocket) can cause frostbite, which is a very serious injury. Do not pick it up, do not hold it in your hand, and most especially, do not put it in your mouth.

Store dry ice in a cooler, preferably a lightweight styrofoam type. Handle it with heavy cotton oven mitts, and only briefly. A good surface to rest dry ice upon is a towel or stack of folded towels.

A secondary risk of dry ice is that it is a concentrated form of carbon dioxide, which can cause asphyxiation (death by suffocation), if allowed to evaporate in an enclosed area. Always ensure excellent ventilation when working with dry ice.

Note 15. Coin Weights

The data table of known weights ("Metric Weights" on page 31) includes weights for the US dime, penny, nickel, quarter, and half dollar. The values in that table have been updated with current values from the US mint (*http://www.usmint.gov*) as of 2014. The modern penny is copper-plated zinc (2.5% copper), while the modern dime, quarter, and half dollar are all 8.33% nickel, with the balance copper. The modern (gold-tone) dollar coin is made of manganese brass: 88.5% Cu, 6% Zn, 3.5% Mn, and 2% Ni.

The original 1963 data from that table was actually quite different, because the dime, quarter, and half-dollar were made from 90% silver and 10% copper until 1964. Additionally, the penny was 95% copper and 5% zinc until 1982. The original weights from that table are given below:

Dime (90% silver)	2.5 g
Penny (95% copper)	3.25 g
Nickel	5 g (unchanged)

Quarter (90% silver)	6.25 g
Half Dollar (90% silver)	12.35 g

Note 16. Acids and the Corner Drugstore

While the proverbial "corner drugstore" once served the role of both pharmacy and chemical supply house, those roles have diverged. In most places, it is difficult for individuals to purchase most types of concentrated acids. They are, of course, "readily" available from chemical supply houses, which most commonly will only sell to schools, institutions, and businesses. Additionally, hazardous material restrictions and local regulations can make shipping difficult and expensive. You should also keep in mind that they are hazardous substances. They carry health risks that you should understand, and they need to be disposed of safely, and in compliance with your local laws.

You can purchase low-concentration sulfuric acid in the form of "battery acid" at an auto parts store. It is possible to increase the concentration by leaving a container of sulfuric acid solution out for several days with the lid off, to allow some of the water to evaporate. (If you do so, ensure that there is ample fresh air flow and no possibility of spillage.) While it is possible to remove acid from a lead-acid storage battery, it is better not to use that as a source for various reasons, including lead contamination.

As an alternative to directly using sulfuric acid (e.g., in battery projects), you can also use a small amount of sodium bisulfate dissolved in water, ½ teaspoon per 9 fluid ounces of water.

Hydrochloric acid, also known as muriatic acid, is more easily available at home improvement stores and hardware stores. It is used as a chemical for cleaning and etching concrete and for changing the pH of swimming pools.

Note 17. Safety: Working with Chemicals

Basic precautions for working with chemicals:

- *Always* wear safety glasses that have side shielding and offer splash protection.

- Wear sturdy, closed-toe shoes, long pants, and a shirt with long sleeves.

- Secure any loose hair or clothing. Do not wear jewelry such as necklaces, rings, or watches.

- A dedicated lab coat—that you leave behind in your chemical work area—can help to prevent accidental transfer of chemicals outside of the area.

- Nitrile rubber gloves are resistant to many kinds of chemicals, and are good general-purpose laboratory gloves. Leave them behind when you leave the lab, to avoid contaminating other areas.

- Do not pipette by mouth. (See "**Modern Safety Practice**" on page 10 for more.)

- Do not directly smell chemicals, whether to identify them or otherwise.

- Do not store or consume food or drink in your work area. And do not store or use chemicals in a food preparation area or refrigerator that is used for food.

- Wash your hands regularly throughout the day with soap and warm water, and again when you leave your chemical work area.

- Keep your work areas free of clutter.

- Have a type ABC fire extinguisher on hand and **know how to use it**.

- Have a planned escape route (or better, *more than one*) from your work area.

- Have a plan for what to do in case of a chemical spill. Certain types of chemicals (such as mercury) require very specialized spill kits. What can you do to make an acid spill kit (for example) should you need it?

- Have easy access to a sink or fountain where you can wash your eyes in cold running water—and make sure that you can actually wash your eyes in it. Supplemental or permanent eye wash stations can be purchased as well.

- Do not work alone. Have a responsible human on hand, who can help in case of emergency.

- Make sure that children and pets do not enter your work area.

Note 18. Mercurochrome

Mercurochrome is a brightly colored topical antiseptic that contains mercury; it is no longer used in the US and some other countries because of its mercury content. Mercurochrome is not necessary to the project; it is only used here as one of several possible *coloring agents*. What other kinds of brightly colored substances do you have access to that might be a better substitute?

Note 19. Density Versus Specific Gravity

The *density* (mass per unit volume) of a substance is a more common figure than specific gravity. The specific gravity of a substance is the ratio of its density to that of a reference substance, usually water (with a density of 1 g/cm^3). Thus if the density of a substance (say, water ice) is 0.92 g/cm^3, then its specific gravity is just 0.92 — a unitless (dimensionless) quantity.

Note 20. Mothballs

Traditional mothballs are made from naphthalene. They are still available if you care to look for them, but there are also now many other types of mothballs. Check the label to make sure that they are pure naphthalene.

Note 21. Carbon Tetrachloride

Carbon tetrachloride has a number of adverse health and environmental effects, and its use is not recommended. What kind of solvent can you find that will dissolve fat, to replace it in the experiments where it is used as a solvent? Would alcohol, turpentine, or acetone work?

Note 22. Star Charts (Planispheres)

The type of star chart with a turning wheel in the center is called a *planisphere*. Several excellent ones are available to purchase online, or you can make your own.

Four popular, well-loved planispheres are the Miller Planisphere, Philip's Planisphere, the David H. Levy "Guide to the Stars," and Chandler's "The Night Sky." You can get one starting at about $10 at your local telescope store, at outdoor gear stores such as REI, or from Amazon.com and various other internet retailers.

Good quality planispheres are designed to work in a specific latitude range, say 30-40 degrees or 40-50 degrees, so be sure to pick one appropriate for your location on Earth. For reference, New Orleans and San Diego are close to 30 degrees North latitude, while Portland Maine and Oregon are at about 45 degrees. If you're in the equatorial region or southern hemisphere, be sure to get a planisphere designed for the equatorial region or southern hemisphere.

You can also download and print a planisphere. Uncle Al's Starwheels are maintained by the emeritus director of the Lawrence Hall of Science Planetarium, and are available at *http://www.uncleal.net/uncle-als-starwheels*.

As a modern alternative, you might also consider "plantarium" software. The basic SkySafari (*skysafariastronomy.com*) apps for phones and tablets are excellent companions for stargazing, and feature a "night mode" that won't spoil your night vision. For the desktop computer, Starry Night (*starrynight.com*) and Stellarium (*stellarium.org*) are excellent software packages that also simulate the night sky for you.

Note 23. Planet Charts

Six planets can be seen with the unaided eye: Mercury, Venus, Earth, Mars, Jupiter, and Saturn. You should be able to locate Earth on your own. Some planispheres have a chart to help you locate the others. But even if yours does have a chart, it may not be up to date for the current year. You can download a current "Planet Locator" chart from davidchandler.com or use an interactive map at in-the-sky.org.

Note 24. Lenses

Either glass or plastic lenses may be used for simple optics experiments. In most cases, glass lenses are of higher quality. You can obtain inexpensive lenses at scientific surplus shops like American Science and Surplus (sciplus.com), at scientific supply shops like Arbor Scientific (arborsci.com), and from optics sources like Anchor Optics (anchoroptics.com). You can also find useful lens elements inside camera lenses (from a second-hand store) or an old projector (digital, overhead, or film/slide).

Note 25. Diffraction Gratings

Transmissive diffraction gratings like the ones discussed here are frequently sold as "rainbow" glasses. See Appendix A for sources.

Reflective diffraction gratings are actually a little easier to find: A CD or DVD produces rainbows in the sunlight because it acts as a diffraction grating. How could you build a spectrograph that uses a reflective grating instead? Visit sci-toys.com for instructions on how to build a digital spectrograph from a CD.

Note 26. Mercury for a Pressure Gauge

Mercury is no longer considered safe to work with, even in small quantities. Sure, you can buy a vacuum gauge for $10 at an auto parts store (see Appendix A), but *can you work like a scientist* and figure out how to do this experiment ("Vacuum Jar Pressure Gauge" on page 79), even without using mercury?

Questions to investigate:

- Why was mercury used in the first place?

- Are there new metal alloys available to purchase that are non-toxic and liquid at room temperature?

- What other liquids could you use in the gauge instead? What would you have to change?

- If you used water in your gauge (density 1 g/cm^3) instead of mercury (density 13.5 g/cm^3), how tall would your columns need to be? Could you use transparent plastic tubing (rather than glass) and try it? Hints: Blue food coloring and a stairwell may come in handy.

- How can you convert between the barometric reading from the weather service (which is reported in *inches of mercury*) and inches of water?

- How long of a straw can you drink out of? What if you were an infinitely-strong comic book superhero?

Note 27. A Telescope to Use

There are great reasons to make your own telescope. And while it is certainly possible to make your own telescope (see "Refracting Telescope" on page 67), it should be noted that there are also high-quality, low-cost telescopes available for purchase, such as the Galileoscope, about $50, or (much better) the Orion StarBlast 4.5 Astro Reflector Telescope, about $200.

Note 28. Radioactive Sources

What radioactive sources do *you* have at home? Would you be surprised if you had any? Look at the warning labels on the smoke detectors in your house. Do they contain radioactive sources? Most commonly, a household smoke detector contains a tiny grain or disk of intensely radioactive americium-241. (Why? How does a smoke detector work?) The americium has a half-life of 432 years, and remains active long after the smoke detector has exceeded its useful life. Is it safe to remove the radioactive source? Is it legal to do so? *Is it a good idea?*

If you have a clock or watch with a luminous dial or hands that glows *even when not exposed to light first*, then the glowing paint contains a radioactive source. To be sure, leave the clock in a light-tight drawer for 24 hours, and then open it with the lights off and the room dark. Other common sources include lantern mantles (as used for common gas camping lanterns), which are sometimes made with thorium.

In addition to the usual precautions that you should take when handling chemicals (see Note 17), do not handle radioactive material with your bare hands. Wash thoroughly after working with radioactive materials. And most importantly, note that otherwise "harmless" radioactive materials become very dangerous if swallowed, even in minuscule quantity. Take extreme caution to prevent young children from having access to them.

Questions to investigate:

- Some foods are radioactive. Is there enough potassium in a banana that you can detect the radiation?

- Is there enough radium in brazil nuts that you can detect it?

- Are these foods still safe to eat? Could you build spinthariscopes (see "Spinthariscope" on page 100) with different materials to figure out which ones are radioactive?

- Could you detect the radiation with your cloud chamber (see "Diffusion Cloud Chamber" on page 103)?

Note 29. Diffusion Cloud Chamber

Use pure alcohol (99% or higher) when building the diffusion cloud chamber. (See Note 6 for more about alcohols.)

The copper strips are specified as 24 gauge (0.020", or 0.5 mm) thick, but that dimension is not critical. For the radioactive source, two excellent choices are a thoriated tungsten welding electrode or a uranium glass bead or marble. Both of these sources are essentially free of dangerous dust, so long as you do not intentionally cut into them. (See also note 28.)

The 240-volt photoflash battery is long obsolete. The cloud chamber can still be impressive without the copper, wires, and battery. However, there will be lower contrast, thanks to a more visible background of condensing "fog."

If you would like to wire up the voltage, you'll need to come up with an alternate voltage source and a way to manage it safely. 240 V DC is not quite "high voltage," but it is a much higher voltage than common batteries produce. Certainly enough to give you a painful shock, and potentially enough to injure or kill, depending on how powerful the voltage source is.

Questions to investigate:

- What safety procedures are required when working with high voltage (i.e., voltage higher than household voltage)?

- One method to consider would be using a set of 9 V batteries connected in series. 27 of them could give a full 240 V. Perhaps surprisingly, they can also supply enough current *to be potentially lethal*. Can you understand *why* that is dangerous? Are there other types of batteries that could be "stacked" well to supply a higher voltage?

- What other sources of DC voltage are available to you?

- Do you need the full 240 V? Is there a minimum voltage that you need, or will using a lower voltage simply produce a weaker effect? Could you compensate for having a lower voltage by moving the lower copper band lower? If so, what does that mean?

Note 30. The No. 6 Dry Cell

The No. 6 dry cell was a type of battery that came with convenient clips or screw terminals on the top. That's not so different from a modern lantern battery, which has large springs on the top.

Most commonly, a No. 6 dry cell was electrically configured with 1.5 V output, much like an overgrown D cell. However, some versions had two cells internally, for a total of 3 V output.

Replicas are made today by connecting a D cell (sometimes, two) to a pair of screw terminals (www.kensclockclinic.com/catalog/batteries/).

A convenient modern interpretation is a regular battery holder, adapted to fit one or two C or D cells. You can make one, starting by harvesting the battery compartment from an old or broken toy that uses C or D cells.

Note 31. Electronic Electroscopes

The electronic electroscope is a vacuum tube circuit based upon a 6AU6 pentode; it's old technology. While it still works just as it used to, none but the rarest enthusiast designs a circuit this way today. A vacuum tube is now an uncommon component, and it requires more current and voltage than modern solid-state electronics.

You may be able to find a 6AU6 (or 6AU6A) at an electronics or ham radio swap meet, on eBay, or elsewhere online (e.g., *http://www.tubesandmore.com)*.

You can find the design for a solid-state 6AU6 replacement online at *http://collinsra.org/home/album/solid-state-6au6/*.

Although it may be a hassle to obtain or substitute for the 6AU6 tube, a bigger concern is that the circuit uses mains (line/wall) current and leaves line voltage exposed. Act accordingly, with extreme care. You may wish to consider using an isolation transformer or an isolated variac to minimize risk.

As an alternative, consider building the clever and simple electroscope demonstrated by Alan Yates (vk2zay.net) in his video about making electrets: youtu.be/1DR-tTU8uIM. Watch the video to see how it can be built and used. This electroscope is only sensitive to *changes* in electric charge, rather than absolute charge. However, it is safe, easy to build, and runs from a single 9 V battery. You'll also need a few common transistors and resistors, as well as red and blue LEDs as output indicators.

Note 32. Building a "Mercury" Switch

As in Note 26, bare mercury is no longer considered safe to work with, nor is it necessarily easy to purchase in bulk. As the text suggests, you can use salt water as a substitute, but it is not as satisfactory to play with.

Questions to investigate:

- Some modern thermometers are made with a non-toxic liquid metal mercury substitute. Could you build a "non-mercury" switch by removing that liquid from a thermometer?

- Modern "tilt switches" (which might be used in a digital camera to sense the camera's orientation) use a steel ball that rolls from one position to another. Could you use a single conductive steel ball as a substitute for the mercury, otherwise using the same design?

- Could you use a multitude of tiny steel balls as a substitute for mercury in a switch like this? (Consider, for example McMaster-Carr PN 9291k41, 1/16" stainless steel balls, pkg of 100.)

- Many hardware stores still sell mercury switches. You might find them in the electronics, automotive, or HVAC sections of the store. (Why are they found in those contexts?) A commercially made mercury switch usually has a clear glass envelope so that you can see the mercury move. So long as you do not actually drop and break the glass envelope, this is a relatively safe way to "play" with mercury and get a feel for how it behaves.

Note 33. A Transformer Versus a Load

An electrical transformer is a magnetic device that uses electromagnetic induction to transfer electrical power between two circuits. In most (but not all) cases, the two circuits are electrically isolated from one another. Transformers are used for electrical isolation, or to step up or step down a given AC voltage.

A key characteristic of transformers is that they *transform* power: A "step down" transformer (like the filament transformer in the power supply project in the "AC or DC Power Supply" on page 126) reduces the voltage, but also supplies a higher amount of current at that lower voltage. A "step up" transformer (which could be that exact same transformer but put into the circuit backwards) produces a higher amount of voltage, but at lower current. In general, for a perfect transformer, $I_P V_P = I_S V_S$, where I_P and V_P are the the current and voltage on the primary side of the transformer, and I_S and V_S are the current and voltage on the secondary side of the transformer. Of course, transformers are not perfect, and the actual power out will always be slightly lower in.

The Variable Load project ("Variable Load" on page 129) was titled "Power Supply Transformer" at the original time of publication. The operation is similar to a transformer in the sense that it can reduce the amount of voltage available, but it is *not* actually a transformer. Why? Because it simply *reduces the available voltage*, rather than transforming the power such that it produces higher current at that voltage.

Questions to investigate:

- Can a transformer be constructed that works for DC electricity, rather than AC electricity? Why or why not?

- Can a transformer be constructed that is based on *electric* fields, rather than magnetic fields?

- Is it more energy efficient to use a transformer or a variable load?

Note 34. A Filament Transformer

A transformer like the one used here is sometimes called a "filament transformer" because it was used to power the incandescent filament of vacuum tubes.

A specific model that you can use is Hammond Mfg part number 166L6, which can supply up to 2 A of current. Note that this transformer is designed for 115 V 60 Hz (US wall current), and you'll need a different transformer for different wall power in some other countries. See Appendix A for sources.

Note 35. A Modern Bridge Rectifier

The rectifier bridge shown in the illustration of the power supply project is an old-school *selenium* rectifier, which was the type used until silicon versions became more popular in the late 1960s. A suitable modern component is the GBU8D rectifier bridge. See *Rectifier bridge* in Appendix A for sources.

The GBU8D is a simpler-looking block, with 4 pins, all of which come off the same side. Hook the output ("secondary") wires from the transformer to the two center pins (marked as either "AC" or "~"). The output is the "+" terminal, which is the pin on the side that has a notched corner.

The "DC" output that this power supply can produce is "full wave rectified." That means that the voltage does not go *negative*, but it is also not the constant voltage that is usually expected from a DC voltage source. What could be done to smooth this "full wave rectified" output into more of a true DC output?

The GBU8D is actually more capable than the original selenium rectifier in that it provides both a positive output and a negative output (the pin of the rectifier that is not yet in use).

Questions to investigate:

- What is the configuration of the diodes inside the selenium rectifier and in the GBU8D that would make this possible?
- Is there a way to get DC by using the rectifier bridge but *without* the center tap of the transformer?
- What would the output voltage be in that case?

Note 36. Current Flow Indicator

The *Current Flow Indicator* project ("Current Flow Indicator" on page 137) is an adaptation of the "Simple Oscilloscope" project that was in this book at the time of its original publication. The original project has been moved here to the appendix; it follows after some discussion.

There are two reasons that we have handled it this way rather than just adding a footnote. First, the original project is based around a General Electric NE-34 neon glow lamp, which is now a genuinely rare component. Second, from all indications, this project *never actually worked as described*.

The illustration from the original project shows the neon bulb hooked up to a regular dry cell, and claims that it will light up with very low voltage. However, neon bulbs typically require upwards of 70 V DC to "strike" (light up). Digging in further, the original GE specifications for this bulb specify a strike voltage of 65 V. Electrical engineer Jonathan Foote (rotormind.com) was able to perform tests on a vintage NE-34 for us, and confirmed that a working bulb would not light up with even 60 V DC.

Two other quibbles that one might point out: A neon bulb has *electrodes*, not filaments. More significantly, the term "oscilloscope" is not an apt description of the apparatus. An oscilloscope is typically a precision instrument that graphs how an electrical signal changes as a function of time. It is further distinguished from a chart recorder in that it is capable of displaying very fast and oscillating signals with great "magnification" of time. What this project can do well is to indicate the direction of current flow, and we have retitled the updated project accordingly.

The updated project is based on LEDs and serves the same intended function of the "Simple Oscilloscope," while working safely at low voltages. The "Can you work like a scientist?" section there has also been updated with a few new questions that might come up when using LED circuits.

The original project is reprinted below.

SIMPLE OSCILLOSCOPE

Purpose: An oscilloscope is used to analyze electrical current to determine whether the current is direct or alternating, the direction of flow, the frequency or times the electricity changes direction, and the strength of the current. Many things such as light and sound can be changed into electrical energy, and thus a picture of them can be formed on an oscilloscope.

Materials: Neon bulb, socket. A neon bulb costs between $1.50 and $2. They may be purchased at most electrical shops.

What to Do: Screw the bulb into the socket. The wires from the object you are going to test are connected to the socket.

Operation of Equipment: The filament of the neon bulb is divided. When current is flowing in one direction, one side of the filament lights up. When the current flows in the other direction, the other half of the filament lights up. If the current alternates or changes directions,

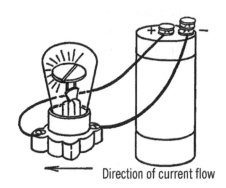
Direction of current flow

the two halves of the filament alternate, producing a blinking effect. The neon bulb operates on very low voltage and yet will still operate on normal house current.

Note 37. The Digital Computer

The dial-based digital computer ("Digital Computer" on page 141) is of the type popularized by the famous computer scientist Edmund C. Berkeley in the 1950s and 1960s. Educational toy computer kits based on this basic design were sold under the names TYNIAC, GENIAC, and BRAINIAC. They were capable of combinational logic, including the simulation of state machines and solving somewhat complex problems, with the help of a human operator to perform sequencing.

While the section in this book on building a digital computer will get you started, it leaves most of the programming up to your ingenuity. Fortunately, there is a wealth of documentation available online, including the original user and wiring guides from several of the kits.

The following resources have scans of one or more of the original manuals. You can also find these and additional resources, as they are found linked online from biyscience.com.

- At vintagecomputer.net: vintagecomputer.net/Geniac/

- At computercollector.com: computer-collector.com/archive/geniac/

- Berkeley, E. Callis. (1956). *Tyniacs: tiny electric brain machines, and how to make them. Also: Manual for Tyniac Electric Construction Kit (K2).* New York: Berkeley Enterprises.

  Viewable online at *http://bit.ly/1z3Kl6J*

- Berkeley, E. Callis. (1955). *Geniacs: simple electric brain machines and how to make them: also, Manual for Geniac Electric Brain Construction Kit No. 1.* 2nd printing. New York: Berkeley Enterprises, Inc.

  Viewable online at: *http://catalog.hathitrust.org/Record/007902253* (bit.ly/ 1szUip5)

Note 38. Centrifugal Force?

If you were in a large bucket — or a space station — being spun around, you would feel an apparent force (an "artificial gravity") pulling you towards the base of the bucket or the wall of the space station. That apparent force that you would feel is called "centrifugal force," (centrifugal meaning "fleeing the center"), and you experience it pushing you directly away from the center of rotation.

However, a formal analysis of the forces involved in a situation like that will always show that there is *not* actually a force acting in that direction. The actual force acting on your body is exerted by the bucket or space station, pushing you directly towards the center: it is called a *centripetal* force.

This is somewhat counterintuitive, so consider a simpler example: riding in an elevator. When you are in an elevator that is accelerating upwards, you feel "heavier" for a moment. That is to say, you feel an *apparent force* that pushes you down towards the floor — just as if gravity had momentarily increased in strength. However, there is *not* actually any new force pushing you down. The actual force being applied to you is *by the floor*, pushing you upwards.

The case of the bucket or space station is slightly more subtle (because of rotation), but the basic idea is the same: the force that holds you to the wall is not an outward force, but of the bucket/spaceship wall pushing you "upwards" towards the pivot point. That force is inwards, and hence *centripetal*. In the case of the spinning bucket, that inwards force is applied by the person spinning the bucket, in order to keep it spinning in a circle. In the case of a space station, it is supplied by the structural elements of the station.

Another useful perspective on the subject may be found at: *http://xkcd.com/123/*

Note 39. Parts for the Motorized Stroboscope

The materials suggested for the project include using a phonograph (record player) record and motor. A piece of cardboard or masonite will make a fine substitute for the record, but there is no single device known as a "phonograph motor"; many different types of motors have been used to turn turntables over the years.

A small battery-driven DC motor is a fine choice for this project. You can commonly find motors like these sold as "hobby motors" or inside of an old RC car or boat. A geared-down version, e.g., the type in a battery-powered hand drill (or even a battery-powered hand drill itself) may be a better choice. Motors like these should be given DC power from batteries, not connected to AC wall power.

The rheostat in this context could be a high-power potentiometer wired up as a variable resistor, or (for example) the pencil rheostat.

Note 40. Phonograph Stop Clock

Timekeeping is easier than it used to be. Most of us carry around devices that function as instant and accurate millisecond timers, with long-term accuracy linked to atomic clocks through precision synchronization via a network of satellites. Nonetheless, it is interesting to consider methods of measuring short time intervals with more basic materials.

The Phonograph Stop Clock project ("Stop Clock" on page 159), as described, does have a flaw in that the record player only begins spinning up (to 45 rpm) when you release the clothespin. Thus, its speed of rotation is *not constant* as it spins up, and the elapsed time per revolution is not constant.

Questions to investigate:

- Can you devise an alternate means of using the stop clock such that you begin counting time only once it is fully up to speed?

- Can you accurately measure the elapsed time of the first, second, and third revolutions (perhaps using your digital device), to see how large of an effect it is? Could you measure the effect well enough to compensate for it?

- How would you improve the precision of an apparatus like this? Would a digital camera be helpful?

Note 41. Soldering to a Tin Can

Common "tin cans" (which are actually tin-plated steel cans) do accept solder but rapidly conduct heat away from the point that you are trying to heat up. A modern soldering iron designed for electronics (usually 15–35 W) has barely enough power to overcome this, and it may be difficult or impossible to solder a can shut. A heavy duty soldering gun or iron designed for plumbing or stained glass (with at least 75–100 W of power) is recommended. A soldering torch may be another avenue worth exploring.

Note that some tin cans are lined (coated) with plastic on their inside surfaces. This plastic coating will not accept solder, and may emit fumes when you solder to the exterior surface. Also, take care that the can may become *very hot* during the soldering operation. Work on a fire-proof surface, have a fire extinguisher handy, wear safety glasses, and don't touch the can while it's hot.

Note 42. Water Barometers

As noted earlier (in Note 26), elemental mercury is no longer considered safe to work with, even in small quantities, and building your own mercury barometer ("Mercury Barometer" on page 185) is no longer recommended. In some locations, it is not even *legal* today. One alternative is to build a *water barometer.*

Water is much less dense than mercury (1 g/cm^3 rather than 13.5 g/cm^3), so air pressure can push a column of water much higher. How high is that? And how can you build something that takes that much vertical space? A stairwell, window on a multistory building, water well, tree, crane, ladder, or balloon may provide the answer. Use semi-rigid translucent plastic tubing (e.g., HDPE, nylon, or teflon tubing) instead of glass, and color the water with food coloring to make it visible through the tubing wall.

A related but easier (and more common) instrument to build is a *weather glass*, also known as a "Goethe barometer." Start with a transparent (glass or clear plastic) teapot or gravy separator of the geometry shown, with a pouring spout that stems from the base.

Fill the main body of the weather glass halfway with water. Seal shut the top of the teapot or gravy separator. If it came with a rigid lid (as

Seal Top

most teapots do), use silicone caulking (from a hardware store) or epoxy to seal it in place. The seal needs to be airtight, to make sure that you locate and fill any air holes in the top. If there is not a rigid lid (but soft plastic or mesh), make or improvise one, perhaps using a ceramic tile or coaster from a hardware or home improvement store. Again, seal it shut with silicone or epoxy.

When you first seal the weather glass, the water level in the spout/arm is equal to the water level inside the body of the glass. As atmospheric pressure changes over a period of days, the level in the spout will be higher or lower than that of the body. To make readings of the value, use a ruler and measure the height of the column (at the spout).

Questions to investigate:

- Can you calibrate your weather glass by comparing the reading on different days to the barometer reading from the weather report?

- What would happen if the weather glass lid were not airtight?

- What would happen if the weather glass lid were flexible, not rigid?

- How is it that this barometer works while being so short, compared to the water barometer described previously that might require a multistorey building? Hint: Compare this apparatus with both the mercury barometer and the "Air Barometer" on page 191. How does the principle of operation compare between the three?

- Does the evaporation of water affect your readings? If so, how much, and can you compensate or correct for it?

- Does the temperature affect your readings? If so, how much, and can you compensate or correct for it?

- If you forgot to fill the weather glass before sealing the lid, how could you go about filling it? Can you refill the weather glass if the water evaporates?

Note 43. The Chemical Weather Glass

The chemical weather glass or storm glass is a traditional Victorian weather forecasting instrument developed by Robert FitzRoy, the captain of Darwin's *Beagle*. It can be a *strikingly beautiful* instrument, with fine and feathery crystals that change character daily. It is a wonderful conversation piece, and elegant modern versions are still constructed with large blown glass globes and brass housings.

The text asks the question "How accurate is the chemical weather glass?" Before you build this to rely upon it, here are some hints: Studies as far back as 1863 have shown that the crystal growth is correlated chiefly with temperature changes. Modern anecdotal and informal tests do not seem to indicate any better-than-chance ability of the storm glass to predict the weather.

Questions to investigate:

- Can you experimentally confirm that the crystal growth is correlated with changes in temperature?

- Is the crystal status determined by the actual temperature, or by the rate of temperature change? Can you create a calibrated scale for the chemical weather glass, indicating your results?

- Find (online) and read Charles Tomlinson's report "An Experimental Examination of the so-called Storm-glass" from 1863. What conclusions were drawn?

- The text makes the claim that the chemical weather glass "helps you predict the weather." If you were to assume for argument that the instrument is no different than a crude thermometer, how can a crude thermometer help you to predict the weather? How well could it predict stormy versus fair weather?

Note 44. Tethered Weather Balloon

Spectra cord, available from kite stores, is an exceptionally light and strong cord that might be helpful for this project.

It is worth noting that above a certain height, a balloon becomes a hazard for aircraft. Check your local laws (FAA regulation, in the US) about tethered balloons before going for any substantial altitude. In some places (close enough to an airport) or weather conditions (when clouds are low enough), tethered balloons may not be allowed at all.

Note 45. On Dissections

The idea of killing a living animal (perhaps especially a vertebrate) merely to dissect it is morally questionable—even repugnant—to many. If you're interested in performing dissections and are resourceful, it is possible to find animal specimens that have died from natural causes or accidents. Consider talking to (for example) the managers of independent pet stores in your town or biology teachers at your local high schools and colleges.

Although it is not the same thing, there are software packages available for download that simulate dissection. You can also now download and 3D print anatomical models, some of which are even "dissectable." As a starting point, search for "anatomy" on *http://www.thingiverse.com*.

Index

wind speed, anemometer for, 175
wind vane, 177
wiring, exposed, safety guidelines for, 281
workbench, science laboratory, 3
worksheet, rock and mineral, 164

X

X connector, 37

XKCD, 292

Y

yardsticks, 30
Yates, Alan, 288

Z

zinc sulfate, 42

About the Authors

Raymond E. Barrett (1926–2011) was the education director of the Oregon Museum of Science and Industry (OMSI) for 22 years.

Windell Oskay is the co-founder of Evil Mad Scientist Laboratories, a Silicon Valley company that has designed and produced specialized electronics and robotics kits since 2007. Evil Mad Scientist Laboratories also runs a popular DIY project blog, and many of its projects have been featured at science and art museums and in *Make*, *Wired*, and *Popular Science* magazines. Windell was also a founding board member of OSHWA, the Open Source Hardware Association. Previously, Windell has worked as a hardware design engineer at Stanford Research Systems and as a research physicist in the Time and Frequency Division of the National Institute of Standards and Technology. He holds a B.A. in Physics and Mathematics from Lake Forest College and a Ph.D. in Physics from the University of Texas at Austin.

Colophon

The cover images are from the original edition of this book, which was illustrated by Joan Metcalf. The cover fonts are URW Typewriter, Guardian Sans, and Heroic Condensed by Silas Dilworth. The text font is Adobe Minion Pro and the heading font is Adobe Myriad Condensed.

CPSIA information can be obtained at www.ICGtesting.com
Printed in the USA
BVOW10n0833300715

411110BV00001B/1/P